MW00760885

DIGITAL PROCESSING

Optical Transmission and Coherent Receiving Techniques

Optics and Photonics

Series Editor

Le Nguyen Binh

Huawei Technologies, European Research Center, Munich, Germany

1. Digital Optical Communications, *Le Nguyen Binh*
2. Optical Fiber Communications Systems: Theory and Practice with MATLAB® and Simulink® Models, *Le Nguyen Binh*
3. Ultra-Fast Fiber Lasers: Principles and Applications with MATLAB® Models, *Le Nguyen Binh and Nam Quoc Ngo*
4. Thin-Film Organic Photonics: Molecular Layer Deposition and Applications, *Tetsuzo Yoshimura*
5. Guided Wave Photonics: Fundamentals and Applications with MATLAB®, *Le Nguyen Binh*
6. Nonlinear Optical Systems: Principles, Phenomena, and Advanced Signal Processing, *Le Nguyen Binh and Dang Van Liet*
7. Wireless and Guided Wave Electromagnetics: Fundamentals and Applications, *Le Nguyen Binh*
8. Guided Wave Optics and Photonic Devices, *Shyamal Bhadra and Ajoy Ghatak*

DIGITAL PROCESSING

Optical Transmission and Coherent Receiving Techniques

Le Nguyen Binh

CRC Press
Taylor & Francis Group
Boca Raton   London   New York

CRC Press is an imprint of the
Taylor & Francis Group, an **informa** business

CRC Press
Taylor & Francis Group
6000 Broken Sound Parkway NW, Suite 300
Boca Raton, FL 33487-2742

© 2014 by Taylor & Francis Group, LLC
CRC Press is an imprint of Taylor & Francis Group, an Informa business

No claim to original U.S. Government works

Printed on acid-free paper
Version Date: 20130916

International Standard Book Number-13: 978-1-4665-0670-1 (Hardback)

Library of Congress Cataloging-in-Publication Data

Binh, Le Nguyen.
  Digital processing : optical transmission and coherent receiving techniques / Le Nguyen Binh.
      pages cm -- (Optics and photonics)
    Includes bibliographical references and index.
    ISBN 978-1-4665-0670-1 (hardback)
    1. Digital communications. 2. Laser communication systems. 3. Optical fiber communication. I. Title.

TK5103.7.B537 2013
621.382'7--dc23   2013034927

Visit the Taylor & Francis Web site at
http://www.taylorandfrancis.com

and the CRC Press Web site at
http://www.crcpress.com

Contents

Preface xv
Author xix
Abbreviations xxi

**1 Overview of Optical Fiber Communications and DSP-Based
    Transmission Systems** 1
  1.1   Introduction 1
  1.2   From Few Mb/s to Tb/s: Transmission and Receiving
        for Optical Communications Systems 3
        1.2.1 Guiding Lightwaves over the Last 40 Years 3
        1.2.2 Guiding Lightwaves: Single Mode, Multimode, and
               Few Mode 8
        1.2.3 Modulation Formats: Intensity to Phase Modulation,
               Direct to External Modulation 8
        1.2.4 Coherent and Incoherent Receiving Techniques 9
        1.2.5 Digital Processing in Advanced Optical
               Communication Systems 10
  1.3   Digital Modulation Formats 11
        1.3.1 Modulation Formats 11
        1.3.2 Pulse Shaping and Modulations for High Spectral
               Efficiency 13
               1.3.2.1 Partial Response 13
               1.3.2.2 Nyquist Pulse Shaping 15
  1.4   Optical Demodulation: Phase and Polarization
        Diversity Technique 18
  1.5   Organization of the Book Chapters 23
  References 24

2 Optical Fibers: Guiding and Propagation Properties 25
  2.1   Optical Fibers: Circular Optical Waveguides 25
        2.1.1 General Aspects 25
        2.1.2 Optical Fiber: General Properties 26
               2.1.2.1 Geometrical Structures and Index Profile ... 26
        2.1.3 Fundamental Mode of Weakly Guiding Fibers 29
               2.1.3.1 Solutions of the Wave Equation for
                       Step-Index Fiber 30
               2.1.3.2 Single and Few Mode Conditions 31
               2.1.3.3 Gaussian Approximation: Fundamental
                       Mode Revisited 36
               2.1.3.4 Cut-Off Properties 38

　　　　2.1.3.5　Power Distribution..40
　　　　2.1.3.6　Approximation of Spot-Size r_0 of a
　　　　　　　　　Step-Index Fiber..41
　　2.1.4　Equivalent-Step Index Description 41
2.2　Nonlinear Optical Effects 42
　　2.2.1　Nonlinear Self-Phase Modulation Effects 42
　　2.2.2　Self-Phase Modulation 43
　　2.2.3　Cross-Phase Modulation.......................................44
　　2.2.4　Stimulated Scattering Effects 45
　　　　2.2.4.1　Stimulated Brillouin Scattering.............46
　　　　2.2.4.2　Stimulated Raman Scattering.................47
　　　　2.2.4.3　Four-Wave Mixing Effects..................48
2.3　Signal Attenuation in Optical Fibers...............................49
　　2.3.1　Intrinsic or Material Absorption Losses..............49
　　2.3.2　Waveguide Losses...50
　　2.3.3　Attenuation Coefficient 52
2.4　Signal Distortion in Optical Fibers.................................53
　　2.4.1　Material Dispersion 55
　　2.4.2　Waveguide Dispersion 58
　　　　2.4.2.1　Alternative Expression for Waveguide
　　　　　　　　　Dispersion Parameter 61
　　　　2.4.2.2　Higher-Order Dispersion........................62
　　2.4.3　Polarization Mode Dispersion 63
2.5　Transfer Function of Single-Mode Fibers 65
　　2.5.1　Linear Transfer Function 65
　　2.5.2　Nonlinear Fiber Transfer Function 72
　　2.5.3　Transmission Bit Rate and the Dispersion Factor 77
2.6　Fiber Nonlinearity Revisited 78
　　2.6.1　SPM, XPM Effects 78
　　2.6.2　SPM and Modulation Instability 80
　　2.6.3　Effects of Mode Hopping......................................81
　　2.6.4　SPM and Intra-Channel Nonlinear Effects.............81
　　2.6.5　Nonlinear Phase Noises...86
2.7　Special Dispersion Optical Fibers....................................87
2.8　SMF Transfer Function: Simplified Linear and Nonlinear
　　　Operating Region...88
2.9　Numerical Solution: Split-Step Fourier Method 95
　　2.9.1　Symmetrical Split-Step Fourier Method...............95
　　　　2.9.1.1　Modeling of Polarization Mode Dispersion 97
　　　　2.9.1.2　Optimization of Symmetrical SSFM 98
2.10　Nonlinear Fiber Transfer Functions and Compensations
　　　in Digital Signal Processing 99
　　2.10.1　Cascades of Linear and Nonlinear Transfer
　　　　　　Functions in Time and Frequency Domains 101

2.10.2 Volterra Nonlinear Transfer Function and Electronic
        Compensation 103
2.10.3 Inverse of Volterra Expansion and Nonlinearity
        Compensation in Electronic Domain........................... 104
        2.10.3.1 Inverse of Volterra Transfer Function............... 106
        2.10.3.2 Electronic Compensation Structure 108
        2.10.3.3 Remarks 111
2.10.4 Back-Propagation Techniques for Compensation of
        Nonlinear Distortion 111
2.11 Concluding Remarks 114
References 115

3 External Modulators for Coherent Transmission and Reception..... 121
3.1   Introduction 121
3.2   External Modulation and Advanced Modulation Formats 122
      3.2.1 Electro-Absorption Modulators.............................. 122
      3.2.2 Electro-Optic Modulators...................................... 124
             3.2.2.1 Phase Modulators.................................... 125
             3.2.2.2 Intensity Modulators 125
             3.2.2.3 Phasor Representation and Transfer
                      Characteristics 127
             3.2.2.4 Bias Control.. 128
             3.2.2.5 Chirp-Free Optical Modulators 129
             3.2.2.6 Structures of Photonic Modulators............. 130
             3.2.2.7 Typical Operational Parameters................. 131
      3.2.3 ASK Modulation Formats and Pulse Shaping............ 131
             3.2.3.1 Return-to-Zero Optical Pulses 131
             3.2.3.2 Phasor Representation.............................. 134
             3.2.3.3 Phasor Representation of CSRZ Pulses......... 135
             3.2.3.4 Phasor Representation of RZ33 Pulses 136
      3.2.4 Differential Phase Shift Keying............................. 137
             3.2.4.1 Background 137
             3.2.4.2 Optical DPSK Transmitter 138
3.3   Generation of Modulation Formats 140
      3.3.1 Amplitude Modulation ASK-NRZ and ASK-RZ............ 140
      3.3.2 Amplitude Modulation Carrier-Suppressed RZ
             Formats.. 141
      3.3.3 Discrete Phase Modulation NRZ Formats 141
             3.3.3.1 Differential Phase Shift Keying 141
             3.3.3.2 Differential Quadrature Phase Shift Keying..... 143
             3.3.3.3 Non Return-to-Zero Differential Phase
                      Shift Keying 143
             3.3.3.4 Return-to-Zero Differential Phase Shift
                      Keying.. 143

3.3.3.5   Generation of M-Ary Amplitude Differential Phase Shift Keying (M-Ary ADPSK) Using One MZIM 144

3.3.3.6   Continuous Phase Modulation PM-NRZ Formats 146

3.3.3.7   Linear and Nonlinear MSK 147

3.4   Photonic MSK Transmitter Using Two Cascaded Electro-Optic Phase Modulators 151

3.4.1   Configuration of Optical MSK Transmitter Using Mach–Zehnder Intensity Modulators: I–Q Approach.... 153

3.4.2   Single-Side Band Optical Modulators 155

3.4.3   Optical RZ-MSK 156

3.4.4   Multi-Carrier Multiplexing Optical Modulators 156

3.4.5   Spectra of Modulation Formats 159

3.5   I–Q Integrated Modulators 164

3.5.1   Inphase and Quadrature Phase Optical Modulators 164

3.5.2   IQ Modulator and Electronic Digital Multiplexing for Ultra-High Bit Rates 167

3.6   DAC for DSP-Based Modulation and Transmitter 168

3.6.1   Fujitsu DAC 168

3.6.2   Structure 170

3.6.2.1   Generation of I and Q Components 171

3.7   Remarks 173

References 176

4 **Optical Coherent Detection and Processing Systems** 179

4.1   Introduction 179

4.2   Coherent Receiver Components 181

4.3   Coherent Detection 182

4.3.1   Optical Heterodyne Detection 185

4.3.1.1   ASK Coherent System 187

4.3.1.2   PSK Coherent System 189

4.3.1.3   Differential Detection 190

4.3.1.4   FSK Coherent System 191

4.3.2   Optical Homodyne Detection 192

4.3.2.1   Detection and OPLL 193

4.3.2.2   Quantum Limit Detection 194

4.3.2.3   Linewidth Influences 195

4.3.3   Optical Intradyne Detection 200

4.4   Self-Coherent Detection and Electronic DSP 201

4.5   Electronic Amplifiers: Responses and Noises 203

4.5.1   Introduction 203

4.5.2   Wideband TIAs 205

4.5.2.1   Single Input/Single Output 205

4.5.2.2 Differential Inputs, Single/Differential Output 205

4.5.3 Amplifier Noise Referred to Input 206

4.6 Digital Signal Processing Systems and Coherent Optical Reception 208

4.6.1 DSP-Assisted Coherent Detection 208

4.6.1.1 DSP-Based Reception Systems 209

4.6.2 Coherent Reception Analysis 211

4.6.2.1 Sensitivity 211

4.6.2.2 Shot-Noise-Limited Receiver Sensitivity 215

4.6.2.3 Receiver Sensitivity under Nonideal Conditions 216

4.6.3 Digital Processing Systems 217

4.6.3.1 Effective Number of Bits 218

4.6.3.2 Impact of ENOB on Transmission Performance 226

4.6.3.3 Digital Processors 228

4.7 Concluding Remarks 228

4.8 Appendix: A Coherent Balanced Receiver and Method for Noise Suppression 231

4.8.1 Analytical Noise Expressions 233

4.8.2 Noise Generators 235

4.8.3 Equivalent Input Noise Current 236

4.8.4 Pole-Zero Pattern and Dynamics 238

4.8.5 Responses and Noise Measurements 242

4.8.5.1 Rise-Time and 3 dB Bandwidth 242

4.8.5.2 Noise Measurement and Suppression 244

4.8.5.3 Requirement for Quantum Limit 245

4.8.5.4 Excess Noise Cancellation Technique 246

4.8.5.5 Excess Noise Measurement 247

4.8.6 Remarks 248

4.8.7 Noise Equations 249

References 252

5 **Optical Phase Locking** 255

5.1 Overview of Optical Phase Lock Loop 255

5.2 Optical Coherent Detection and Optical PLL 258

5.2.1 General PLL Theory 258

5.2.1.1 Phase Detector 259

5.2.1.2 Loop Filter 260

5.2.1.3 Voltage-Controlled Oscillator 261

5.2.1.4 A Second-Order PLL 261

5.2.2 PLL 263

5.2.3 OPLL 265

5.2.3.1 Functional Requirements 265

                         5.2.3.2   Nonfunctional Requirements............................265
              5.2.4   Digital LPF Design...266
                         5.2.4.1   Fixed-Point Arithmetic...............................266
                         5.2.4.2   Digital Filter 268
                         5.2.4.3   Interface Board..270
                         5.2.4.4   FPGA Implementation................................272
                         5.2.4.5   Indication of Locking State 272
                         5.2.4.6   OPLL Hardware Details............................273
      5.3   Performances: Simulation and Experiments...........................274
              5.3.1   Simulation...274
              5.3.2   Experiment: Digital Feedback Control 275
                         5.3.2.1   Noise Sources..278
                         5.3.2.2   Quality of Locking State 278
                         5.3.2.3   Limitations 280
              5.3.3   Simulation and Experiment Test Bed: Analog
                         Feedback Control 281
                         5.3.3.1   Simulation: Analog Feedback Control Loop 281
                         5.3.3.2   Laser Beating Experiments........................288
                         5.3.3.3   Loop Filter Design......................................289
                         5.3.3.4   Closed-Loop Locking of LO and Signal
                                       Carrier: Closed-Loop OPLL 290
                         5.3.3.5   Monitoring of Beat Signals........................291
                         5.3.3.6   High-Resolution Optical Spectrum Analysis.....293
                         5.3.3.7   Phase Error and LPF Time Constant.................293
                         5.3.3.8   Remarks 295
      5.4   OPLL for Superchannel Coherent Receiver.............................296
      5.5   Concluding Remarks...298
      References 299

6   **Digital Signal Processing Algorithms and Systems Performance** 301
      6.1   Introduction 301
      6.2   General Algorithms for Optical Communications Systems........304
              6.2.1   Linear Equalization 305
                         6.2.1.1   Basic Assumptions.....................................306
                         6.2.1.2   Zero-Forcing Linear Equalization (ZF-LE).......307
                         6.2.1.3   ZF-LE for Fiber as a Transmission Channel.....308
                         6.2.1.4   Feedback Transversal Filter 310
                         6.2.1.5   Tolerance of Additive Gaussian Noises 310
                         6.2.1.6   Equalization with Minimizing MSE in
                                       Equalized Signals......................................312
                         6.2.1.7   Constant Modulus Algorithm for Blind
                                       Equalization and Carrier Phase Recovery........314
              6.2.2   Nonlinear Equalizer or DFEs...319
                         6.2.2.1   DD Cancellation of ISI 319
                         6.2.2.2   Zero-Forcing Nonlinear Equalization..............321

6.2.2.3 Linear and Nonlinear Equalization of a
Factorized Channel Response 323
6.2.2.4 Equalization with Minimizing MSE in
Equalized Signals 324
6.3 MLSD and Viterbi 324
6.3.1 Nonlinear MLSE 325
6.3.2 Trellis Structure and Viterbi Algorithm 326
6.3.2.1 Trellis Structure 326
6.3.2.2 Viterbi Algorithm 327
6.3.3 Optical Fiber as a Finite State Machine 328
6.3.4 Construction of State Trellis Structure 328
6.3.5 Shared Equalization between Transmitter and
Receivers 329
6.3.5.1 Equalizers at the Transmitter 329
6.3.5.2 Shared Equalization 332
6.4 Maximum a Posteriori Technique for Phase Estimation 333
6.4.1 Method 333
6.4.2 Estimates 334
6.5 Carrier Phase Estimation 339
6.5.1 Remarks 339
6.5.2 Correction of Phase Noise and Nonlinear Effects 340
6.5.3 Forward Phase Estimation QPSK Optical Coherent
Receivers 341
6.5.4 CR in Polarization Division Multiplexed Receivers:
A Case Study 342
6.5.4.1 FO Oscillations and Q-Penalties 343
6.5.4.2 Algorithm and Demonstration of Carrier
Phase Recovery 345
6.6 Systems Performance of MLSE Equalizer–MSK Optical
Transmission Systems 348
6.6.1 MLSE Equalizer for Optical MSK Systems 348
6.6.1.1 Configuration of MLSE Equalizer in Optical
Frequency Discrimination Receiver 348
6.6.1.2 MLSE Equalizer with Viterbi Algorithm 349
6.6.1.3 MLSE Equalizer with Reduced-State
Template Matching 351
6.6.2 MLSE Scheme Performance 351
6.6.2.1 Performance of MLSE Schemes in 40 Gb/s
Transmission Systems 351
6.6.2.2 Transmission of 10 Gb/s Optical
MSK Signals over 1472 km
SSMF Uncompensated Optical Link 352
6.6.2.3 Performance Limits of Viterbi–MLSE
Equalizers 355
6.6.2.4 Viterbi–MLSE Equalizers for PMD Mitigation ... 359

            6.6.2.5   On the Uncertainty and Transmission
                      Limitation of Equalization Process...................364
   References 365

7 **DSP-Based Coherent Optical Transmission Systems** 369
   7.1   Introduction 369
   7.2   QPSK Systems...371
         7.2.1   Carrier Phase Recovery......................................371
         7.2.2   112 G QPSK Coherent Transmission Systems.................371
         7.2.3   I–Q Imbalance Estimation Results 374
         7.2.4   Skew Estimation...375
         7.2.5   Fractionally Spaced Equalization of CD and PMD.........377
         7.2.6   Linear and Nonlinear Equalization and Back-
                 Propagation Compensation of Linear and
                 Nonlinear Phase Distortion................................377
   7.3   16 QAM Systems...381
   7.4   Tera-Bits/s Superchannel Transmission Systems 385
         7.4.1   Overview..385
         7.4.2   Nyquist Pulse and Spectra 386
         7.4.3   Superchannel System Requirements 388
         7.4.4   System Structure 389
                 7.4.4.1   DSP-Based Coherent Receiver 389
                 7.4.4.2   Optical Fourier Transform-Based Structure 394
                 7.4.4.3   Processing..395
         7.4.5   Timing Recovery in Nyquist QAM Channel 398
         7.4.6   128 Gb/s 16 QAM Superchannel Transmission 399
         7.4.7   450 Gb/s 32 QAM Nyquist Transmission Systems.........401
         7.4.8   DSP-Based Heterodyne Coherent Reception Systems 403
   7.5   Concluding Remarks 406
   References 407

8 **Higher-Order Spectrum Coherent Receivers** 409
   8.1   Bispectrum Optical Receivers and Nonlinear Photonic Pre-
         Processing 409
         8.1.1   Introductory Remarks..409
         8.1.2   Bispectrum..411
         8.1.3   Bispectrum Coherent Optical Receiver 412
         8.1.4   Triple Correlation and Bispectra.......................412
                 8.1.4.1   Definition..412
                 8.1.4.2   Gaussian Noise Rejection...................413
                 8.1.4.3   Encoding of Phase Information 413
                 8.1.4.4   Eliminating Gaussian Noise...............413
         8.1.5   Transmission and Detection...............................414
                 8.1.5.1   Optical Transmission Route and Simulation
                           Platform 414

        8.1.5.2   Four-Wave Mixing and Bispectrum
Receiving 415

        8.1.5.3   Performance 415

8.2   NL Photonic Signal Processing Using Higher-Order Spectra 419

   8.2.1   Introductory Remarks 419

   8.2.2   FWM and Photonic Processing 420

        8.2.2.1   Bispectral Optical Structures 420

        8.2.2.2   The Phenomena of FWM 422

   8.2.3   Third-Order Nonlinearity and Parametric
FWM Process 424

        8.2.3.1   NL Wave Equation 424

        8.2.3.2   FWM Coupled-Wave Equations 425

        8.2.3.3   Phase Matching 427

        8.2.3.4   Coupled Equations and Conversion
Efficiency 427

   8.2.4   Optical Domain Implementation 428

        8.2.4.1   NL Wave Guide 428

        8.2.4.2   Third-Harmonic Conversion 429

        8.2.4.3   Conservation of Momentum 429

        8.2.4.4   Estimate of Optical Power Required for
FWM 429

   8.2.5   Transmission Models and NL Guided Wave Devices 430

   8.2.6   System Applications of Third-Order Parametric
Nonlinearity in Optical Signal Processing 431

        8.2.6.1   Parametric Amplifiers 431

        8.2.6.2   Wavelength Conversion and NL Phase
Conjugation 436

        8.2.6.3   High-Speed Optical Switching 437

        8.2.6.4   Triple Correlation 442

        8.2.6.5   Remarks 448

   8.2.7   NL Photonic Pre-Processing in Coherent Reception
Systems 449

   8.2.8   Remarks 455

References 456

Index 459

Preface

Optical communication technology has been extensively developed over the last 50 years, since the proposed idea by Kao and Hockham [1]. However, only during the last 15 years have the concepts of communication foundation, that is, the modulation and demodulation techniques, been applied. This is possible due to processing signals using real and imaginary components in the baseband in the digital domain. The baseband signals can be recovered from the optical passband region using polarization and phase diversity techniques, as well as technology that was developed in the mid-1980s.

The principal thrust in the current technique and technology differs distinctively in the processing of baseband signals in the discrete/sampled digital domain with the aid of ultra-high-speed digital signal processors and analog-to-digital and digital-to-analog converters. Hence, algorithms are required for such digital processing systems.

Over the years, we have also witnessed intensive development of digital signal processing algorithms for receivers in wireless transceivers, and especially in band-limited transmission lines to support high-speed data communications [2] for the Internet in its early development phase.

We have now witnessed applications and further development of the algorithms from wireless and digital modems to signal processing in lightwave coherent systems and networks. This book is written to introduce this new and important development direction of optical communication technology. Currently, many research groups and equipment manufacturers are attempting to produce real-time processors for practical deployment of these DSP-based coherent transmission systems. Thus, in the near future, there will be new and significant expansion of this technology due to demands for more effective and memory-efficient algorithms in real time. Therefore, the author believes that there will be new books addressing these coming techniques and technological developments.

This book is organized into seven chapters. Chapter 1 gives an introduction and overview of the development of lightwave communication techniques from intensity modulation direct detection, to coherent modulation and detection in the early stage (1980s), to self-homodyne coherent in the last decade of the 20th century, and then current digital signal processing techniques in coherent homodyne reception systems for long-haul nondispersion compensating multispan optical links. Thus, a view of the fiber transmission property is given in Chapter 2. Chapter 3 then discusses the optical modulation technique using external modulators, especially the modulation of the inphase and quadrature phase components of the quadrature amplitude modulation scheme.

Chapters 4 and 5 then introduce optical coherent reception techniques and technological development in association with digital signal processors. Optical phase locking of the local oscillator and the channel carrier is also important for performance improvement of reception sensitivity, and is described in Chapter 5.

Chapters 6 and 7 present digital processing algorithms and their related performances of some important transmission systems, especially those employing quadrature modulation schemes, which are considered to be the most effective ones for noncompensating fiber multispan links.

Further, the author would like to point out that the classical term "synchronization systems" employed some decades ago can now be used in its true sense to refer to DSP-based coherent reception transmission systems. Synchronization refers to processing at the receiver side of a communications systems link, in order to recover optimal sampling times and compensate for frequency and phase offsets of the mixing of the modulated channel and local oscillator carriers, induced by the physical layers and transmission medium. In digital optical communications, designs of synchronization algorithms are quite challenging due to their ultrahigh symbol rates, ultrahigh sampling rates, minimal memory storage and power consumptions, stringent latency constraints, and hardware deficiencies. These difficulties are coupled with the impairments induced by other nonlinear physical effects in the linearly polarized guided modes of the single-mode fibers, when more wavelength channels are multiplexed to increase transmission capacity. Recently, there have been published works on synchronization algorithms for the digital coherent optical communications systems, with some of these aspects touched upon in this book, but still, little is known about the optimal functionality and design of these DSP-based algorithms due to impacts of synchronization error. We thus expect extensive research on these aspects in the near future.

Finally, higher-order spectra techniques, a multidimensional spectrum of signals with amplitude and phase distribution for processing in optical coherent receivers, are introduced.

The author wishes to thank his colleagues at Huawei Technologies Co. Ltd. for discussions and exchanges of processing techniques in analytical and experimental works during the time that he worked in several fruitful projects of advanced optical transmission systems. He also acknowledges the initial development phases of major research projects funded by the Australian Industry R&D Grant, involving development of DSP-based algorithms with colleagues of CSIRO Australia and Ausanda Pty. Ltd. of Melbourne, Australia. A number of his former PhD and undergraduate students of Monash University of Melbourne, Australia have also contributed to discussions and learning about processing algorithms for the minimum shift keying self-heterodyne reception.

Last but not least, the author thanks his family for their understanding during the time that he spent compiling the chapters of the book. He also thanks Ashley Gasque of CRC Press for her encouragement during the time of writing each chapter of this book.

Le Nguyen Binh
Huawei Technologies, European Research Center
Munich, Germany

_____

References

1. K.C. Kao and G.A. Hockham, Dielectric-fibre surface waveguides for optical frequencies, *Proc. IEE*, 113(7), 1151–1158, 1966.
2. A.P. Clark, *Equalizers for Digital Modems*, Pentech Press, London, 1985.

MATLAB® and Simulink® are registered trademarks of The MathWorks, Inc. For product information, please contact:

The MathWorks, Inc.
3 Apple Hill Drive
Natick, MA 01760-2098 USA
Tel: 508 647 7000
Fax: 508-647-7001
E-mail: info@mathworks.com
Web: www.mathworks.com

Author

Le Nguyen Binh earned his BE(Hons) and PhD degrees in electronic engineering and integrated photonics in 1975 and 1978, respectively, both from the University of Western Australia, Nedlands, Western Australia. In 1980, he joined the Department of Electrical Engineering of Monash University after spending 3 years as a research scientist with CSIRO Australia.

Dr. Binh has worked on several major advanced world-class projects, especially the Multi-Tb/s and 100G-400G DWDM optical transmission systems and transport networks, employing both direct and coherent detection techniques. He has worked for the Department of Optical Communications of Siemens AG Central Research Laboratories in Munich, Germany, and the Advanced Technology Centre of Nortel Networks in Harlow, England. He was the Tan-Chin Tuan Professorial Fellow of Nanyang Technological University of Singapore. He was also a professorial fellow at the Faculty of Engineering of Christian Albretchs University in Kiel, Germany.

Dr. Binh has authored and coauthored more than 250 papers in leading journals and refereed conferences, and 7 books in the field of photonics and digital optical communications. His current research interests are in advanced modulation formats for superchannel long-haul optical transmission, electronic equalization techniques for optical transmission systems, ultrashort pulse lasers and photonic signal processing, optical transmission systems, and network engineering.

Dr. Binh has developed and taught several courses in engineering, such as Fundamentals of Electrical Engineering, Physical Electronics, Small-Signal Electronics, Large-Signal Electronics, Signals and Systems, Signal Processing, Digital Systems, Micro-Computer Systems, Electromagnetism, Wireless and Guided Wave Electromagnetics, Communications Theory, Coding and Communications, Optical Communications, Advanced Optical Fiber Communications, Advanced Photonics, Integrated Photonics, and Fiber Optics. He has also led several courses and curriculum development in electrical and electronic engineering (EEE), and joint courses in physics and EEE.

In January 2011, he joined the European Research Center of Huawei Co. Ltd., working on several engineering aspects of 100G, Tb/s long-haul, and metro-coherent optical transmission networks. He is also the series editor for *Optics and Photonics* for CRC Press and chairs Commission D (Electronics and Photonics) of the National Committee of Radio Sciences of the Australian Academy of Sciences.

Abbreviations

4QAM	equivalent with QPSK
16QAM	QAM with 16 point constellation
32QAM	QAM with 32 point constellation
256QAM	QAM with 256 point constellation
ADC	analog-to-digital conversion/converter
Algo	algorithms
α_{NL}	scaling factor of nonlinear phase noises (NLPN)
ASE	amplification stimulated emission
ASK	amplitude shift keying
Balanced detectors	BalD
BalORx	balanced optical receiver
BP	back propagation
CD	chromatic dispersion
CMA	constant modulus amplitude
Co-OFDM	coherent optical orthogonal frequency division multiplexing
Co-ORx	coherent optical receiver
CoQPSK	coherent QPSK
$D(\lambda)$	waveguide dispersion factor
DAC	digital-to-analog conversion/converter
DCF	dispersion compensating fiber
DD	direct detection
DFE	decision feedback equalizer
DQPSK	differential QPSK
DTIA	differential transimpedance amplifier
DuoB	duobinary
DWDM	dense wavelength division multiplexing
ECL	external cavity laser
EDFA	erbium-doped fiber amplifier/amplification
ENOB	effective number of bits of ADC or DAC
EQ	equalization
ESA	electrical spectrum analyzer
ETDM	electrical time division multiplexing
FC	fiber coupler
FDEQ	frequency domain equalization
FF	feedforward
FFE	feedforward equalizer
FIR	finite inpulse response
FMF	few mode fiber
FSK	frequency shift keying

FTVF	feedback transversal filter
FWM	four-wave mixing
Ga/s	Giga samples per second
GB or GBaud	Giga-baud or giga symbols/s
Gb/s	Giga-bits per second
GBaud	Giga symbols/s—see also Gsy/s
GSa/s	Giga samples per seconds
GSy/s	Giga symbols per second, i.e., Giga bauds
HD	hard decision
IM	intensity modulation
IM/DD	intensity modulation/direct detection
IQ	inphase quadrature
ISI	intersymbol interference
iTLA	integrated tunable laser assembler
IXPM	intrachannel cross-phase modulation
LD	laser diode
LE	linear equalizer
LMS	least-mean-square
LO	local oscillator
LPF	low-pass filter
LSB	least significant bit
LUT	look-up table
$M(\lambda)$	material dispersion factor
MAP	maximum a posteriori probability
M-ary QAM	M-order QAM, $M = 4, 8, 166, 32,\ldots,256,\ldots$
MIMO	multiple input/multiple output
MLD	multilane distribution
MLSE	minimum likelihood sequence estimation/estimator
MLSE	maximum likelihood sequence estimation
MMF	multimode fiber
MSB	most significant bit
MSE	mean square error
MSK	minimum shift keying
MZDI	Mach–Zehnder delay interferometer
MZIM	Mach–Zehnder interferometric/intensity modulator
MZM	Mach–Zender modulator
NL	nonlinear
NLE	nonlinear equalizer
NLPE	nonlinear performance enhancement
NLPN	nonlinear phase noise
NLSE	nonlinear Schrödinger equation
NRZ	nonreturn to zero
Nyquist QPSK	Nyquist pulse-shaping modulation QPSK
NZDSF	nonzero dispersion-shifted fiber
OA	optical amplifier

OFDM	orthogonal frequency division multiplexing
OFE	optical front-end
OOK	on-off keying
OPLL	optical phase lock loop
ORx	optical receiver
OSA	optical spectrum analyzer
OSNR	optical signal-to-noise ratio
OTDM	optical time domain multiplexing
PAM	pulse amplitude modulation
PD	photodetector
PDF	probability density function
PDL	polarization-dependent loss
PDM	polarization division multiplexing
PDM-QAM	polarization multiplexed QAM
PDM-QPSK	polarization multiplexed QPSK
PDP	photo-detector pair—see also balanced detection
pi/2 HC	pi/2 hybrid coupler—optical domain
PLL	phase-locked loop
PMD	polarization mode dispersion
PMF	polarization-maintaining fiber
PSK	phase shift keying
QAM	quadrature amplitude modulation
QPSK	quadrature phase shift keying
RF	radio frequency
RFS	recirculating frequency shifting
RFSCG	RFS comb generator
RIN	relative intensity noise
ROA	Raman optical amplification
Rx	receiver
RZ	return-to-zero
RZ33, RZ50, RZ63	RZ pulse shaping with full-width half mark of the "1" pulse equal to 33%, 50%, and 63% of the pulse period, respectively
SMF	single-mode fiber
SoftD	soft decision
SOP	state of polarization
SPM	self-phase modulation
SSFM	split-step Fourier method
SSMF	standard single-mode fiber (G.652)
TDEQ	time domain equalization
TDM	time division multiplexing
TF	transfer function
TIA	transimpedance amplifier
TIA gain	transimpedance gain, Z_T
TVE	transversal equalizer

TVF	transversal filter
Tx	transmitter
VOA	variable optical attenuator
VSTF	Volterra series transfer function
XPM	cross-phase modulation

1

Overview of Optical Fiber Communications and DSP-Based Transmission Systems

1.1 Introduction

Since the proposed "dielectric waveguides" in 1966 by Charles Kao and George Hockham of Standard Telephone Cables (STC) Ltd. in Harlow of England [1], optical fiber communication systems have used lightwaves as carriers to transmit information from one place to the other. The distance of a few meters in laboratory, to a few kilometers, to hundreds of kilometers, and now thousands of kilometers in the first decade of this century with bit rates reaching from few tens of Mb/s in late 1960s to 100 Gb/s and Tb/s at present. Tremendous progress has been made though nearly the last 50 years due to two main significant phenomena, the guiding of lightwaves in optical fibers and transmission, and modulation and detection techniques. The progress of long-haul transmission with extremely high capacity is depicted in Figure 1.1, with transmission distance reaching several thousands of kilometers of one SMF (single mode fiber). We can see since the invention of optical amplifiers, the Erbium doped fiber amplifier (EDFA), in the late 1980s, the demonstration in experimental platform has reached 2.5 Gb/s, that is when the attenuation of the transmission medium can be overcome and only the dispersion remains to be resolved, hence the dispersion management technique developed to push the bit rate to 10 Gb/s. Thence the transmission capacity is further increased with multiplexing of several wavelength channels in the C-band as well as L- and S- bands. With gain equalization, the total capacity was able to reach 40–100 Gbps in 1995. The exploitation of spectral efficiency would then be exploded with further R&D and experimental demonstration by deploying channels over the entire C-band and then over L- and S-bands using hybrid amplifiers to reach 2000 Gb/s at the turn of this century. So over the first decade of this century we have witnessed further progresses to push the capacity with high spectral efficiency, coherent detection, and digital signal processing (DSP) techniques employed at both the transmitter and receivers to achieve 64 Tbps and even higher in the near future. Coherent detection allows further gain in the receiver sensitivity and DSP, overcoming several

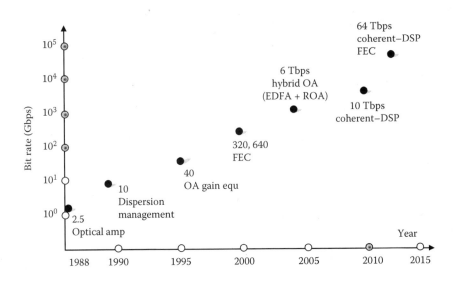

FIGURE 1.1
Experimentally demonstrated single-mode, single-fiber transmission capacity.

difficulties in coherent receiving and recovery of signals. Furthermore, the modulation techniques such as QPSK, M-ary QAM, and spectral shaping such as Nyquist and orthogonality have assisted in the packing of high symbol rates in the spectrum whose bandwidth would be the same as that of the symbol rate of the transmitted channels. DSP algorithms have also been employed to tackle problems of dispersion compensation, clock recovery, compensation on nonlinear effects, polarization dispersion, and cycle slip in walk-off over transmission. It is noted that the transmission is a multi-span optically amplified line and no dispersion compensation is used, unlike the dispersion-managed transmission systems developed and deployed in several terrestrial and undersea networks currently installed, as depicted in Figures 1.2 and 1.3.

This chapter is thus organized as follows: the next section gives an overview of the fiber development over the last few decades and the important features of the single- or even few-mode fibers currently attracting much interest for increasing the transmission capacity. Section 1.3 then gives an overview of advanced modulation techniques, and Section 1.4 gives an overview of coherent detection and DSP process. Details of DSP algorithms will be discussed briefly, then later chapters will provide more details, along with a description of improvement of transmission performance.

This introduction chapter is organized in the following sequence: The historical aspects of optical guiding and transmission techniques over the last 50 years are outlined, followed by an introduction of present progress in the processing of received signals as well as a brief explanation of the generation of modulated signals in the digital domain.

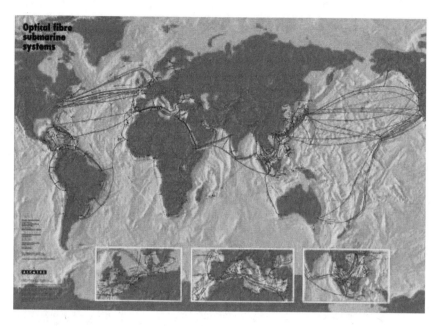

FIGURE 1.2
Global submarine cable systems.

1.2 From Few Mb/s to Tb/s: Transmission and Receiving for Optical Communications Systems

1.2.1 Guiding Lightwaves over the Last 40 Years

Since the proposed dielectric waveguides, the idea of transmission via an optical waveguide was like a lightening stroke through the telecommunication engineering, physics, and material engineering communities, alike. The physicists, mathematicians, and electrical engineers were concerned with the design of the optical waveguides, the guiding conditions, and the formation of the wave equations employed Maxwell's equations and their solutions for guiding and propagation, as well as the eigenvalue equations subject to the boundary conditions, among others. Material engineers played a very important role in determining the combination of elements of the guided medium so that the scattering loss was minimal and, even more important, that the fabrication of such optical waveguides and hence the demonstration of the guided waves through such waveguides.

In 1970, the propagation and fabrication of circular optical waveguides were then successfully demonstrated with only 16 dB/km at red line of 633 nm wavelength, thus the name optical fibers, which consist of a circular core and cladding layer. Don Keck's research alongside two other Corning scientists,

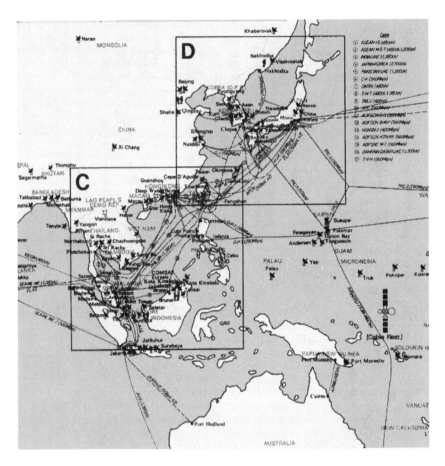

FIGURE 1.3
Optical fiber cable networks in Southeast Asia and the Australia Oceana Region.

Maurer and Schultz, transformed the communications industry to the forefront of the communications revolution, from narrow bandwidth with electromagnetic radiation to guiding lightwaves. Compared to the attenuation of 0.2 dB/km today, that attenuation factor was not ideal, but it did serve as a pivotal point in time for the current revolution of global information systems [2,3].

However, the employment of guided lightwaves in transmission for communication purposes has evolved over the last four decades from multimode to single mode, and then once again in the first decade of this century when the few mode fibers attracted once again the "multimode" used in transmission to increase the total capacity per fiber. The detection of lightwaves also evolves from direct to coherent, then direct self-homodyne, and then coherent homodyne with analog-to-digital processing. The structure of transmission systems over the decades is depicted in Figure 1.4, which shows the evolution of such system through the end of the twentieth century when

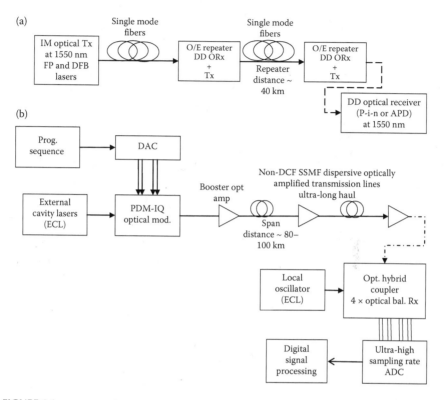

FIGURE 1.4
Schematic structures of the first and recent single-mode optical transmission systems: (a) single-mode non-DCF optically amplified transmission system with DSP-based coherent detection, the fiber can be a single-mode or few-mode type; (b) nonoptically amplified repeated link; (b) optically repeated transmission line with coherent reception. Note the optical transmission line is non-DCF (dispersive) and hence a dispersive optically amplified link.

direct detection, in fact self-homodyne detection, with external modulation of the lightwaves emitted from an external cavity laser whose line width is sufficiently narrow and various modulation formats employed to exploit the combat of dispersion and sensitivities of the optical receivers.

The transmission systems were limited due to attenuation or losses of the fibers and associated components as well as the receiver sensitivity. The transmission was at first operating in the 810 nm near the infrared region due to availability of the source developed in GaAs. This wavelength is then shifted to 1310 nm where the dispersion of the fiber is almost zero, thus limited only by the attenuation factor. This loss can be further reduced when the wavelength is moved to 1550 nm at which the Rayleigh scattering for silica-based fiber is lowest with a value of 0.2 dB/km. This is about half of the attenuation factor at 1310 nm spectral region. However, at this wavelength the dispersion is not zero. The attenuation was further eliminated by the invention of optical amplification by erbium-doped fiber amplifiers.

Optical amplification has changed design considerations for long-haul transmission. With 30 dB gain in the optical domain the fiber attenuation becomes negligible, and over 100 km or 80 km can be equalized without much difficulty as far as the power to the optical channels can satisfy the amplification minimum input level. Furthermore, the insertion loss of integrated modulators would pose no obstacle for their uses as external modulators, which would preserve the linewidth of the laser and hence further reduce the dispersion effects and then the pulse broadening. The schematic structure of this optically amplified transmission system is shown in Figure 1.5e. Note also that distributed optical amplification such as Raman amplifiers are also commonly used in transmission link, in which the distance between spans is longer than the maximum optical gain provided by EDFA. Such requirement would normally be faced by the designer in an overseas environment. For example, the optical link between Melbourne Australia and Hobart of Tasmania, the large island in the southern-most location of Australia. The coast-to-coast link distance is about 250 km, and thus the EDFA is employed as a power booster and optical pre-amplification of the receiver, and Raman pumping from both sides (i.e., co- and contra pumping with respect to the signal propagation direction) of the link located on shore to provide a further 30 dB gain.

During the last decade of the twentieth century we witnessed an explosion of research interests in pushing the bit rates and transmission distance with the dispersion of the standard optical fibers compensated by dispersion compensating fiber; that is, management of the dispersion of the transmission link either by DCF or by distributed dispersion optical compensators such as fiber Bragg gratings (FBG). However, the detection was still by direct detection, or by self-homodyne detection and the processing was still in analog domain.

Another way of compensating the dispersion of the fiber link can be by pre-distortion or chirping the phase of the lightwave source at the transmitter. The best technique is to use the digital-to-analog conversion (DAC) and to tailor the phase distortion of the lightwaves. This is done by modulating the optical modulator by the analog outputs of the DAC, which can be programmable and provide the flexibility that the sampling rate of DAC can meet the Nyquist criteria. This can be met due to significant progresses in the development of digital signal processors for wireless communication systems and computing systems. Under such digital processing, the optical signals can be pre-distorted to partially compensate the dispersion as well as post compensation at the receiver DSP sub-systems.

The DSP field could then be combined with the opto-electronic detection to advance the technology for optical communication systems.

Coherent detection has then been employed with DSP to overcome several hurdles that were met by the development and research of coherent communications in the early 1980s, when single-mode fibers were employed. The limited availability of narrow linewidth sources that would meet the requirement for receiver sensitivities by modulation formats such as DPSK (differential phase shift keying), DQPSK, FSK, MSK, and the recovery of clock for sampling, among

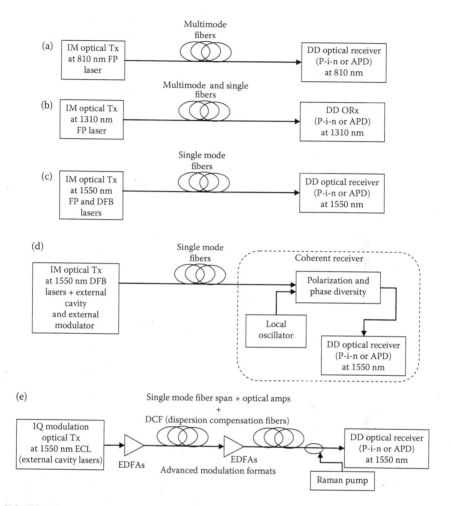

FIGURE 1.5
Schematic structures of optical transmission systems over the decades: (a) earliest multimode systems, (b) single-mode fiber transmission, (c) single-mode fiber as the transmission medium with 1550 nm wavelength, (d) first optical coherent systems with external modulator and cavity lasers and homodyne or heterodyne with polarization and phase diversity detection in analog domain, and (e) optically amplified single-mode fiber links with lumped EDFA and distributed Raman amplification.

others, can be resolved without much difficulty if ultra-high-speed analog-to-digital converter (ADC) is available and combined with ultra-high-speed DSP.

Thus, we have witnessed once again significant progress in the DSP of advanced modulated lightwaves and detection for ultra-long-haul, ultra-sensitive optical fiber communications, without the management of dispersion. A generic schematic of the most advanced transmission is shown in Figure 1.4b, in which both DAC and ADC at the transmitter receive analog signals produced from conventional coherent optical receivers. In contrast to the coherent and DSP-based optical transmission, Figure 1.4a shows the first single-mode optical fiber transmission system with several opto-electronic repeaters, where the distance between them is about 40 km, deployed in the 1980s. In these systems the data sequence must be recovered back into the electrical domain, which is then used to modulate the lasers for further transmission. The distance between these repeaters is about 40 km for a wavelength of 1550 nm. It is at this distance that several housing infrastructures were built and remain to be the housing for present-day optical repeaters, hence the span length of 80 km, with optical attenuation at about 22 dB that fits well into the optical amplification using lumped amplifiers such as the EDFAs for the C-band region of 1550 nm.

1.2.2 Guiding Lightwaves: Single Mode, Multimode, and Few Mode

Lightwaves are coupled into the circular dielectric waveguide, the optical fiber whose refractive index profile consists of a core region and a circular covering on the outside. The refractive index difference between the core and the cladding regions would normally be very small, in order of less than 1%, 0.3% typically. The main principles of operation of such guiding lightwaves are due to the condition that would satisfy the boundary conditions and the guiding such that the interface between the core and cladding would not contribute much to the scattering of the guided waves. Thus, typically the dimension for standard single-mode optical fiber (SSMF) is a core diameter ~8.0 µm, with a cladding of about 125 µm to assure mechanical strength and distribution of the tails of the guided waves in the core. The refractive index is about ~0.3% and a mode spot size of about 4.2 µm. The operational parameters of the SSMF are a dispersion factor of 17 ps/nm km at 1550 nm with a dispersion slope of 0.01 ps/µm^2, and a nonlinear coefficient of 2.3×10^{-23} µm^{-2} with GeO$_2$:doped silica as the core materials. The cutoff wavelength of the SSMF is in the 1270–1290 nm range, above which only one single mode, the fundamental mode LP_{01}, can be guided. This linearly polarized mode consists of two polarized modes, the EH_{11} and HE_{11}, or the field distribution is nearly the same but the polarizations of these modes are spatially orthogonal. Under the nonuniformity of the core of the fiber, these two polarized modes travel at different propagation velocity due to the difference in their propagation constant and hence the delay difference. This delay difference is termed as the polarization mode dispersion.

1.2.3 Modulation Formats: Intensity to Phase Modulation, Direct to External Modulation

The invention and availability of optical amplification in the 1550 nm with EDF has allowed integrated community reconsideration of the employment of integrated optical modulators, especially the LiNbO$_3$-based components

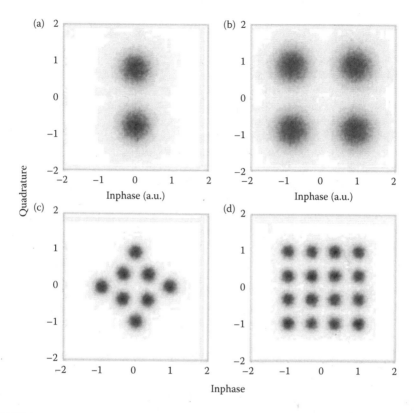

FIGURE 1.6
Constellation of M-ary QAM with $M = 2$ (a), 4 (b), 8 (c), and 16 (d).

due to reasonably high insertion loss, about 3–4 dB for a single Mach–Zehnder interferometric modulator (MZIM). The MZIM offers significant features in terms of bandwidth and extinction ration, defined as the difference in intensity, or field between the "on" and "off." The bandwidth of LiNbO$_3$ can be up to 50 GHz if the travelling wave electrode can be fabricated with the thickness sufficiently high.

Thus we have seen in recent years several modulation formats, especially the quadrature amplitude modulation (QAM) techniques in which both the real and imaginary or inphase and quadrature components are used to construct the constellation in the complex plane, as shown in Figure 1.6 for $M = 2$, 4, 8, and 16. The phase shift keying modulation formats were employed in the first-generation optical communications in guided wave systems in the 1980s.

1.2.4 Coherent and Incoherent Receiving Techniques

Coherent, incoherent, or direct reception of the modulated and transmitted lightwave modulated signals are currently considered, but depending on applications and whether they are in the long haul (carrier side), or metropolitan access (client side, or access networks). Direct modulation should also be considered as this solution for offering significantly inexpensive deployment in metro networks, while coherent solution would offer significant advantages to long-haul transmission systems in terms of reach and symbol rates or baud rates. Both incoherent and coherent systems can employ digital processing techniques to improve the receiver sensitivity and error coding to achieve coding gain, thus gaining longer transmission distance. We have witnessed the development of chirp-managed lasers by taking advantage of the biasing of distributed feedback laser (DFB) about 4 to 5 times the level of the laser threshold, so that the inverse NRZ driving of the DFB would produce chirp and thence the phase difference between the "1" and "0" about π_rads. Thus, any dispersion due to these pulses over long distances of fiber would be cancelled out, hence the dispersion tolerance of such management of the chirp by laser direct modulation.

1.2.5 Digital Processing in Advanced Optical Communication Systems

A generic block diagram of the digital coherent receiver and associate DSP techniques is shown in the flow chart presented in Figure 1.6. Obviously the reception of the modulated and transmitted signals is conducted via an optical receiver in coherent mode. Commonly known in coherent reception techniques are homodyne, heterodyne, and now intradyne, which are dependent on the frequency difference between the local oscillator and that of the carrier of the received channel. For homodyne-coherent detection, the frequency difference is nil, thus locking the local oscillator frequency to that of the carrier of the channel is essential, while with heterodyne coherent detection there is a frequency difference that is outside the 3 dB bandwidth of the channel. When the frequency difference is less than the 3 dB and can be close to the carrier, then the coherent reception is of intra-dyne type. Indeed this difference has degraded the first-generation coherent reception systems for optical fiber communications in the mid-1980s. With DSP, the phase carrier recovery techniques can be developed and overcome these difficulties. Heterodyne reception would require an electrical filter to extract the beating channel information outside the signal band and may become troublesome, with cross talks between received channels. With the bit rate and symbol rate now expected to reach several tens of GHz, as well as due to its complexity, heterodyne detection is not the preferred technique.

For a DSP-based coherent receiver, the availability of a high-speed sampling rate ADC is a must. However, with tremendous progress in digital technology, ADC at 56–64 GSa/s is available and the sampling speed is expected to rise when 28 nm SiGe technologies are employed. In addition, significant progress in the development of algorithms for processing the received sampled data sequence in real time must be made, so that real-time recovery of data sequences can be realized. Currently, offline processing has been done to ensure the availability of processing algorithms.

1.3 Digital Modulation Formats

1.3.1 Modulation Formats

In this book we concentrate on digital modulation formats as a way of carrying information over long distance via the use of the optical carrier. These modulation formats have been developed over the last 50 years and are now well known. However, for completeness we will provide a brief revision of the concepts, as these will lead to further detailed understanding of the modulation of the lightwaves in the optical domain.

The modulation of the lightwave carrier can be in the following forms:

The optical signal filed has the ideal form during the duration of one bit period, given as

$$E_s(t) = E_P(t)a(t)\cos\left[\omega(t)\cdot t + \theta(t)\right] \qquad 0 \le t \le T \tag{1.1}$$

where $E_s(t)$, $E_P(t)$, $a(t)$, $\omega(t)$, and $\theta(t)$ are the signal optical field, the polarized field coefficient as a function of time, the amplitude variation, the optical frequency change with respect to time, and the phase variation with respect to time, respectively. Depending on the modulation of the carrier by amplitude, frequency, or phase, as follows:

- For amplitude shift keying (ASK), the amplitude $a(t)$ takes the value $a(t) > 0$ for a "1" symbol and the value of 0 for a "0" symbol. Other values such as the angular frequency and the phase parameter remain unchanged over one bit period.

- For phase shift keying (PSK), the phase angle $\theta(t)$ takes a value of π rad for a "1" symbol, and zero rads for the symbol "0" so that the distance between these symbols on the phase plane is at maximum, and hence minimum interference or error can be obtained. These values are changed accordingly if the number of phase states is increased, as shown in Figure 1.7. The values of $a(t)$, $\omega(t)$, and $E_p(t)$ remain unchanged.

- For frequency shift keying (FSK), the value of $\omega(t)$ takes the value ω_1 for the "1" symbol and ω_2 for the "0" symbol. The values of $a(t)$, $\theta(t)$, and $E_p(t)$ remain unchanged. Indeed, FSK is a form of phase modulation provided that the phase is continuous. Sometimes continuous phase modulation is also used as the term for FSK. In the case that the frequency spacing between ω_1 and ω_2 equals to a quarter of the bit rate, then the FSK is called minimum shift keying.

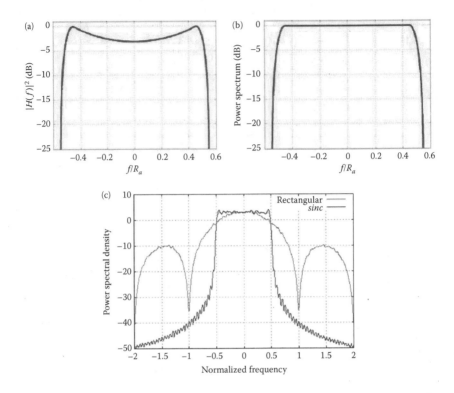

FIGURE 1.7
(a) Desired Nyquist filter for spectral equalization; (b) output spectrum of the Nyquist filtered QPSK signal; (c) spectra of pulse sequences with *sinc* function and rectangular pulse shape.

- For polarization shift keying (PolSK), we have $E_p(t)$ taking one direction for the "1" symbol and the other for the "0" symbol. Sometimes continuous polarization of light waves is used to multiplex two optically modulated signal sequences to double the transmission capacity.

- Furthermore, to increase the transmission capacity there is a possibility to increase the number of bits per symbol by using M-ary QAM, such as 16 QAM, 32 QAM, or 64 QAM, for which constellations are as shown in Figures 1.8 and 1.9. However, the limitation is that the required OSNR would be increased accordingly. For example, an extra 6–7 dB would be required for 16 QAM as compared to QPSK, which is a 4 QAM. The estimated theoretical BER versus SNR is depicted in Figure 1.10 by using the bertoool.m in MATLAB®.

Clearly we can observe that at a BER of 1e–4 the required energy per bit of 16 QAM is about 5 dB above that required for QPSK. So where can we get this energy for a symbol in the optical domain? We can naturally increase the carrier power to achieve this, but this will hit the threshold level of nonlinear effects, thus further penalty. This can be resolved by a number of

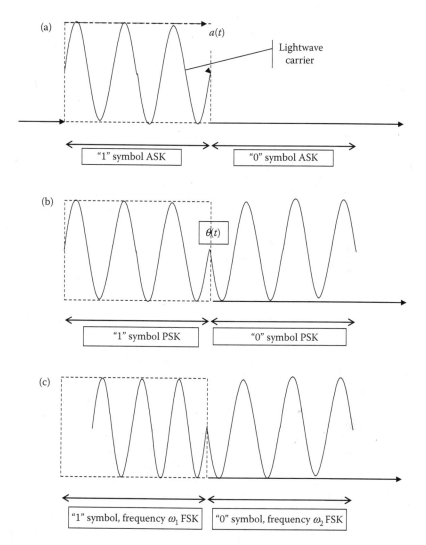

FIGURE 1.8
Illustration of ASK, PSK, and FSK with the symbol and variation of the optical carrier (a) amplitude, (b) phase, and (c) frequency.

techniques that will be explained in detail in the corresponding chapters related to transmission systems.

1.3.2 Pulse Shaping and Modulations for High Spectral Efficiency

1.3.2.1 Partial Response

The M-ary-QAM digital modulation formats form the basis of modulation for digital optical fiber communications systems due to the availability of

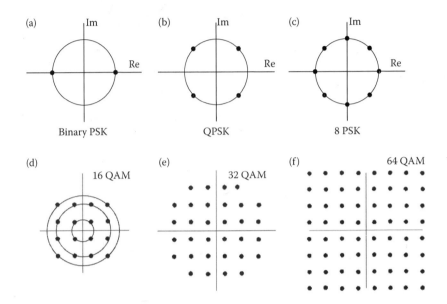

FIGURE 1.9
Constellations of the inphase and quadrature phases of lightwave carrier under modulation formats (a) with π phase shift of the BPSK at the edge of the pulse period, (b) QPSK, (c) 8 PSK, (d) 16 QAM with three rings, (e) 32 QAM, and (f) 64 QAM.

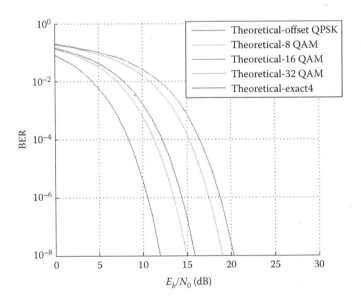

FIGURE 1.10
BER versus SNR for multi-level M-ary-QAM.

the integrated PDM IQ-modulator, which can be fabricated on the LiNbO$_3$ substrate for multiplexing the polarized modes and modulating both the inphase and quadrature phase components. We have witnessed tremendous development of transmission using such modulators over the last decade. Besides these formats, the pulse shaping does also play an important part in these advanced systems; the need to pack more channels for a given limited C-band motivates several research groups in the exploitation of the employment of partial signal technique, such as the duobinary or vestoigial single-side band and Nyquist pulse shaping.

They include nonreturn-to-zero (NRZ), return-to-zero (RZ), and duobinary (DuoB). RZ and NRZ are of binary-level format, taking two levels "0" and "1," while DuoB is a tri-level format, taking the values of "-1 0   1." The -1 in optical waves can be taken care of by an amplitude of "1" and a phase of π phase shift with respect to the "+1," which means a differential phase is used to distinguish between the +1 and -1 states.

The modulated lightwaves at the output of the optical transmitter are then fed into the transmission fibers and fiber spans, as shown in Figure 1.7.

1.3.2.2 Nyquist Pulse Shaping

One way to shape the pulse sequence is to employ the Nyquist pulse-shaping techniques; that is, the pulse spectrum must satisfy the three Nyquist criteria. Considering the rectangular spectrum with a *sinc*, that is (($\sin x$)/x), time-domain impulse response, at the sampling instants $t = kT$ ($k = 1,2..., N$ as nonzero integer) its amplitudes reach zero, implying that at the ideal sampling instants, the ISI from neighboring symbols is thus negligible, or free of intersymbol interference (ISI). Figure 1.11 depicts such Nyquist pulse and its spectrum for either a single channel or multiple channels. Note that the maximum of the next pulse raise is the minimum of the previous impulse of the consecutive Nyquist channel.

Now considering one sub-channel carrier 25 GBaud PDM-DQPSK signal, then the resulting capacity is 100 Gbps for a sub-channel, hence to reach 1 Tbps, 10 sub-channels would be required. To increase the spectral efficiency, the bandwidth of these 10 sub-channels must be packed densely together. The most likely technique for packing the channel as close as possible in the frequency with minimum ISI is the Nyquist pulse shaping. Thus the name Nyquist-WDM system is used. However, in practice, such "brick-wall-like" spectrum shown in Figure 1.11 is impossible to obtain, and hence a nonideal solution for non-ISI pulse shape should be found so that the raise cosine pulse with some roll-off property condition can be met.

The raised-cosine filter is an implementation of a low-pass Nyquist filter, that is, one that has the property of vestigial symmetry. This means that its spectrum exhibits odd symmetry, about $1/2T_s$, where T_s is the symbol-period. Its frequency-domain representation is a brick-wall-like function, given by

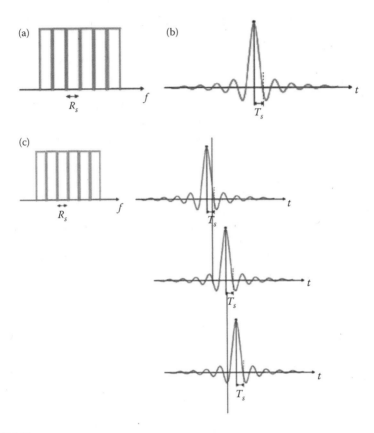

FIGURE 1.11
A super-channel Nyquist spectrum and its corresponding "impulse" response (a) spectrum, (b) impulse response in time domain of a single channel, and (c) sequence of pulse to obtain consecutive rectangular spectra. A superposition of these pulse sequences would form a rectangular "brick wall-like" spectrum.

$$H(f) = \begin{cases} T_s & |f| \leq \dfrac{1-\beta}{2T_s} \\[2ex] \dfrac{T_s}{2}\left[1 + \cos\left(\dfrac{\pi T_s}{\beta}\left\{ |f| - \dfrac{1-\beta}{2T_s} \right\} \right) \right] & \dfrac{1-\beta}{2T_s} < |f| \leq \dfrac{1+\beta}{2T_s} \\[2ex] 0 & \text{otherwise} \end{cases} \quad (1.2)$$

with $0 \leq \beta \leq 1$

This frequency response is characterized by two values: β, the *roll-off factor*, and T_s, the reciprocal of the symbol rate in Sym/s, that is $1/2T_s$, which is the half bandwidth of the filter. The impulse response of such a filter can

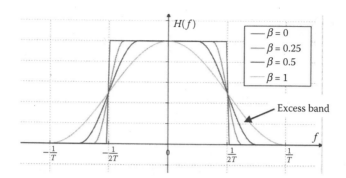

FIGURE 1.12
Frequency response of a raised-cosine filter with various values of the roll-off factor β.

be obtained by analytically taking the inverse Fourier transformation of Equation 1.2, in terms of the normalized *sinc* function, as

$$h(t) = sinc\left(\frac{t}{T_s}\right)\frac{\cos(\pi\beta t/T_s)}{1-(2(\pi\beta t/T_s))^2} \tag{1.3}$$

where the roll-off factor, β, is a measure of the *excess bandwidth* of the filter, that is, the bandwidth occupied beyond the Nyquist bandwidth as from the amplitude at $1/2T$. Figure 1.12 depicts the frequency spectra of a raised cosine pulse with various roll-off factors. Their corresponding time domain pulse shapes are given in Figure 1.13.

When used to filter a symbol stream, a Nyquist filter has the property of eliminating ISI, as its impulse response is zero at all nT (where n is an integer), except when $n=0$. Therefore, if the transmitted waveform is correctly sampled at the receiver, the original symbol values can be recovered completely. However, in many practical communications systems, a matched

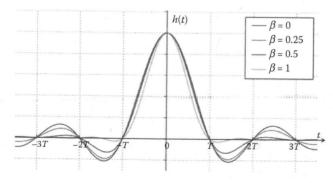

FIGURE 1.13
Impulse response of a raised-cosine filter with the roll-off factor β as a parameter.

filter is used at the receiver, so as to minimize the effects of noises. For zero ISI, the net response of the product of the transmitting and receiving filters must equate to $H(f)$, thus we can write:

$$H_R(f)H_T(f) = H(f) \qquad (1.4)$$

Or alternatively, we can rewrite that

$$|H_R(f)| = |H_T(f)| = \sqrt{|H(f)|} \qquad (1.5)$$

The filters that can satisfy the conditions of Equation 1.5 are the root-raised-cosine filters. The main problem with root-raised-cosine filters is that they occupy larger frequency bands than that of the Nyquist *sinc*-pulse sequence. Thus, for the transmission system we can split the overall raised cosine filter with the root-raise cosine filter at both the transmitting and receiving ends, provided the system is linear. This linearity is to be specified accordingly. An optical fiber transmission system can be considered linear if the total power of all channels is under the nonlinear SPM threshold limit. When it is over this threshold, a weakly linear approximation can be used.

The design of a Nyquist filter influences the performance of the overall transmission system. Oversampling factor, selection of roll-off factor for different modulation formats, and FIR Nyquist filter design are key parameters to be determined. If taking into account the transfer functions of the overall transmission channel, including fiber, WSS, and the cascade of the transfer functions of all O/E components, the total channel transfer function is more Gaussian-like. To compensate this effect in the Tx-DSP, one would thus need a special Nyquist filter to achieve the overall frequency response equivalent to that of the rectangular or raised cosine with roll-off factors shown in Figure 1.13. The spectra of data sequences for which pulse shapes follow a rectangle and a *sinc* function are shown in Figure 1.14a and b. The spectrum of a pulse sequence of the raised cosine function shows its close approximation to a *sinc* function. This will allow effective packing of adjacent information channels and transmission.

1.4 Optical Demodulation: Phase and Polarization Diversity Technique

A generic schematic of the transmission is depicted in Figure 1.7. The output-transmitted signals that are normally distorted are then detected by a digital optical receiver. The main function of this optical receiver is to recognize

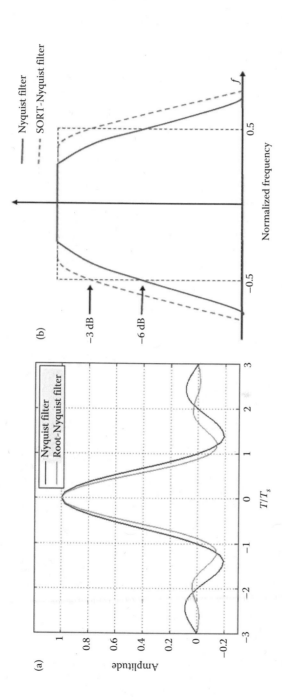

FIGURE 1.14

(a) Impulse and (b) corresponding frequency response of *sinc* Nyquist pulse shape or root-raise cosine (RRC) Nyquist filters.

whether the current received and therefore the "bit symbol" voltage at the output of the amplifiers following the detector is a "1" or "0." The modulation of amplitude, phase, or frequency of the optical carrier requires an optical demodulation. That is, the demodulation of the optical carrier is implemented in the optical domain. This is necessary due the extremely high frequency of the optical carrier (in order, or nearly 200 THz for 1550 nm wavelength); it is impossible to demodulate in the electronic domain by direct detection using a single photo-detector. The second most common technique is coherent detection by mixing the received signals with a local oscillator laser. The beating signal in photodetection with square law application would result in three components: one is the DC component, and the other two located at the summation and the difference of the two lightwave frequencies. Tone, thus, is very far away in the electrical domain and only the difference component would be detected in the electrical domain provided that this difference is within the bandwidth of the electronic detection and amplification. Indeed, it is quite straightforward to demodulate in the optical domain using optical interferometers to compare the phases of the carrier in two consecutive bits for the case of differential coding, which is commonly used to avoid demand on absolute stability of the lightwave carrier.

However, the phase and frequency of the lightwave signals can be recovered via an intermediate step by mixing the optical signals with a local oscillator, a narrow linewidth laser, to beat it to the baseband or an intermediate frequency region. This is known as the coherent detection technique. Figure 1.15 shows the schematics of optical receivers using direct detection and coherent detection. If both polarization modes of the fiber line are employed, then a 90° hybrid coupler would be used to split and mix the polarization of both the received channels and the local oscillator. Further, the optical frequency regions of the lightwaves employed for optical communications are indicated in Figure 1.16. In this case, the terms polarization and phase diversity coherent detection can be used. As we can see, 4 pairs of photodetectors are connected back to back as balanced detectors. They are required for detection of two polarized channels and two pairs of the in-phase and quadrature components of the QAM modulated channels. The received signals are sampled by an ultra-high speed analog to digital converter (ADC) and then processed in realtime by algorithms stored a DSP. Figure 1.17 shows generic flow diagram of the algorithm which are commonly employed in the digital processing of transmitted signals. Figure 1.18 shows the schematic diagram of a DSP-based coherent optical receiver in which both the analog and digital processing parts are included.

The main difference between these detection systems and those presented in several other textbooks is the electronic signal processing sub-system following the detection circuitry. In the first decade of this century we have witnessed tremendous progress in the speed of electronic ultra-large-scale integrated circuits, with the number of samples per second now reaching a few tens of GSa/s. This has permitted considerations for applications of DSP of distorted received

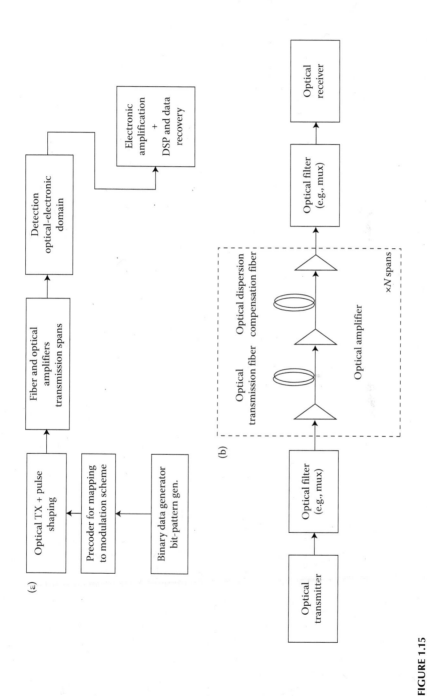

FIGURE 1.15

(a) Generalized diagram of dispersion-managed optical transmission systems; (b) more details of the dispersion management of multi-span link.

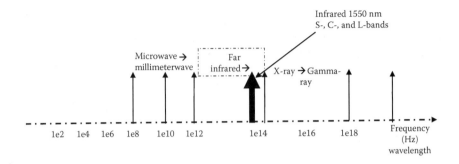

FIGURE 1.16
Electromagnetic spectrum of waves for communications, and lightwaves region for silica-based fiber optical communications.

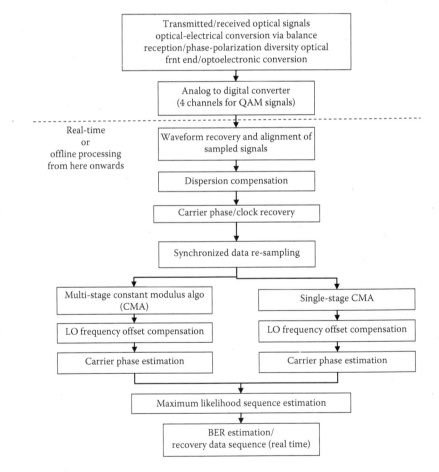

FIGURE 1.17
Flow chart of block schematic of the optical digital receiver and DSP.

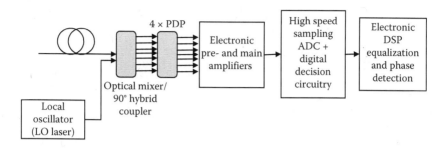

FIGURE 1.18
Schematics of optical receivers employing coherent detection and DSP. PDP = photodetector pair connected back to back.

optical signals in the electronic domain. Thus, flexibility in the equalization of signals in transmission systems and networks is very attractive.

1.5 Organization of the Book Chapters

The chapters of this book are dedicated to the latest development in research and practical systems to date. The presentation of this book follows the integration of optical components and digital modulation and DSP techniques in coherent optical communications in the following manner.

Chapter 2 briefly summarizes the fundamental properties of the waveguiding phenomena, especially the polarization modes and few mode aspects in optical fibers and essential parameters of such waveguides that would influence the transmission and propagation of optical modulated signals through the circular optical waveguide. This chapter presents the static parameters, including the index profile distribution and the geometrical structure of the fiber, the mode spot size and mode field diameter of optical fibers, and thence the estimation of the nonlinear self-phase modulation effects. Operational parameters such as group velocity, group velocity dispersion, and dispersion factor and dispersion slope of single-mode fibers as well as attenuation factors are also given. The frequency responses, including impulse and step responses, of optical fibers are also given, so that the chirping of an optically modulated signal can be understood from the point of view of phase evolution when propagating through an optical fiber, a quadratic phase modulation medium. The propagation equation, the nonlinear Schroedinger equation (NLSE) that represents the propagation of the complex envelope of the optical signals, is also described so that the modeling of the signal propagation can be related.

Chapter 4 describes the optical receiver configurations based on principles of coherent reception and the concepts of polarization, phase diversity, and

DSP technique. A local oscillator (LO) is required for mixing with the spectral components of the modulated channel to recover the signals back to the base band. Any jittering of the central frequency of the LO would degrade the system performance. The DSP algorithms in real time will recover the carrier phase, but only within a certain limit or tolerance of the carrier frequency. Thus, an optical phase locking may be required. This technique is presented in Chapter 5.

Chapter 6 outlines the principles of DSP and associated algorithms for dispersion compensation, carrier phase recovery, and nonlinear equalization by Volterra transfer functions and back propagation, Nyquist post filtering, and pre-filtering.

Chapter 7 then gives detailed designs, experimental and field demonstrations, and transmission performance of optical transmission systems employing DSP technique.

Chapter 8 introduces processing techniques in frequency domain, employing higher-order spectral techniques for both DSP-based coherent receivers and photonic processing, incorporating nonlinear optical waveguides for optical multi-dimensional spectrum identification.

References

1. C. Kao and G. Hockham, Dielectric-fibre surface waveguides for optical frequencies, *Proc. IEE*, 113(7), 1151–1158, 1966.
2. Maurer et al., *Fused silica optical waveguide.* U.S. Patent 3,659,915 (1972–05).
3. Keck, *IV Method of producing optical waveguide fibers.* U.S. Patent 3,711,262 (1973–01).

2

Optical Fibers: Guiding and Propagation Properties

2.1 Optical Fibers: Circular Optical Waveguides

2.1.1 General Aspects

Planar optical waveguides compose a guiding region, a slab imbedded between a substrate and a superstrate having identical or different refractive indices. The lightwaves are guided by the confinement of the lightwaves with oscillation solution. The number of oscillating solutions that satisfy the boundary constraints is the number of modes that can be guided. The guiding of lightwaves in an optical fiber is similar to that of the planar waveguide, except the lightwaves are guiding through a circular core embedded in circular cladding layer.

Within the context of this book, optical fibers would be most relevant as the circular optical waveguides that can support single mode with two polarized modes or few modes with different polarizations. We should point out the following development in optical fiber communications systems.

- Step-index and graded index multimode optical fibers find very limited applications in systems and networks for long-haul applications.

- Single-mode optical fibers have structured with very small difference in the refractive indices between the core and cladding regions. Thus, the guiding in modern optical fiber for telecommunications is called "weakling" guiding. This development was intensively debated and agreed upon by the optical fiber communications technology community during the late 1970s.

- The invention of optical amplification in rare-earth doped, single-mode optical fibers in the late 1980s has transformed the design and deployment of optical fiber communications systems and networks in the last decade and the coming decades of the twenty-first century. The optical loss of the fiber and the optical components in the optical networks can be compensated for by using these fiber in-line optical amplifiers.

- Therefore, the pulse broadening of optical signals during transmission and distribution in the networks becomes much more important for system design engineers.

- Recently, due to several demonstrations of the use of digital signal processing of coherently received modulated lightwaves, multiple input-multiple output (MIMO) techniques can be applied to enhance significantly the sensitivity of optical receivers and thus the transmission distance and the capacity of optical communication systems [1]. MIMO techniques would offer some possibilities of the uses of different guided modes through a single fiber, for example, few mode fibers that can support more than one mode but not too many, as in the case of multimode types. Thus the conditions under which circular optical waveguides can operate as a few-mode fibers are also described in this chapter.

Owing to the above development we shall focus the theoretical approach to the understanding of optical fibers on the practical aspects for designing optical fibers with minimum dispersion or for a specified dispersion factor. This can be carried out by, from practical measurements, the optical field distribution that would follow a Gaussian distribution. Knowing the field distribution, one would be able to obtain the propagation constant of the single guided mode, the spot size of this mode, and thus the energy concentration inside the core of the optical fiber. The basic concept of optical dispersion by using the definition of group velocity and group delay we would be able to derive the chromatic dispersion in single-mode optical fibers. After arming ourselves with the basic equations for dispersion we would be able to embark on the design of optical fibers with a specified dispersion factor.

2.1.2 Optical Fiber: General Properties

2.1.2.1 Geometrical Structures and Index Profile

An optical fiber consists of two concentric dielectric cylinders. The inner cylinder, or core, has a refractive index of $n(r)$ and radius a. The outer cylinder, or cladding, has index n_2 with $n(r) > n_2$ and a larger outer radius. A core of about 4–9 µm and a cladding diameter of 125 µm are the typical values for silica-based single-mode optical fiber. A schematic diagram of the structure of a circular optical fiber is shown in Figure 2.1. Figure 2.1a shows the core and cladding region of the circular fiber, while Figures 2.1b and 2.1c show the figure of the etched cross sections of a multimode and single mode, respectively. The silica fibers are etched in a hydroperoxide solution so that the core region doped with impurity would be etched faster than that of pure silica, thus the exposure of the core region as observed. Figure 2.2 shows the index

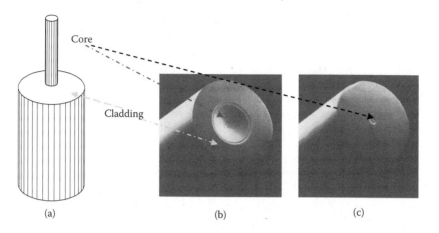

FIGURE 2.1

(a) Schematic diagram of the step-index fiber: coordinate system, structure. The refractive index of the core is uniform and slightly larger than that of the cladding. For silica glass, the refractive index of the core is about 1.478 and that of the cladding about 1.47 at 1550 nm wavelength region. (b) Cross-section of an etched fiber—multimode type—50 micrometer diameter. (c) Single-mode optical fiber etched cross-section.

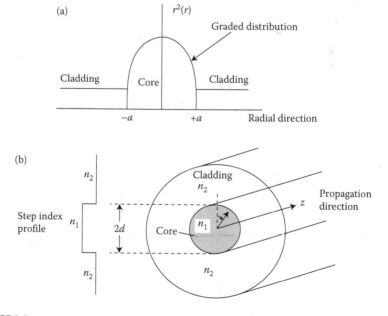

FIGURE 2.2

(a) Refractive index profile of a graded index profile; (b) fiber cross-section and step index profile with a as the radius of fiber.

profile and the structure of circular fibers. The refractive index profile can be step or graded.

The refractive index $n(r)$ of a circular optical waveguide is usually changed with radius r from the fiber axis ($r = 0$) and is expressed by

$$n^2(r) = n_2^2 + NA^2 s\left(\frac{r}{a}\right) \tag{2.1}$$

where NA is the numerical aperture at the core axis, while $s(r/a)$ represents the profile function that characterizes any profile shape ($s = 1$ at maximum) with a scaling parameter (usually the core radius).

For a step-index profile, the refractive index remains constant in the core region, thus

$$s\left(\frac{r}{a}\right) = \begin{cases} 1 & r \leq a \\ 0 & r > a \end{cases} \xrightarrow[\substack{hence \\ ref\_index}]{} n^2(r) = \begin{cases} n_1^2 & r \leq a \\ n_2^2 & r > a \end{cases} \tag{2.2}$$

For a graded-index profile, we can consider the two most common types of graded-index profiles: power-law index and the Gaussian profile.

For power-law index profile, the core refractive index of optical fiber is usually following a graded profile. In this case, the refractive index rises gradually from the value n_2 of the cladding glass to value n_1 at the fiber axis. Therefore, $s(r/a)$ can be expressed as

$$s\left(\frac{r}{a}\right) = \begin{cases} 1 - \left(\frac{r}{a}\right)^\alpha & \text{for } r \leq a \\ 0 & \text{for } r < a \end{cases} \tag{2.3}$$

with α as power exponent. Thus, the index profile distribution $n(r)$ can be expressed in the usual way, by using Equations 2.3 and 2.2, and substituting $NA^2 = n_1^2 - n_2^2$.

$$n^2(r) = \begin{cases} n_1^2 \left[1 - 2\Delta \left(\frac{r}{a}\right)^\alpha \right] & \text{for } r \leq a \\ n_2^2 & \text{for } r > a \end{cases} \tag{2.4}$$

$\Delta = (NA^2 / n_1^2)$ is the relative refractive difference with small difference between that of the cladding and the core regions. The profile shape given in Equation 2.4 offers three special distributions: (i) $\alpha = 1$: the profile function $s(r/a)$ is linear and the profile is called a triangular profile; (ii) $\alpha = 2$: the profile is a quadratic function with respect to the radial distance and the profile is called the parabolic profile; and (iii) $\alpha = \infty$; then the profile is a step type.

For Gaussian profile, the refractive index changes gradually from the core center to a distance very far away from it, and $s(r)$ can be expressed as

$$s\left(\frac{r}{a}\right) = e^{-\left(\frac{r}{a}\right)^2} \tag{2.5}$$

2.1.3 Fundamental Mode of Weakly Guiding Fibers

The electric and magnetic fields $E(r,\phi,z)$ and $H(r,\phi,z)$ of the optical fibers in cylindrical coordinates can be found by solving Maxwell's equations. Only the lower-order modes of ideal step index fibers are important for digital optical communication systems. The fact is that for $\Delta < 1\%$, the optical waves are confined very weakly and are thus gently guided. Thus, the electric and magnetic fields E and H can then take approximate solutions of the scalar wave equation in a cylindrical coordinate system (x,θ,ϕ),

$$\left[\frac{\delta^2}{\delta r^2} + \frac{1}{r}\frac{\delta}{\delta r} + k^2 n_j^2\right]\varphi(r) = \beta^2\varphi(r) \tag{2.6}$$

where $n_j = n_1, n_2$, and $\varphi(r)$ is the spatial field distribution of the nearly transverse EM waves.

$$E_x = \psi(r)e^{-i\beta z}$$

$$H_y = \left(\frac{\varepsilon}{\mu}\right)^{1/2}E_x = \frac{n_2}{Z_0}E_x \tag{2.7}$$

With E_y, E_z, H_x, H_z negligible, $\varepsilon = \varepsilon_0 n_2^2$ and $Z_0 = (\varepsilon_0\,\mu_0)^{1/2}$ is the vacuum impedance. We can assume that the waves can be seen as a plane wave travelling down along the fiber tube. This plane wave is reflected between the dielectric interfaces; in another word, it is trapped and guided along the core of the optical fiber. Note that the electric and magnetic components are spatially orthogonal with each other. Thus, for a single mode there are always two polarized components that are then the polarized modes of single-mode fiber. It is further noted that the Snell's law of reflection would not be applicable for single-mode propagation, but Maxell's equations. However, we will see in the next section that the field distribution of single-mode optical fibers follows closely to that of a Gaussian shape. Hence the solution of the wave Equation 2.6 can be assumed, and hence the eigenvalue or the propagation constant of the guided wave can be found or optimized to achieve the best fiber structure. However, currently due to the potentials of digital signal processing, the uses of the modes of multimode fibers can be beneficial, so few mode optical fibers are intensively investigated.

Thus, in the next section we give a brief analysis of the wave equations subject to the boundary conditions, so that the eigenvalue equation can be found, the propagation or wave number of the guided modes can be found, and hence the propagation delay of these group of lightwaves along the fiber transmission line. Then we will revisit the single-mode fiber with a Gaussian mode field profile to give insight into the weakly guiding phenomenon that is so important for the understanding of the guiding of lightwaves over very long distance with minimum loss and optimum dispersion, the group delay difference.

2.1.3.1 Solutions of the Wave Equation for Step-Index Fiber

The field spatial function $\phi(r)$ would have the form of Bessel functions given by Equation 2.6 as

$$\varphi(r) = \begin{cases} A\dfrac{J_0(ur/a)}{J_0(u)} & 0 < r < a\text{---core} \\[3mm] A\dfrac{K_0(vr/a)}{K_0(v)} & r > a\text{---cladding} \end{cases} \tag{2.8}$$

where J_0; K_0 are the Bessel functions of the first kind and modified of second kind, respectively, and u,v are defined as

$$\frac{u^2}{a^2} = k^2 n_1^2 - \beta^2 \tag{2.9}$$

$$\frac{v^2}{a^2} = -k^2 n_2^2 + \beta^2 \tag{2.10}$$

Thus, following the Maxwell's equations relation, we can find that E_z can take two possible orthogonal solutions,

$$E_z = -\frac{A}{kan_2}\begin{pmatrix} \sin\phi \\ \cos\phi \end{pmatrix}\begin{cases} \dfrac{uJ_1\left(u\dfrac{r}{a}\right)}{J_0(u)} & \text{for } 0 \le r < a \\[5mm] \dfrac{vK_1\left(\dfrac{vr}{a}\right)}{K_0(v)} & \text{for } r > a \end{cases} \tag{2.11}$$

The terms u and v must simultaneously satisfy two equations

$$u^2 + v^2 = V^2 = ka(n_1^2 - n_2^2)^{1/2} = kan_2(2\Delta)^{1/2} \qquad (2.12)$$

$$u\frac{J_1(u)}{J_0(u)} = v\frac{K_1(v)}{K_0(v)} \qquad (2.13)$$

where Equation 2.13 is obtained by applying the boundary conditions at the interface $r = a$ (E_z is the tangential component and must be continuous at this dielectric interface). Equation 2.13 is commonly known as the eigenvalue equation of the wave equation, bounded by the continuity at the boundary of the two dielectric media, hence the condition for guiding in the transverse plane such that the maximum or fastest propagation velocity in the axial direction. The solution of this equation would give specific discrete values of β, the propagation constants of the guided lightwaves.

2.1.3.2 Single and Few Mode Conditions

Over the years, since the demonstration of guiding in circular optical wave-guides, the eigenvalue Equation 2.13 is employed to find the number of modes supported by the waveguide and their specific propagation constants. Hence, the Gaussian mode spatial distribution can be approximated for the fundamental mode based on experimental measurement of the mode fields, and the eigenvalue equation would no longer needed when single-mode fiber (SMF) is extensively used. However, under current extensive research interests in the spatial multiplexing in DSP-based coherent optical communication systems, few mode fibers have attracted quite a lot of interest due to their potential supporting of many channels, with their modes and their related filed polarizations. This section is thus extended to consider both the fundamental mode and higher-order modes.

Equation 2.12 shows that the longitudinal field is in the order of $u/(kan_2)$, with respect to the transverse component. In practice, with $\Delta \ll 1$, by using Equation 2.12, we observe that this longitudinal component is negligible compared with the transverse component. Thus the guided mode is *transversely polarized*. The fundamental mode is then usually denominated as LP_{01} mode (linearly polarized mode) for which the field distribution is shown in Figures 2.5a and 2.5b. The graphical representation of the eigenvalue Equation 2.13 calculated as the variation of $b = (\beta/k)$ as the normalized propagation constant and the V-parameter is shown in Figure 2.6d. There are two possible polarized modes; the horizontal and vertical polarizations that are orthogonal to each other. These two polarized modes can be employed for transmission of different information channels. They are currently exploited, in the first two decades of this century, in optical transmission systems employing polarization division multiplexed so as to offer the transmission bit rate to 100 Gb/s and beyond. Furthermore, when the number of guided modes is higher than two polarized modes, they do form a set of modes over which information channels can be simultaneously carried and spatially demultiplexed at the receiving end, so as to increase the transmission capacity as illustrated in Figures 2.5a and 2.5b [2,3]. Such few mode fibers are employed in the most modern optical transmission system, for which the schematic is shown in Figure 2.3. Mode multiplexer acts as mode spatial mixing and likewise, the demultiplexer splits the modes LP_{11}, LP_{01} into individual modes. The demultiplexer then converts them to LP_{01} mode field and then injects them into SMF to feed into the coherent receiver. Obviously there must be mode spatial demultiplexing and then modulation and then multiplexing back into the transmission fiber for transmission. Similar structures would be available at the receiver to separate and detect the channels. Note the two possible polarizations of the mode LP_{11}. Note that there are 4 polarized modes of the LP_{11}

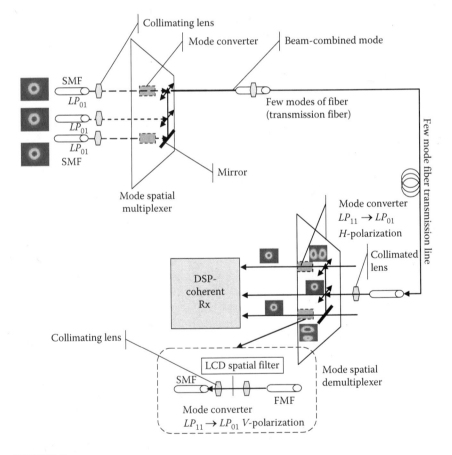

FIGURE 2.3
Mode fiber employed as a spatial multiplexing and demultiplexing in DSP-based coherent optical transmission systems operating at 100 Gb/s and higher bit rate.

mode; only two polarized modes are shown in this diagram. The delay due to the propagation velocity, from the difference in the propagation constant, can be easily compensated in the DSP processing algorithm, similar to that due to the polarization mode dispersion (PMD). The main problem to resolve in this spatial mode multiplexing optical transmission system is the optical amplification for all modes so that long-haul transmission can be achieved; that is, the amplification in multimode fiber structure (Figure 2.4).

The number of guided modes is determined by the number of intersecting points of the circle of radius V and the curves representing the eigenvalue solutions (2.13). Thus, for a single-mode fiber the V-parameter must be less than 2.405 and for few-mode fiber this value is higher. For example, if $V = 2.8$ we have three intersecting points between the circle of radius V and three curves, then the number of modes would be LP_{01}, LP_{11}, and their corresponding alternative polarized modes as shown in Figures 2.5b and 2.5c. The optic arrangement shown in Figure 2.5d is used to split the LP modes and their corresponding polarizations. For single mode there are two polarized modes whose polarizations can be vertical or horizontal. Thus, an SMF

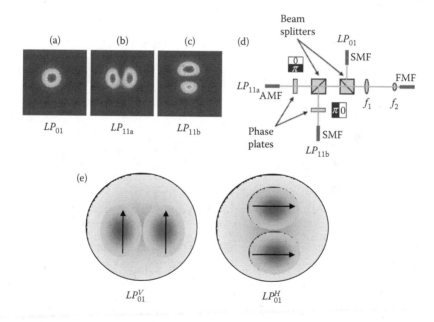

FIGURE 2.4

(a), (b), and (c) Intensity profiles of first few order modes of a few-mode optical fiber employed for 5×65 Gbps optical transmission system [4] and optical system arrangement for spatially demux and mux of modal channels. (d) Setup of optical system for spatial demultiplexing the LP modes and their polarizations. (e) Horizontal [H] and vertical [V] polarized modes $LP_{01}^{V,H}$. Amplitude mode field distribution for LP_{11} with H and V as horizontal and vertical polarization directions, respectively. Polarization directions indicated by arrows. (Adapted from S. Randel, *Optics Express*, 19(17), 16697, Aug 2011.)

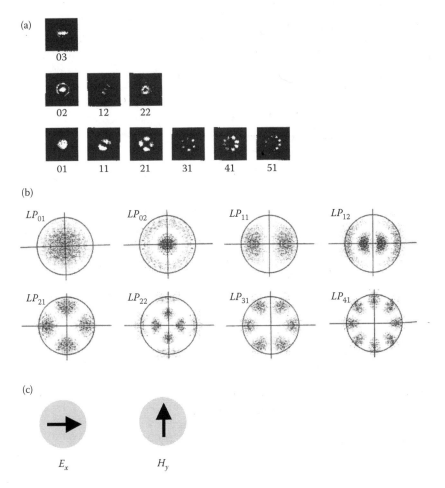

FIGURE 2.5
(a) Spectrum of guided modes in a few/multi-mode fiber, numbers indicate order of modes.
(b) Calculated intensity distribution of LP-guided modes in a step-index of optical fibers with
$V = 7$. (c) Electric and magnetic field distribution of an LP_{01} mode polarized along Ox (H-mode)
and Oy (V-mode) of the fundamental mode of a single-mode fiber.

is not a monomode but supports two polarized modes! The main issues are
also on the optical amplification gain for the transmission of modulated signals in such few mode fibers. This remains to be the principal obstacle.

We can illustrate the propagation of the fundamental mode and higher-
order modes as in Figures 2.6a and 2.6b. The rays of these modes can be
axially straight or skewed and twisted around the principal axis of the fiber.
Thus, there are different propagation times between these modes. This
property can be employed to compensate for chromatic dispersion effect [5].
Figure 2.6 shows a spectrum of the graphical solution of the modes of optical
fibers. In Figure 2.6d, the regions of single operation and then a higher order,

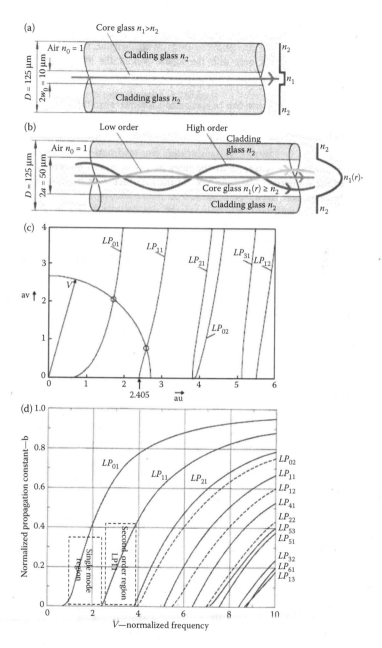

FIGURE 2.6

(a) Guided modes as seen by "a ray" in the transverse plane of a circular optical fiber; "Ray" model of lightwave propagating in single-mode fiber. (b) Ray model of propagation of different modes guided in a few/multi-mode graded-index fiber. (c) Graphical illustration of solutions for eigenvalues (propagation constant—wave number of optical fibers). (d) b-V characteristics of guided fibers.

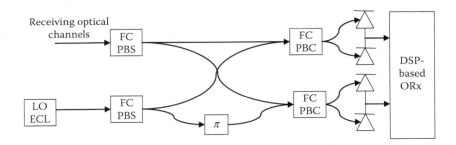

FIGURE 2.7
$\pi/2$ hybrid coupler for polarization demultiplexing and mixing with local oscillator in a coherent receiver of modern DSP-based optical receiver for detection of phase-modulated schemes.

second-order mode regions as determined by the value of the *V*-parameter are indicated. Naturally, due to manufacturing accuracy, the mode regions would be variable from fiber to fiber (Figure 2.7).

2.1.3.3 Gaussian Approximation: Fundamental Mode Revisited

We note again that the **E** and **H** are approximate solutions of the scalar wave equation and the main properties of the fundamental mode of weakly guiding fibers that can be observed as follows:

- The propagation constant β (in *z*-direction) of the fundamental mode must lie between the core and cladding wave numbers. This means the effective refractive index of the guided mode lies within the range of the cladding and core refractive indices.
- Accordingly, the fundamental mode must be a nearly transverse electro-magnetic wave as described by Equation 2.7.

$$\frac{2\pi n_2}{\lambda} < \beta < \frac{2\pi n_1}{\lambda} \tag{2.14}$$

- The spatial dependence $\psi(r)$ is a solution of the scalar wave Equation 2.6.

The main objectives are to find a good approximation for the field $\psi(r)$ and the propagation constant β. This can be found though the eigenvalue equation and the Bessel's solutions as shown in previous section. It is desirable if we can approximate the field to a good accurate number to obtain simple expressions to have a clearer understanding of light transmission on single-mode optical fiber without going through graphical or numerical methods. Furthermore, experimental measurements and numerical solutions for step and power-law profiles show that $\psi(r)$ is approximately

Gaussian in appearance. We thus approximate the field of the fundamental mode as

$$\varphi(r) \cong Ae^{-\frac{1}{2}\left(\frac{r}{r_0}\right)^2}$$ (2.15)

where r_0 is defined as the spot size, that is, at which the intensity equals to e^{-1} of the maximum. Thus, if the wave Equation 2.6 is multiplied by $r\psi(r)$ and using the identity

$$r\varphi\frac{\delta^2\varphi}{\delta r^2} + \varphi\frac{\delta\varphi}{\delta r} = \frac{\delta}{\delta r}\left(r\varphi\frac{\delta\varphi}{\delta r}\right) - r\left(\frac{\delta\varphi}{\delta r}\right)^2$$ (2.16)

then, by integrating from 0 to infinitive and using $[r\Psi(d\varphi/dr)]_0^\infty = 0$ we have

$$\beta^2 = \frac{\int_0^\infty \left[-\left(\frac{\delta\varphi}{\delta r}\right)^2 + k^2 n^2(r)\varphi^2\right] r\delta r}{\int_0^\infty r\varphi^2\delta r}$$ (2.17)

The procedure to find the spot size is then followed by substituting $\psi(r)$ (Gaussian) in Equation 2.15 into 2.17, then differentiating and setting $(\delta^2\beta/\delta r)$ evaluated at r_0 to zero; that is, the propagation constant β of the fundamental mode *must* give the largest value of r_0. Therefore, knowing r_0 and β, the fields E_x and H_y (Equation 2.7) are fully specified.

2.1.3.3.1 Step-Index Profile

Substituting the step-index profile given by Equation 2.7 and $\psi(r)$ into Equation 2.15 and then Equation 2.17 leads to an expression for β in terms of r_0 given by

$$V = NA \cdot k \cdot a = NA\frac{2\pi}{\lambda}a$$ (2.18)

The spot size is thus evaluated by setting

$$\frac{\delta^2\beta}{\delta r_0} = 0$$ (2.19)

and r_0 is then given by

$$r_0^2 = \frac{a^2}{\ln V^2}$$ (2.20)

Substituting Equation 2.20 into Equation 2.18 we have

$$(a\beta)^2 = (akn_1)^2 - \ln V^2 - 1 \qquad (2.21)$$

This expression is physically meaningful only when $V > 1$; that is, when r_0 is positive which is naturally feasible.

2.1.3.3.2 Gaussian Index Profile Fiber

Similarly, for the case of a Gaussian index profile, by following the procedures for step-index profile fiber we can obtain

$$(a\beta)^2 = (an_1 k)^2 - \left(\frac{a}{r_0}\right)^2 + \frac{V^2}{\left(\frac{a}{r_0} + 1\right)} \qquad (2.22)$$

and

$$r_0^2 = \frac{a^2}{V - 1} \text{ by using } \frac{\delta^2 \beta}{\delta^2 r_0} = 0 \qquad (2.23)$$

That is maximizing the propagation constant of the guided waves. The propagation constant is at maximum when the "light ray" is very close to the horizontal direction. Substituting Equation 2.23 into Equation 2.22 we have

$$(a\beta)^2 = (akn_1)^2 - 2V + 1 \qquad (2.24)$$

Thus, Equations 2.23 and 2.24 are physically meaningful only when $V > 1$ ($r_0 > 0$).

It is obvious from Equation 2.25 that the spot size of the optical fiber with a V-parameter of 1 is extremely large. This is very important; one must not design the optical fiber with a near-unit value of the V-parameter, hence under this scenario all the optical field is distributed in the cladding region. In practice, we observe that the spot size is large but finite (observable). In fact, if V is smaller than 1.5, the spot size becomes large. This occurrence will be investigated in detail in the next chapter.

2.1.3.4 Cut-Off Properties

Similar to the case of planar dielectric waveguides, in Figure 2.6 we observe that when we have $V < 2.405$, only the fundamental LP_{01} exists. Thus, we have the field and intensity distribution of this guided mode across the fiber cross section as shown in Figure 2.8a. Figure 2.8b also shows the variation of this fundamental mode as a function of the V-parameter. Obviously, the

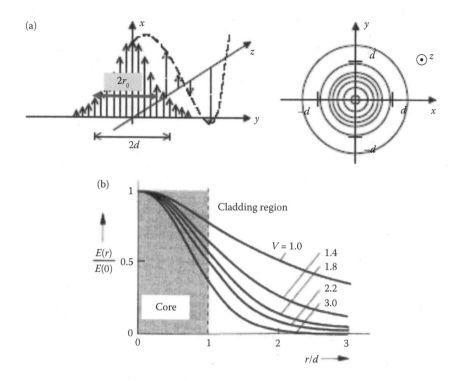

FIGURE 2.8
(a) Intensity distribution of the LP01 mode in radial direction (left) and in contour profile in step of 20% (right). (b) Variation of the spot size–field distribution with radial distance r with V as a parameter.

smaller the V-parameter, this narrower the mode distribution. There must be an optimum mode distribution so that the transmission of the modulated optical signals can suffer the least dispersion and attenuation effects.

It is noted that for single-mode operation the V parameter must be less than or equal to 2.405. However, in practice $V < 3$ may be acceptable for single-mode operation. Indeed, the value 2.405 is the first zero of the Bessel function $J_0(u)$. In practice, one cannot really distinguish between the V value between 2.3 and 3.0. Experimental observation also shows that the optical fiber can still support only one mode. Thus, designers do usually take the value of V as 3.0 or less to design a single-mode optical fiber.

The V parameter is inversely proportional with respect to the optical wavelength, which is directly related to the operating frequency. Thus, if an optical fiber is launched with lightwaves with optical wavelengths smaller than the operating wavelength at which the optical fiber is single mode, then the optical fiber is supporting more than one mode. The optical fiber is said to be operating in a few regions and then multi-mode region when the total number of modes reaches few hundreds.

Thus, one can define the cut-off wavelength for optical fibers as follows: the wavelength (λ_c), *above which* only the fundamental mode is guided in the fiber, is called the *cut-off wavelength* λ_c. This cut-off wavelength can be found by using the *V*-parameter of $V_C = V|_{cut\text{-}off} = 2.405$, thus

$$\lambda_c = \frac{2\pi a NA}{V_c} \tag{2.26}$$

In practice, the fibers tend to be effectively single mode for larger values of *V*, say $V < 3$, for the step profile, because the higher-order modes suffer radiation losses due to fiber imperfections. Thus if $V = 3$, from Equation 2.12 we have a $< 3\lambda/2\ NA$, in this case that $\lambda = 1\ \mu m$ and the numerical aperture *NA* must be very small ($\ll 1$) for radius *a* to have some reasonable dimension. Usually Δ is about 1% or less for standard single-mode optical fibers (SSMF) employed in long-haul optical transmission systems, so as to minimize the loss factor and the dispersion.

2.1.3.5 Power Distribution

The axial power density or intensity profile $S(r)$, the *z*-component of the Poynting's vector, is given by

$$S(r) = \frac{1}{2} E_x H_y^* \tag{2.27}$$

Substituting Equation 2.7 into Equation 2.27 we have

$$S(r) = \frac{1}{2} \left(\frac{\varepsilon}{\mu} \right)^{1/2} e^{-\left(\frac{r}{r_0} \right)^2} \tag{2.28}$$

The total power is then given by

$$P = 2\pi \int_0^\infty r S(r) dr = \frac{1}{2} \left(\frac{\varepsilon}{\mu} \right)^{1/2} r_0^2 \tag{2.29}$$

and hence the fraction of power $\eta(r)$ within $0 \to r$ across the fiber cross section is given by

$$\eta(r) = \frac{\int_0^r r S(r) dr}{\int_0^\infty r S(r) dr} = 1 - e^{-\left(\frac{r^2}{r_0^2} \right)} \tag{2.30}$$

Thus, given a step profile or approximated profile one can either analytically or numerical estimate the power confined in the core region and the propagation constant, hence the phase velocity of the guided mode. Experimentally or in production, the mode spot size can be obtained from the digital image from an infrared camera, and then using Equations 2.29 and 2.30 the power of the mode confined inside the core can be obtained.

2.1.3.6 Approximation of Spot-Size r_0 of a Step-Index Fiber

As stated above, spot-size r_0 would play a major role in determining the performance of single-mode fiber. It is useful if we can approximate the spot size as long as the fiber is operating over a certain wavelength. When a single-mode fiber is operating above the cut-off wavelength, a good approximation (>96% accuracy) for r_0 is given by

$$\frac{r_0}{a} = 0.65 + 1.619\, V^{-3/2} + 2.879\, V^{-6} = 0.65 + 0.434 \left(\frac{\lambda}{\lambda_c} \right)^{+3/2} + 0.0419 \left(\frac{\lambda}{\lambda_c} \right)^{+6}$$

$$\text{for} \quad 0.8 \leq \frac{\lambda}{\lambda_c} \leq 2.0 \ \text{single mode} \tag{2.31}$$

2.1.4 Equivalent-Step Index Description

As we can observe, there are two possible orthogonally polarized modes (E_x, Hy) and (E_y, H_x) that can be propagating *simultaneously*. The superposition of these modes can usually be approximated by a single linearly polarized (LP) mode. These modes' properties are well known and well understood for step-index optical fibers, and analytical solutions are also readily available.

Unfortunately, practical SM optical fibers never have perfect step-index profile due to the variation of the dopant diffusion and polarization. These nonstep index fibers can be approximated, under some special conditions, by the *equivalent-step index (ESI)* profile technique.

A number of index profiles of modern single-mode fibers, for example, non-zero dispersion shifted fibers, is shown in Figure 2.9. The ESI profile is determined by approximating the fundamental mode electric field spatial distribution $\psi(r)$ by a Gaussian function as described above. The electric field can thus be totally specified by the e^{-1} width of this function or mode spot size (r_0). Alternatively, the term mode field diameter (MFD) is also used and equivalent to twice the size of the mode spot size r_0. The ESI method is important in practice since the measured refractive index profile of manufactured fibers would never follow the ideal geometrical profile, so the ESI is applied to reconstruct it to an equivalent ideal distribution so that the analytical estimation can be used to confirm with that obtained by experimental measurement (Figure 2.9).

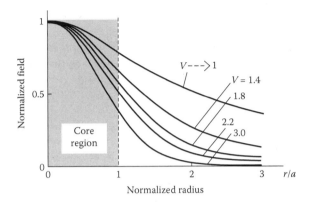

FIGURE 2.9
Index profiles of a number of modern fibers, for example, dispersion-shifted SMFs.

2.2 Nonlinear Optical Effects

In this section, the nonlinear effects on the guided lightwaves propagating through a long length of optical fibers, the single-mode type, are described. These effects play important roles in the transmission of optical pulses along single-mode optical fibers as distortion due to the modification of the phase of the lightwaves. The nonlinear effects can be classified into three types: the effects that change the refractive index of the guided medium due to the intensity of the pulse, the self-phase modulation; the scattering of the lightwave to other frequency-shifted optical waves when the intensity reaches over a certain threshold, the Brillouin and Raman scattering phenomena; and the mixing of optical waves to generate a fourth wave, the degenerate four-wave mixing. Besides these nonlinear effects there is also photorefractive effect, which is due to the change of refractive index of silica due to the intensity of ultra-violet optical waves. This phenomenon is used to fabricate grating with spacing between dark and bright regions satisfying the Bragg diffraction condition. These are fiber Bragg gratings and would be used as optical filters and dispersion compensators when the spacing varies or chirps.

In modern coherent optical communication systems incorporating digital signal processors at the receiver, the compensation can be done in the electronic domain and back propagation (BP) of the lightwaves can be implemented to reverse the nonlinear effects imposing on the phase of the guided mode using the frequency domain transfer function [6–8].

2.2.1 Nonlinear Self-Phase Modulation Effects

All optical transparent materials are subject to the change of the refractive index with the intensity of the optical waves, the optical Kerr effect. This physical phenomenon is originated from the harmonic responses of electrons of optical fields leading to the change of the material susceptibility. The modified refractive index $n_{1,2}^K$ of the core and cladding regions of the silica-based material can be written as

$$n_{1,2}^K = n_{1,2} + \overline{n_2} \frac{P}{A_{\text{eff}}} \tag{2.32}$$

where n_2 is the nonlinear index coefficient of the guided medium, the average typical value of n_2 is about 2.6×10^{-20} m²/W. P is the average optical power of the pulse and A_{eff} is the effective area of the guided mode. The nonlinear index changes with the doping materials in the core. Although this nonlinear index coefficient is small, but the effective area is also very small, about 50–70 µm², and the fiber transmission length is very long, so according to Equation 2.32 the accumulated phase change is not negligible over this distance. This leads to the self-phase modulation (SPM) and cross-phase modulation (XPM) effects in the optical channels.

2.2.2 Self-Phase Modulation

The origin of the SPM is due to the phase variation of the guided lightwaves exerted by the intensity of its own power or field accumulated along the propagation path which is quite long, possibly a few hundreds to thousands of kilometers. Under a linear approximation we can write the modified propagation constant of the guided LP mode in a single-mode optical fiber as

$$\beta^K = \beta + k_0 \overline{n_2} \frac{P}{A_{\text{eff}}} = \beta + \gamma P \quad \text{where } \gamma = \frac{2\pi \overline{n_2}}{\lambda A_{\text{eff}}} \tag{2.33}$$

where n_2 and γ are the nonlinear coefficient and parameter of the guided medium taking, respectively, and effective values of 2.3×10^{-23} m⁻² and from 1 to 5 (kmW)⁻¹, depending on the effective area of the guided mode and the operating wavelength. Thus, the smaller the mode spot size or MFD, the larger the nonlinear SPM effect. For dispersion-compensating fiber, the effective area is about 15 µm² while for SSMF and NZ-DSF the effective area ranges from 50 to 80 µm². Thus, the nonlinear threshold power of DCF is much lower than that of SSMF and NZ-DSF. The maximum launch power into DCF would be limited at about 0 dBm or 1.0 mW in order to avoid nonlinear distortion effect while about 5 dBm for SSMF.

The accumulated nonlinear phase changes due to the nonlinear Kerr effect over the propagation length L are given by

$$\phi_{NL} = \int_0^L (\beta^K - \beta)dz = \int_0^L \gamma P(z)dz = \gamma P_{in} L_{eff} \quad \text{with } P(z) = P_{in}e^{-\alpha z} \tag{2.34}$$

This equation represents the phase change under nonlinear effects over a length L along the propagation direction z. When considering that the nonlinear SPM effect is small compared to the linear chromatic dispersion effect, one can set $\phi_{NL} \gg 1$ or $\phi_{NL} = 0.1$ rad. and the effective length of the propagating fiber is set at $L_{eff} = 1/\alpha$ with optical losses equalized by cascaded optical amplification sub-systems. Then the maximum input power to be launched into the fiber can be set at

$$P_{in} < \frac{0.1\alpha}{\gamma N_A} \tag{2.35}$$

For $\gamma = 2$ (W·km)$^{-1}$ and $N_A = 10$, $\alpha = 0.2$ dB/km (or 0.0434×0.2 km^{-1}), then $P_{in} < 2.2$ mW or about 3 dBm. Similarly, this threshold level is about 1 mW for DCF, with an effective area of 15 μm^2. In practice, due to the randomness of the arrival "1" and "0," this nonlinear threshold input power can be set at about 10 dBm, as the total average power of all wavelength multiplexed channels of WDM transmission systems launched into the fiber link.

2.2.3 Cross-Phase Modulation

The change of the refractive index of the guided medium as a function of the intensity of the optical signals can also lead to the phase of optical channels in different spectral regions close to that of the original channel. This is known as cross phase modulation effect (XPM), which is critical in wavelength-division multiplexed (WDM) channels, and even more critical in dense WDM when the frequency spacing between channels is 50 GHz or even narrower, as the cross-interference between channels can generate unwanted noises in the optical domain, leading to the detected electronic signals at the receiver. In such systems, the nonlinear phase shift of a particular channel depends not only on its power but also those of other multiplexed channels. The phase shift of the ith channel can be written as [9]:

$$\phi_{NL}^i = \gamma L_{eff} \left(P_{in}^i + 2 \sum_{j \neq i}^{M} P_j \right) \quad \text{with } M = \text{number of multiplexed channels} \tag{2.36}$$

The factor of 2 in Equation 2.36 is due to the bipolar effects of the susceptibility of silica materials and the total phase noises that are integrated over both sides of the channel spectrum. The XPM thus depends on the bit pattern and the randomness of the synchronous arrival of the "1." It is hard to estimate analytically, so numerical simulations would normally be employed to obtain the XPM distortion effects using the nonlinear Schroedinger wave propagation equation involving the signal envelopes of all channels. The

FIGURE 2.10
Illustration of XPM effects—phase modulation conversion to amplitude modulation and hence interference between adjacent channel.

evolution of slow-varying complex envelopes $A(z,t)$ of optical pulses along a single-mode optical fiber is governed by the nonlinear Schroedinger equation (NLSE) [7]:

$$\frac{\partial A(z,t)}{\partial z} + \frac{\alpha}{2} A(z,t) + \beta_1 \frac{\partial A(z,t)}{\partial t} + \frac{j}{2} \beta_2 \frac{\partial^2 A(z,t)}{\partial t^2} - \frac{1}{6} \beta_3 \frac{\partial^3 A(z,t)}{\partial t^3}$$

$$= -j\gamma |A(z,t)|^2 A(z,t) \tag{2.37}$$

where z is the spatial longitudinal coordinate, α accounts for fiber attenuation, β_1 indicates DGD, β_2 and β_3 represent second- and third-order factors of fiber CD, and γ is the nonlinear coefficient. This equation is derived from Maxwell's equations under external perturbation.

The phase modulation due to nonlinear phase effects is then converted to amplitude modulation and therefore the cross-talk to other adjacent channels. This is shown in Figure 2.10.

2.2.4 Stimulated Scattering Effects

Scattering of lightwaves by impurities can happen due to the absorption and vibration of the electrons and dislocation of molecules in silica-based materials. The back scattering and absorption is commonly known as Raleigh scattering losses in fiber propagation, in which phenomena of the frequency of the optical carrier does not change. Other scattering processes in which the frequency of the lightwave carrier is shifted to another spectral regions are commonly known as inelastic scattering, Raman scattering, or Brillouin scattering. In both cases, the scattering of photons to a lower energy-level photon with energy difference between these levels is fallen with the energy of phonons. Optical phonons are resulted from the electronic vibration for Raman scattering, while acoustic phonons or mechanical vibration of the linkage between molecules lead to Brillouin scattering. At high power, when the intensity reaches over a certain threshold, and the number of scattered photons is exponentially grown, then the phenomena is a simulated process. Thus, the phenomena can be called stimulated Brillouin scattering (SBS) and stimulated Raman scattering (SRS). SRS and SBS were first observed in the 1970s [10–12].

2.2.4.1 Stimulated Brillouin Scattering

Brillouin scattering comes from the compression of the silica materials in the presence of an electric field, the electrostriction effect. Under the pumping of an oscillating electric field of frequency, f_p, an acoustic wave of frequency F_a is generated. Spontaneous scattering is an energy transfer from the pump wave to the acoustic wave, and then a phase matching to transfer a frequency-shifted optical wave of frequency as a sum of the optical signal waves and the acoustic wave. This acoustic wave frequency shift is around 11 GHz with a bandwidth of around 50–100 MHz (due to the gain coefficient of the SBS) and a beating envelope would be modulating the optical signals. Thus, jittering of the received signals at the receiver would be formed, hence the closure of the eye diagram in the time domain.

Once the acoustics wave is generated, it beats with the signal waves to generate the side band components. This beating beam acts as a source and further transfers the signal beam energy into the acoustic wave energy, amplifying this wave to generate further jittering effects. The Brillouin scattering process can be expressed by the following coupled equations [13]:

$$\frac{dI_p}{dz} = -g_B I_p I_s - \alpha_p I_p$$

$$-\frac{dI_s}{dz} = +g_B I_p I_s - \alpha_s I_s$$

(2.38)

The SBS gain g_B is frequency–dependent, with a gain bandwidth of around 50–100 MHz for pump wavelength at around 1550 nm. For silica fiber, g_B is about 5e–11 mW^{-1}. The threshold power for the generation of SBS can be estimated (using Equation 2.38) as

$$g_B P_{th\_SBS} \frac{L_{eff}}{A_{eff}} \approx 21 \quad \text{with the effective length} \quad L_{eff} = \frac{1 - e^{-\alpha L}}{\alpha}$$

(2.39)

where
  I_p = intensity of pump beam
  I_s = intensity of signal beam
  g_B = Brillouin scattering gain coefficient
  α_s, α_p = losses of signal and pump waves

For the standard single-mode optical fiber (SSMF), this SBS power threshold is about 1.0 mW. Once the launched power exceeds this power threshold level, the beam energy is reflected back. Thus the average launched power is usually limited to a few dBm due to this low-threshold power level.

2.2.4.2 Stimulated Raman Scattering

Stimulated Raman scattering (SRS) occurs in silica-based fiber when a pump laser source is launched into the guided medium, and the scattering light from the molecules and dopants in the core region are shifted to a higher energy level and then jump down to a lower energy level, hence amplification of photons in this level. Thus a transfer of energy from different frequency- and energy-level photons occurs. The stimulated emission happens when the pump energy level reaches above the threshold level. The pump intensity and signal beam intensity are coupled via the coupled equations:

$$\frac{dI_p}{dz} = -g_R I_p I_s - \alpha_p I_p$$

$$-\frac{dI_s}{dz} = +g_R I_p I_s - \alpha_s I_s$$

(2.40)

where

I_p = intensity of pump beam
I_s = intensity of signal beam
g_R = Raman scattering gain coefficient
α_s, α_p = losses of signal and pump waves

The spectrum of the Raman gain depends on the decay lifetime of the excited electronic vibration state. The decay time is in the range of 1 ns and Raman gain bandwidth is about 1 GHz. In single-mode optical fibers the bandwidth of the Raman gain is about 10 THz. The pump beam wavelength is usually about 100 nm below the amplification wavelength region. Thus, in order to extend the gain spectra, a number of pump sources of different wavelengths are used. Polarization multiplexing of these beams is also used to reduce the effective power launched in the fiber so as to avoid the damage of the fiber. The threshold for stimulated Raman gain is given by

$$g_R P_{th\_SRS} \frac{L_{eff}}{A_{eff}} \approx 16 \quad \text{with the effective length } L_{eff} = \frac{1 - e^{-\alpha L}}{\alpha} \quad \text{or}$$

(2.41)

$$\approx 1/\alpha \quad \text{for long length}$$

For SSMF with an effective area of 50 μm², $g_R \sim$ 1e–13 m/W, then the threshold power is about 570 mW near the C-band spectral region. This would require at least two pump laser sources, which should be polarization multiplexed. The distributed amplification of SRS offers significant advantages as compared with lumped amplifiers such EDFA and SRS is used frequently in modern optical communications systems, especially when no undersea optical amplification is required. The broadband gain and low-gain ripple of SRS is also another advantage for DWDM transmission.

2.2.4.3 Four-Wave Mixing Effects

Four-wave mixing (FWM) is considered a scattering process in which three photons are mixed to generate the fourth wave. This happens when the momentum of the four waves satisfies a phase- matching condition. That is the condition of maximum power transfer. Figure 2.11 illustrates the mixing of different wavelength channels to generate inter-channel cross talk. The phase matching can be represented by a relationship between the propagation constant along the z-direction in a single-mode optical fiber, as

$$\beta(\omega_1) + \beta(\omega_2) - \beta(\omega_3) - \beta(\omega_4) = \Delta(\omega) \tag{2.42}$$

with ω_1, ω_2, ω_3, ω_4 representing the frequencies of the 1st to 4th waves, and Δ representing the phase mismatching parameter. In the case that the channels are equally spaced with a frequency spacing of Ω, as in DWDM optical transmission, thus $\omega_1 = \omega_2$; $\omega_3 = \omega_1 + \Omega$; $\omega_4 = \omega_1 - \Omega$. One can use the Taylor-

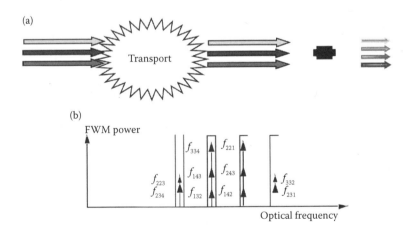

FIGURE 2.11
Illustration of FWM of optical channels; (a) momentum vectors of channels, and (b) frequencies resulted from mixing of different channels.

series expansion around the propagation constant at the center frequency of the guide carrier β_0. Then we can obtain [14]:

$$\Delta(\omega) = \beta_2 \Omega^2 \qquad (2.43)$$

The phase matching is thus optimized when β_2 is null, indicating that in the region where there is no dispersion, FWM is largest, and hence there is maximum inter-channel crosstalk. This is the reason why dispersion-shifted fiber is not commonly used when the zero dispersion wavelength is fallen in the spectral region of operation of channel. Instead, nonzero dispersion-shifted fibers are used in which dispersion-zero wavelengths are shifted away from the spectral region of the active channels so that there exists some dispersion value that would make the FWM condition not satisfied, and result in minimum generation of the fourth waves. In modern transmission fiber, the zero dispersion wavelength is shifted to outside the C-band, say 1510 nm, so that there is a small dispersion factor at 1550 nm and the C-band ranging from 2 to 6 ps/nm km; for example, Corning LEAF or nonzero dispersion-shifted fibers (NZ-DSF). This small amount of dispersion is sufficient to avoid the FWM with a channel spacing of 100 GHz or 50 GHz.

The XPM signal is proportional to instantaneous signal power. Its distribution is bounded by <5 channels and otherwise effectively unbounded. Thus, the link budgets include XPM evaluated at maximum outer bounds.

2.3 Signal Attenuation in Optical Fibers

Optical loss in optical fibers is one of the two main fundamental limiting factors, as it reduces the average optical power reaching the receiver. The optical loss is the sum of three major components: intrinsic loss, microbending loss, and splicing loss.

2.3.1 Intrinsic or Material Absorption Losses

Intrinsic loss consists mainly of absorption loss due to OH impurities and Rayleigh scattering loss. The intrinsic is a function of λ^{-6}. Thus, the longer the operating wavelength, the lower the loss. However it also depends on the transparency of the optical materials that are used to form the optical fibers. For silica fiber the optical material loss is low over the wavelength range 0.8–1.8 μm. Over this wavelength range there are three optical windows in which optical communication is utilized. The first window over the central wavelength 810 nm is about 20.0 nm spectral window over the central wave-length. The second and third windows are most commonly used in present optical communications over 1300 and 1550 nm, with a range of 80 and 40 nm, respectively. The intrinsic losses are about 0.3 and 0.15 dB/km at 1550 and 1300 nm regions, respectively.

This is a few hundred thousand times improvement over the original transmission of signal over 5.0 m, with a loss of about 60 dB/km. Most communication fiber systems are operating at 1300 nm due to its minimum dispersion at this range. For "power-hungry" systems, optical or extra-long systems should operate at 1550 nm.

The absorption loss in silica glass is composed mainly of ultraviolet (UV) and infra-red (IR) absorption tales of pure silica. The IR absorption tale of pure silica has been shown due to the vibration of the basic tetrahedron and thus strong resonance occurs around 8–13 μm, with a loss about 10^{-10} dB/km. This loss is shown in curve IR of Figure 2.1. Overtones and combinations of these vibrations lead to various absorption peaks in the low wavelength range, as shown by curve UV.

Various impurities that also lead to spurious absorption effects in the wavelength range of interest (1.2–1.6 μm) are transition metal ions and water in the form of OH ions. These sources of absorption have been reduced in recent years.

The Raleigh scattering loss, L_R, which is due to microscopic nonhomogeneities of the material, shows a λ^{-4} dependence and is given by

$$L_R = (0.75 + 4.5\Delta)\lambda^{-4} \text{ dB/km} \tag{2.44}$$

where Δ is the relative index difference as defined above and λ is the wavelength in μm. Thus to minimize the loss, Δ should be made as low as possible.

2.3.2 Waveguide Losses

The losses due to waveguide structure arise from power leakage, bending, microbending of the fiber axis, and defects and joints between fibers. The power leakage is significant only for depressed cladding fibers.

When a fiber is bent, the plane wave fronts associated with the guided mode are pivoted at the center of curvature and their longitudinal velocity along the fiber axis increases with the distance from the center of curvature. As the fiber is bent further over a critical curve, the phase velocity would exceed that of plane wave in the cladding, and radiation occurs.

The bending loss L_B for a radius R, the radius of curvature, is given by

$$L_B = -10\log_{10}(1 - 890)\frac{r_0^6}{\lambda^4 R^2} \quad \text{for silica} \tag{2.45}$$

Microbending loss results from power coupling from the guided fundamental mode of the fiber to radiation modes. This coupling takes place when the fiber axis is bent randomly in a high spatial frequency. Such bending can occur during packing of the fiber during the cabling process. The microbending loss of an SM fiber is a function of the fundamental mode spot size r_0. Fibers with large spot size are extremely sensitive to microbending. It is therefore desirable to design the fiber to have as small a spot size as possible to minimize bending loss. The microbending loss can be expressed by the relation (Figure 2.12)

$$L_m = 2.15 \times 10^{-4} r_0^6 \lambda^{-4} L_{mm} \text{ dB/km} \tag{2.46}$$

where L_{mm} is the microbending loss of a 50 μm core multimode fiber having an NA of 0.2.

Ultimately, the fibers will have to be spliced together to form the final transmission link. With fiber cable that averages 0.4–0.6 dB/km, splice loss in excess of 0.2 dB/splice drastically reduces the nonrepeated distance that can be achieved. It is therefore extremely important that the fiber be designed such that splicing loss be minimized.

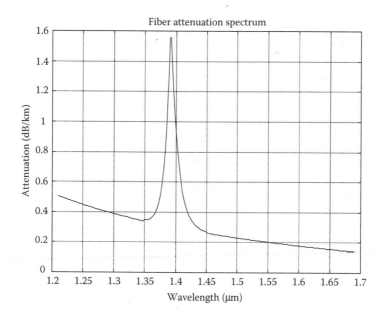

FIGURE 2.12

Attenuation of optical signals as a function of wavelength. The minimum loss at wavelength: at $\lambda = 1.3$ μm about 0.3 dB/km and at $\lambda = 1.5$ μm loss of about 0.13 dB/km. For cabled fibers, the attenuation factor at 1550 nm is 0.25 dB/km.

(a) (b)

FIGURE 2.13
(a) Misalignment in splicing two optical fibers generating losses. (b) Aligned spliced fibers.

Splice loss is mainly due to axial misalignment of the fiber core, as shown in Figure 2.13.

Splicing techniques, which rely on aligning the outside surface of the fibers, require extremely tight tolerances on core to outside surface concentricity. Offsets of the order of 1 μm can produce significant splice loss. This loss is given by

$$L_s = \frac{10}{\ln 10}\left(\frac{d}{r_0}\right)^2 \text{ dB} \tag{2.47}$$

where d is the *axial* misalignment of the fiber cores. It is obvious that minimizing optical loss involves making trade-offs between the different sources of loss. It is advantageous to have a large spot size to minimize both Raleigh and splicing losses, whereas minimizing bending and microbending losses requires a small spot size. In addition, as will be described in the next section, the spot size plays a significant role in the chromatic dispersion properties of SMFs.

2.3.3 Attenuation Coefficient

Under general conditions of power attenuation inside an optical fiber, the attenuation coefficient of the optical power P can be expressed as

$$\frac{dP}{dz} = -\alpha P \tag{2.48}$$

where α is the attenuation factor in linear scale. This attenuation coefficient can include all effects of power loss when signals are transmitted though the optical fibers.

Considering optical signals with an average optical power entering at the input of the fiber length, L is P_{in} and P_{out} is the output optical power, then we have P_{in} and P_{out} related to the attenuation coefficient α as

$$P_{out} = P_{in}e^{(-aL)} \tag{2.49}$$

It is customary to express α in dB/km by using the relation

$$\alpha(\text{dB/km}) = -\frac{10}{L}\log_{10}\left(\frac{P_{out}}{P_{in}}\right) = 4.343\,\alpha \tag{2.50}$$

Standard optical fibers with a small Δ would exhibit a loss of about 0.2 dB/km, that is, that the purity of the silica is very high. Such a purity of a bar of silica would allow us to see a person standing at the other end of a 1 km glass bar without any distortion! The attenuation curve for silica glass is shown in Figure 2.1.

2.4 Signal Distortion in Optical Fibers

Consider a *monochromatic* field given by

$$E_x = A\cos(\omega t - \beta z) \tag{2.51}$$

where A is the wave amplitude, ω is the radial frequency, and β is the propagation constant along the z-direction. If setting $(\omega t - \beta z)$ constant, then the wave phase velocity is given by

$$v_p = \frac{dz}{dt} = \frac{\omega}{\beta} \tag{2.52}$$

Now consider that the propagating wave consists of two monochromatic fields of frequencies $\omega + \delta\omega;\ \omega - \delta\omega$

$$E_{x1} = A\cos[(\omega + \delta\omega)t - (\beta + \delta\beta)z)] \tag{2.53}$$

$$E_{x2} = A\cos[(\omega - \delta\omega)t - (\beta - \delta\beta)z)] \tag{2.54}$$

The total field is then given by

$$E_x = E_{x1} + E_{x2} = 2A\cos(\omega t - \beta z)\cos(\delta\omega t - \delta\beta z) \tag{2.55}$$

If $\omega \gg \delta\omega$, then $\cos(\omega t - \beta z)\cos(\omega t - \beta z)$ varies much faster than $\cos(\delta\omega t - \delta\beta z)$, hence by setting $(\delta\omega t - \delta\beta z)$ *invariant* we can define the group velocity as

$$v_g = \frac{d\omega}{d\beta} \rightarrow v_g^{-1} = \frac{d\beta}{d\omega} \tag{2.56}$$

The group delay t_g per unit length (setting L at 1.0 km) is thus given as

$$t_g = \frac{L(\text{of } 1\,\text{km})}{v_g} = \frac{d\beta}{d\omega} \tag{2.57}$$

The pulse spread $\Delta\tau$ per unit length due to group delay of light sources of spectral width σ_λ, that is, the full-width-half-mark (FWHM) of the optical spectrum of the light source, is

$$\Delta\tau = \frac{dt_g}{d\lambda}\sigma_\lambda \tag{2.58}$$

The spread of the group delay due to the spread of source wavelength can be in ps/km. Thus, the linewidth of the light source contributes significantly to the distortion of optical signal transmitted through the optical fiber due to the fact that the delay differences between the guided modes carried by the spectral components of the lightwaves. Hence, the narrower the source linewidth, the less dispersed the optical pulses. Typical linewidth of Fabry–Perot semiconductor lasers is about 1–2.0 nm while, the DFB (distributed feedback) laser would exhibit a linewidth of 100 MHz. (How many nm is this 100 MHz optical frequency equivalent to?) Later we will see that, under the case that the source linewidth is very narrow, such as the external cavity laser (ECL), then the components of the modulated sources, the bandwidth of the channel would play the principal role in the distortion.

Optical signal traveling along a fiber becomes increasingly distorted. This distortion is a consequence of *intermodal* delay effects and *intramodal* dispersion. Intermodal delay effects are significant in multimode optical fibers due to each mode having different value-of-group velocity at a specific frequency, while intermodal dispersion is pulse spreading that occurs within a single mode. It is the result of the group velocity being a function of the wavelength λ and is therefore referred as chromatic dispersion.

Two main causes of intermodal dispersion are: (i) Material dispersion, which arises from the variation of the refractive index $n(\lambda)$ as a function of wavelengths. This causes a wavelength dependence of the group velocity of any given mode. (ii) Waveguide dispersion, which occurs because the mode propagation constant $\beta(\lambda)$ is a function of wavelength $\beta(\lambda)$ and core radius a and the refractive index difference.

The group velocity associated with the fundamental mode is frequency dependent because of chromatic dispersion. As a result, different spectral components of the light pulse travel at different group velocities, a phenomenon referred to as the *group velocity dispersion* (GVD), intra-modal dispersion, or as material dispersion and waveguide dispersion.

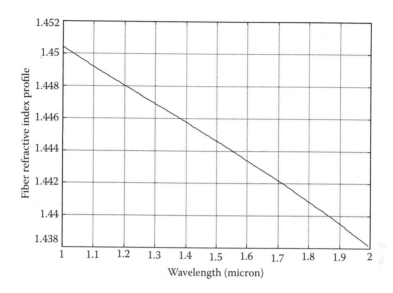

FIGURE 2.14
Variation in the refractive index as a function of optical wavelength of silica.

2.4.1 Material Dispersion

The refractive index of silica as a function of wavelength is shown in Figure 2.14. The refractive index is plotted over the wavelength region of 1.0–2.0 μm, which is the most important range for silica-based optical communications systems, as the loss is lowest at 1300 and 1550 nm windows (Figure 2.14).

The propagation constant β of the fundamental mode guided in the optical fiber can be written as (see Figures 2.15 through 2.17)

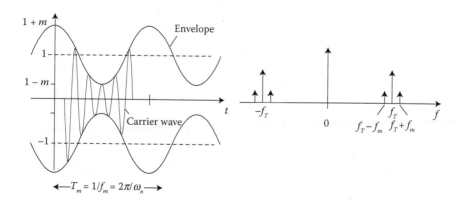

FIGURE 2.15
Time signal and spectrum.

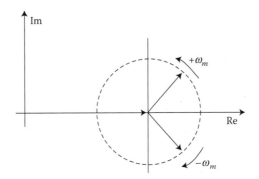

FIGURE 2.16
Vector phasor diagram of the complex envelope.

$$\beta(\lambda) = \frac{2\pi n(\lambda)}{\lambda} \tag{2.59}$$

The group delay t_{gm} per unit length of Equation 3.9 can be obtained

$$t_{gm} = \frac{d\beta}{d\omega} \tag{2.60}$$

where we can use

$$d\omega = d\left(\frac{2\pi c}{\lambda}\right) = -\frac{2\pi c}{\lambda^2} d\lambda \tag{2.61}$$

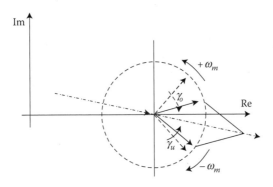

FIGURE 2.17
Magnitude of complex envelope when not sinusoidal; the envelope subject to nonlinear distortions.

thus

$$t_{gm} = -\frac{\lambda^2}{2\pi c}\frac{d\beta}{d\lambda} \tag{2.62}$$

Substituting Equation 2.59 into Equation 2.62 we have

$$t_{gm} = \frac{1}{c}\left[n(\lambda) - \frac{\lambda dn(\lambda)}{d\lambda}\right] \tag{2.63}$$

Thus, the pulse dispersion per unit length $\Delta\tau_m/\Delta\lambda$ due to material (using Equation 2.63) for a source having RMS spectral width σ_λ is

$$\Delta\tau_m = -\frac{\lambda}{c}\frac{d^2 n}{d\lambda^2}\sigma_\lambda \tag{2.64}$$

if setting $\Delta\tau_m = M(\lambda)\sigma_\lambda$, then

$$M(\lambda) = -\frac{\lambda}{c}\frac{d^2 n}{d\lambda^2} \tag{2.65}$$

$M(\lambda)$ is assigned as *the material dispersion factor* or *material dispersion parameter*. Its unit is commonly expressed in ps/(nm km). Thus, if the refractive index can be expressed as a function of the optical wavelength, then the material dispersion can be calculated. In practice, optical material engineers have to characterize all optical properties of new materials. The refractive index $n(\lambda)$ can usually be expressed in Sellmeier's dispersion formula as

$$n^2(\lambda) = 1 + \sum_k \frac{G_k \lambda^2}{(\lambda^2 - \lambda_k^2)} \tag{2.66}$$

where G_k are Sellmeier's constants and k is an integer and normally takes a range of $k = 1$–3. In late 1970s, several silica-based glass materials were manufactured and their properties measured. The refractive indices are usually expressed using Sellmeier's coefficients. These coefficients for several optical fiber materials are given in Table 2.1.

By using curve fitting, the refractive index of pure silica $n(\lambda)$ can be expressed as

$$n(\lambda) = c_1 + c_2\lambda^2 + c_3\lambda^{-2} \tag{2.67}$$

where $c_1 = 1.45084$, $c_2 = -0.00343\ \mu m^{-2}$, and $c_3 = 0.00292\ \mu m^2$. Thus from Table 2.1 and either Equation 2.67, we can use Equation 2.65 to determine the material dispersion factor for certain wavelength ranges.

TABLE 2.1

Sellmeier's Coefficients for Several Optical Fiber Silica-Based Materials
with Germanium Doped in the Core Region

Sellmeiere's Constants	Germanium Concentration, C (mol%)			
	0 (pure silica)	3.1	5.8	7.9
G_1	0.6961663	0.7028554	0.7088876	0.7136824
G_2	0.4079426	0.4146307	0.4206803	0.4254807
G_3	0.8974794	0.8974540	0.8956551	0.8964226
λ_1	0.0684043	0.0727723	0.0609053	0.0617167
λ_2	0.1162414	0.1143085	0.1254514	0.1270814
λ_3	9.896161	9.896161	9.896162	9.896161

For the doped core of the optical fiber, the Sellmeier's expression (2.66) can be approximated by using a curve-fitting technique to approximate it to the form in Equation 2.67. The material dispersion factor $M(\lambda)$ becomes zero at wavelengths around 1350 nm and about −10 ps/(nm km) at 1550 nm. However, the attenuation at 1350 nm is about 0.4 dB/km compared with 0.2 dB/km at 1550 nm, as shown in Table 2.1 (Figure 2.18).

2.4.2 Waveguide Dispersion

The effect of waveguide dispersion can be approximated by assuming that the refractive index of the material is independent of wavelength. Let us now consider the group delay, that is, the time required for a mode to travel along a fiber of length L. This kind of dispersion depends strongly on Δ and V parameters. To make the results of fiber parameters, we define a *normalized propagation constant b* as

$$b = \frac{(\beta^2/k^2) - n_2^2}{n_1^2 - n_2^2} \tag{2.68}$$

for small Δ. We note that the β/k is in fact the *"effective" refractive index* of the guided optical mode propagating along the optical fiber; that is, the guided waves traveling the axial direction of the fiber "see" it as a medium with a refractive index of an equivalent "effective" index.

In case the fiber is a weakly guiding waveguide with the effective refractive index taking a value significantly close to that of the core or cladding index, Equation 2.68 can be approximated by

$$b \cong \frac{(\beta/k) - n_2}{n_1 - n_2} \tag{2.69}$$

solving Equation 2.69 for β, we have

$$\beta = n_2 k(b\Delta + 1)$$

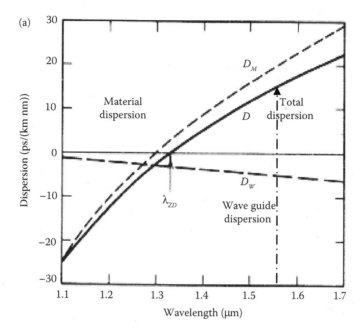

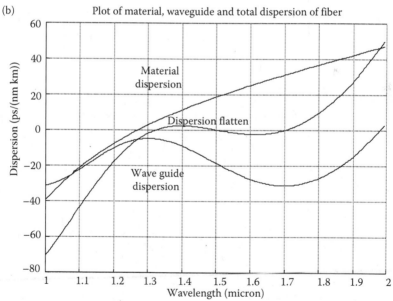

FIGURE 2.18
Total chromatic dispersion factor of (a) SSMF and (b) a dispersion flatten fiber.

the group delay for waveguide dispersion is then given by (per unit length)

$$t_{wg} = \frac{d\beta}{d\omega} = \frac{1}{c}\frac{d\beta}{dk} \tag{2.70}$$

$$t_{wg} = \frac{1}{c}\left[n_1 + n_2\Delta\frac{d(bk)}{dk}\right] = \frac{1}{c}\left[n_1 + n_2\Delta\frac{d(bk)}{dk}\right] = \frac{1}{c}\left[n_1 + n_2\Delta\frac{d(bV)}{dV}\right] \tag{2.71}$$

Equation 2.71 can be obtained from Equation 2.70 by using the expression of V. Thus, the pulse spreading $\Delta\tau_\omega$ due to the waveguide dispersion per unit length by a source having an optical bandwidth (or linewidth σ_λ) is given by

$$\Delta\tau_\omega = \frac{dt_{gw}}{d\lambda}\sigma_\lambda = -\frac{n_2\Delta}{c\lambda}V\frac{d^2(Vb)}{dV^2}\sigma_\lambda \tag{2.72}$$

and similar to the definition of the material dispersion factor, the *waveguide dispersion factor* or *"waveguide dispersion parameter"* can be defined as

$$D(\lambda) = -\frac{n_2(\lambda)\Delta}{c\lambda}V\frac{d^2(Vb)}{dV^2} \tag{2.73}$$

This waveguide factor can take unit of ps/(nm km). In the range of $0.9 < \lambda/\lambda_c <$ 2.6, the factor $V(d^2(Vb)/dV^2)$ can be approximated (to <5% error) by

$$V\frac{d^2(Vb)}{dV^2} \cong 0.080 + 0.549(2.834 - V)^2 \tag{2.74}$$

or, alternatively, using the definition of cut-off wavelength and the expression of the V-parameters we obtain

$$V\frac{d^2(Vb)}{dV^2} \cong 0.080 + 3.175\left(1.178 - \frac{\lambda_c}{\lambda}\right)^2 \tag{2.75}$$

The fiber dispersion factor in the linear operation region is the summation of the material dispersion factor and that due to waveguide dispersion. The curves given in Figure 2.18a and b show the total dispersion factors for SSMF and for a dispersion flattened fiber, respectively. These curves are generated as an example. For SSMF which are currently installed throughout the world, the total dispersion is around +17 ps/(nm km) at 1550 nm and almost zero at 1310 nm. We can estimate the waveguide dispersion curve for the SSMF at around 1300 and 1550 nm windows and hence estimate the pulse broadening after transmission over a link of distance L km at a certain bit rate evaluated in equivalent bandwidth in units of nanometer.

It is not so difficult to prove the equivalent equation of Equations 2.70 through 2.75. Further, the sign assignment of the material and waveguide

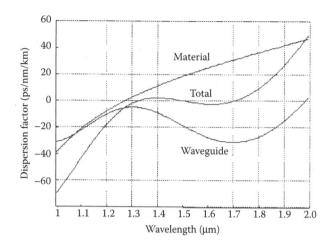

FIGURE 2.19
Variation of the dispersion factors due to material and waveguide group delay effects with respect to wavelength, hence the total dispersion factor of a dispersion flattening fiber.

dispersion factors must be the same. Otherwise, a negative and positive of these dispersion factors would create confusion. Can you explain what would happen to the pulse if it is transmitted through an optical fiber having a total negative dispersion factor?

Thus, from Equations 2.74 and 2.75 we can calculate the waveguide dispersion factor and hence the pulse dispersion factor for a particular source spectral width σ_λ. It is noted that the dispersion considered in this chapter is for step-index fiber only. For grade-index fiber, ESI parameters must be found and the chromatic dispersion can then be calculated. Figure 2.19 shows a design of single-mode optical fibers with the total dispersion factor contributed by material and waveguide effects (Figure 2.19).

2.4.2.1 Alternative Expression for Waveguide Dispersion Parameter

Alternatively, the waveguide dispersion parameter can be expressed as function of the propagation constant β. By using $\omega = (2\pi c/\lambda_g)$ and Equation 2.75, the waveguide dispersion factor can be written as

$$D(\lambda) = -\frac{2\pi c}{\lambda^2}\beta_2 = -\frac{2\pi c}{\lambda^2}\frac{d\beta^2}{d\omega^2} \tag{2.76}$$

Thus the waveguide dispersion factor is directly related to the second-order derivative of the propagation constant with respect to the optical radial frequency. An example of a design of an optical fiber operating in the single-mode region is given in Figure 2.19. The cladding material is pure

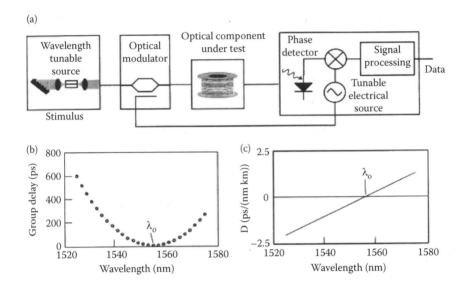

FIGURE 2.20
(a) Chromatic dispersion measurement of two-port optical device; (b) relative group delay versus wavelength; and (c) dispersion parameter versus wavelength.

silica. Shown in this figure are the curves of the material dispersion factor, waveguide dispersion factor, and total dispersion for a single-mode optical fiber with nonuniform refractive index profile in the core.

A typical measurement set-up for determination of the dispersion is the measurement of the differential group delay as shown in Figure 2.20, in which a narrow linewidth source is modulated by an optical modulator, the Mach–Zehnder interferometric modulator (MZIM), through which the RF signal comes from the detected signal at the end of the device under test (DUT). This set-up is thus very similar to an RF network analyzer, but with the RF signal transferred to the optical domain by modulating the optical modulator, then recovered at the output of the optical device, and then tuned to vary the RF signals as feedback to the modulator to scan the excitation frequency. The group delay can then be estimated without much difficulty, hence the dispersion factor (Figure 2.20).

2.4.2.2 Higher-Order Dispersion

We also observe from Figure 2.19 that the bandwidth-length product of the optical fiber can be extended to infinity if the system is operating at the wavelength, at which the total dispersion factor is zero. However, the dispersive effects do not disappear completely at this zero-dispersion wavelength. Optical pulses still experience broadening because of higher-order dispersion effects. It is easily imagined that the total dispersion factor cannot be made zero to "flatten" over the optical spectrum. This is higher-order dispersion, which is governed by the slope of the total dispersion curve, called the dispersion slope $S = d[D(\lambda) + M(\lambda)]/d\lambda$; $S(\lambda)$ can thus be expressed as

$$S(\lambda) = \left(\frac{2\pi c}{\lambda^2}\right)^2 \frac{d^3\beta}{d\lambda^3} + \left(\frac{4\pi c}{\lambda^3}\right)\frac{d^2\beta}{d\lambda^2} \qquad (2.77)$$

$S(\lambda)$ is also known as the differential-dispersion parameter.

2.4.3 Polarization Mode Dispersion

The delay between two PSPs is normally negligible, at a bit rate less than 10 Gb/s (Figures 2.21 and 2.22). However, at a high bit rate and in ultra long-haul transmission, PMD severely degrades the system performance [15–18]. The instantaneous value of DGD ($\Delta \tau$) varies along the fiber and follows a Maxwellian distribution [19,20] (see Figure 2.23).

The Maxwellian distribution is governed by the following expression:

$$f(\Delta\tau) = \frac{32(\Delta\tau)^2}{\pi^2\langle\Delta\tau\rangle^3} \exp\left\{-\frac{4(\Delta\tau)^2}{\pi\langle\Delta\tau\rangle^2}\right\} \quad \Delta\tau \geq 0 \qquad (2.78)$$

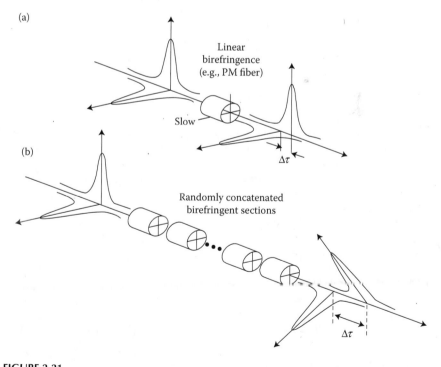

FIGURE 2.21
Conceptual model of PMD: (a) simple birefringence device; (b) randomly concatenated birefringence.

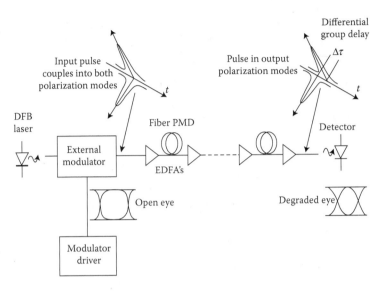

FIGURE 2.22
Effect of PMD in a digital optical communication system, degradation of the received eye diagram.

The mean DGD value $\langle \Delta\tau \rangle$ is commonly termed "fiber PMD" and provided in the fiber specifications. The following expression gives an estimate of the maximum transmission limit L_{max} due to the PMD effect as

$$L_{max} = \frac{0.02}{\langle \Delta\tau \rangle^2 \cdot R^2} \tag{2.79}$$

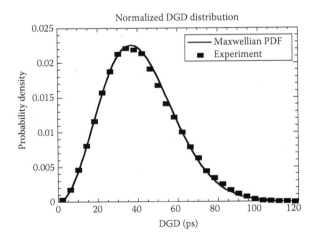

FIGURE 2.23
Maxwellian distribution of PMD random process.

where R is the bit rate. Based on Equation 2.79, L_{max} for both old fiber vintage and contemporary fibers are obtained as follows: (i) $\langle \Delta \tau \rangle = 1$ ps/km (old fiber vintages); for a bit rate of $R = 40$ Gb/s, then the maximum distance $L_{max} = 12.5$ km; for $R = 10$ Gb/s, then $L_{max} = 200$ km; (ii) $\langle \Delta \tau \rangle = 0.1$ ps/km (contemporary fiber for modern optical systems): the if the bit rate $R = 40$ Gb/s then the maximum transmission distance is $L_{max} = 1250$ km; for $R = 10$ Gb/s; $L_{max} = 20{,}000$ km.

2.5 Transfer Function of Single-Mode Fibers

2.5.1 Linear Transfer Function

The treatment of the propagation of modulated lightwaves through single-mode fiber in the linear and nonlinear regimes has been well documented [21–26]. For completeness of the transfer function of single-mode optical fibers, in this section we restrict our study to the frequency transfer function and impulse responses of the fiber to the linear region of the media. Furthermore, the delay term in the NLSE can be ignored, as it has no bearing on the size and shape of the pulses. From NLSE we can thus model the fiber simply as a quadratic phase function. This is derived from the fact that the nonlinear term of NLSE can be removed, the Taylor-series approximation around the operating frequency (central wavelength) can be obtained, and a frequency and impulse responses of the SMF. The input–output relationship of the pulse can therefore be depicted. Equation 2.80 expresses the time-domain impulse response $h(t)$ and the frequency domain transfer function $H(\omega)$ as a Fourier transform pair:

$$h(t) = \sqrt{\frac{1}{j4\pi\beta_2}} \exp\left(\frac{jt^2}{4\beta_2}\right) \quad \leftrightarrow \quad H(\omega) = \exp(-j\beta_2\omega^2) \qquad (2.80)$$

where β_2 is well known as the group velocity dispersion (GVD) parameter. The input function $f(t)$ is typically a rectangular pulse sequence and β_2 is proportional to the length of the fiber. The output function $g(t)$ is the dispersed waveform of the pulse sequence. The propagation transfer function in Equation 2.80 is an exact analogy of diffraction in optical systems (see Item 1, Table 2.1, p. 14, A. Papoulis [27]). Thus, the quadratic phase function also describes the diffraction mechanism in one-dimensional optical systems, where distance x is analogous to time t. The establishment of this analogy affords us to borrow many of the imageries and analytical results that have been developed in the diffraction theory. Thus, we may express the step response $s(t)$ of the system $H(\omega)$ in terms of Fresnel cosine and sine integrals as follows:

$$s(t) = \int_0^t \sqrt{\frac{1}{j4\pi\beta_2}} \exp\left(\frac{jt^2}{4\beta_2}\right) dt = \sqrt{\frac{1}{j4\pi\beta_2}}\left[C\left(\sqrt{1/4\beta_2}\,t\right) + jS\left(\sqrt{1/4\beta_2}\,t\right)\right] \qquad (2.81)$$

with

$$C(t) = \int_0^t \cos\left(\frac{\pi}{2}\tau^2\right)d\tau$$

$$S(t) = \int_0^t \sin\left(\frac{\pi}{2}\tau^2\right)d\tau$$

(2.82)

where $C(t)$ and $S(t)$ are the Fresnel cosine and sine integrals.

Using this analogy, one may argue that it is always possible to restore the original pattern $f(x)$ by refocusing the blurry image $g(x)$ (e.g., image formation, item 5, Table 2.1, [15]). In the electrical analogy, it implies that it is possible to compensate the quadratic phase media perfectly. This is not surprising. The quadratic phase function $H(\omega)$ in Equation 2.80 is an all-pass transfer function, thus it is always possible to find an inverse function to recover $f(t)$. One can express this differently in information theory terminology, that the quadratic phase channel has a theoretical bandwidth of infinity; hence its information capacity is infinite. Shannon's channel capacity theorem states that there is no limit on the reliable rate of transmission through the quadratic phase channel. Figure 2.24 shows the pulse and impulse responses of the fiber. It is noted that only the envelope of the pulse is shown and the phase of the lightwave carrier is included as the complex value of the amplitude. As observed, the chirp of the carrier is significant at the edges of the pulse. At the center of the pulse, the chirp is almost negligible at some limited fiber length, thus the frequency of the carrier remains nearly the same as at its original starting value. One could obtain the impulse response quite easily, but in this work we believe that the pulse response is much more relevant in the investigation of the uncertainty in the pulse sequence detection. Rather, the impulse response is much more important in the process of equalization.

The uncertainty of the detection depends on the modulation formats and detection process. The modulation can be implemented by manipulation of the amplitude, the phase or the frequency of the carrier, or both amplitude and phase or multi-sub-carriers such as the orthogonal frequency division multiplexing (OFDM) [16]. The amplitude detection would be mostly affected by the ripples of the amplitudes of the edges of the pulse. The phase of the carrier is mostly affected near the edge due to the chirp effects. However, if differential phase detection is used then the phase change at the transition instant is the most important and the opening of the detected eye diagram. For frequency modulation the uncertainty in the detection is not very critical, provided that the chirping does not enter into the region of the neighborhood of the center of the pulse in which the frequency of the carrier remains almost constant.

The picture changes completely if the detector/decoder is allowed only a finite time window to decode each symbol. In the convolution coding scheme, for example, it is the decoder's constraint length that manifests due to the finite time window. In an adaptive equalization scheme, it is the number of equalizer

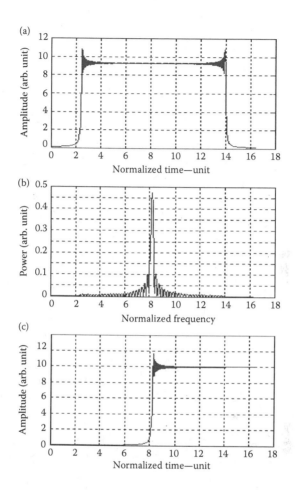

FIGURE 2.24

Rectangular pulse transmission through an SMF: (a) pulse response; (b) frequency spectrum; (c) step response of the quadratic-phase transmittance function. Note the horizontal scale in normalized unit of time.

coefficients that determines the decoder window length. Since the transmitted symbols have already been broadened by the quadratic phase channel, if they are next gated by a finite time window, the information received could be severely reduced. The longer the fiber, the more the broadening of the pulses is widened, hence the more uncertain it becomes in the decoding. It is the interaction of the pulse broadening on one hand, and the restrictive detection time window on the other, that gives rise to the finite channel capacity.

It is observed that the chirp occurs mainly near the edge of the pulses when it is in the near field region, about a few kms for standard single-mode fibers. In this near-field distance the accumulation of nonlinear effects is still very weak and thus the chirp effects dominate the behavior of the single-mode fiber.

The nonlinear Volterra transfer function presented in the next section would thus have minimum influence. This point is important for understanding the behavior of lightwaves circulating in short-length fiber devices in which both the linear and nonlinear effects are to be balanced, such as active-mode locked soliton and multi-bound soliton lasers [28,29]. In the far field, the output of the fiber is Gaussian-like for the square pulse launched at the input. In this region the nonlinear effects would dominate over the linear dispersion effect as they have been accumulated over a long distance.

The linear time-variant system such as the SMF would have a transfer function of

$$H(f) = |H(f)| e^{-j\alpha(f)} \tag{2.83}$$

where $\alpha = \pi^2 \beta_2 L = (-\pi DL\lambda^2/2c)$ is proportional to the length L and the dispersion factor $D(\lambda)$ (ps/nm/km). The phase of the frequency transfer response is a quadratic function of the frequency, thus the group delay would follow a linear relationship with respect to the frequency as observed in Figure 2.25. The frequency response in amplitude terms is infinite and is a constant, while the phase response is a quadratic function with respect to the frequency of the base band signals. The carrier is chirped accordingly as observed in Figures 2.26 and 2.27. The chirping effect is very significant near the edge of the rectangular pulse and almost nil at the center of the pulse, in the near field region of less than 1 km of standard SMF. In the far-field region the pulse becomes Gaussian-like. Thus, the response of the fiber in the linear region can be seen as shown in Figure 2.28 for a Gaussian pulse input to the fiber. The output pulse is also Gaussian by taking the Fourier transform of the input pulse and multiplied by the fiber transfer function. Hence, an inverse Fourier would indicate the output pulse shape follows a Gaussian profile.

This leads to a rule of thumb for consideration of the scaling of the bit rate and transmission distance as *"Given that a modulated lightwave of a bit rate B can be transmitted over a maximum distance L of single-mode optical fiber with a BER of error-free level, than if the bit rate is halved then the transmission distance can be increase by four times."* For example, for a 10 Gb/s amplitude shift, keying modulation format signals can be transmitted over 80 km of standard single-mode optical fiber, then at 40 Gb/s only 5 km can be transmitted for a bit error rate (BER) of 10^{-9}. Figure 2.25 shows a typical frequency response in magnitude phase and the low ps property. Ideally, one could see that the amplitude response of the fiber is constant throughout all frequency if no attenuation or constant attenuation throughout all frequency range. Only the phase of the lightwaves is altered, that is the chirping of the carrier. It is this chirping of the carrier that would then limit the response frequency range. In the case of phase modulation, this chirp would rotate the constellation of signals modulated by the QAMs given in Chapter 1. The digital signal processing could then be used to determine the exact dispersion and hence

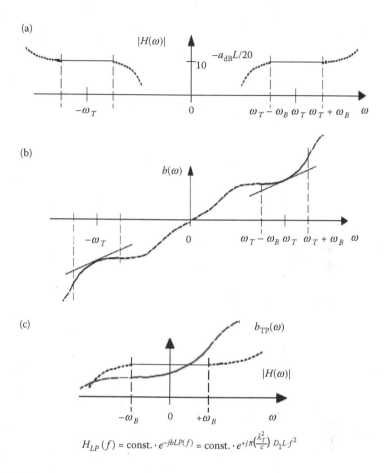

FIGURE 2.25

Frequency response of a single-mode optical fiber: (a) magnitude; (b) phase response in bandpass regime; and (c) baseband equivalence.

the rotation of the constellation by an angle such that it is recovered back to its original position. The chirp of the carrier can be seen in Figure 2.26, in which the chirp is much less at the center but heavily chirped near the edge of the pulse. The step responses shown in Figures 2.27 and 2.28 indicate the damping oscillation of the step pulse, which is due to the chirp of the carrier as we also observe from the calculated impulse and step responses given in Figure 2.24. Figure 2.27 shows the Gaussian-like impulse response when the transmission distance is large. This is the typical pulse shape in long-haul nondispersion compensating transmission. These dispersive pulse sequences are then coherently detected and processed by the digital signal processors. The number of samples must be long enough to cover the dispersive sequence and the number of taps of the finite impulse response (FIR) filter must be high enough to ensure the whole dispersive pulse is covered and fallen within the

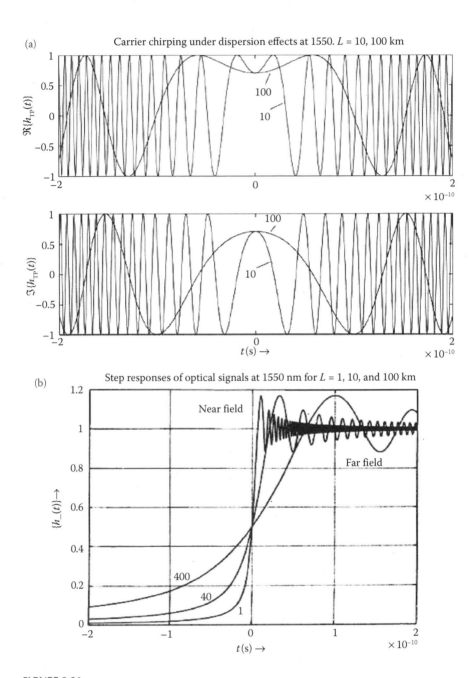

FIGURE 2.26
(a) Carrier chirping effects and (b) step response of a single-mode optical fiber of $L = 1$, 10, and 100 km.

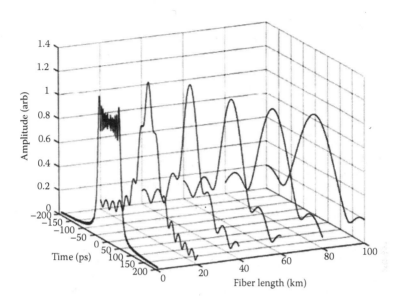

FIGURE 2.27
Pulse response from near field (~<2 km) to far field (>80 km).

filter length. So the longer the transmission length, the higher the number of taps of the FIR. This is the first step in the DSP-based optical receiver, by using the constant modulus amplitude (CMA) algorithm to compensate for the dispersion effects before compensating and recovering the phase constellation of the modulated and transmitted channels. It is noted that the chirp of the pulse envelope is much higher when the pulse in the near-field region than that in the far field, as observed in the step responses given in Figures 2.26a and 2.27. Thus, if a Gaussian pulse is launched and propagating in the

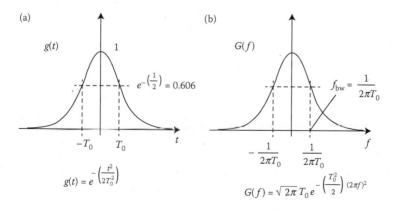

FIGURE 2.28
Fiber response to Gaussian pulse (a) in time domain and (b) in frequency domain.

single-mode fiber, we would obtain Gaussian pulse shape at the output (see Figure 2.27) provided that the fiber length is sufficiently long, typically more than 20 km, which is commonly met in real transmission system.

2.5.2 Nonlinear Fiber Transfer Function

The weakness of most of the recursive methods in solving the NLSE is that they do not provide much useful information to help the characterization of nonlinear effects. The Volterra series model provides an elegant way of describing a system's nonlinearities, and enables the designers to see clearly where and how the nonlinearity affects the system performance. Although Refs. [30,31] have given an outline of the kernels of the transfer function using the Volterra series, it is necessary for clarity and physical representation of these functions that brief derivations are given here on the nonlinear transfer functions of an optical fiber operating under nonlinear conditions.

The Volterra series transfer function of a particular optical channel can be obtained in the frequency domain as a relationship between the input spectrum $X(\omega)$ and the output spectrum $Y(\omega)$, as

$$Y(\omega) = \sum_{n=1}^{\infty} \int_{-\infty}^{\infty} \cdots \int_{-\infty}^{\infty} H_n(\omega_1, \cdots, \omega_{n-1}, \omega - \omega_1 - \cdots - \omega_{n-1}) \times X(\omega_1) \cdots X(\omega_{n-1})$$

$$\times X(\omega - \omega_1 - \cdots - \omega_{n-1}) d\omega_1 \cdots d\omega_{n-1} \qquad (2.84)$$

where $H_n(\omega_1, \ldots, \omega_n)$ is the nth-order frequency domain Volterra kernel, including all signal frequencies of orders 1 to n. The wave propagation inside an SMF can be governed by a simplified version of the NLSE, given above in this chapter with only the SPM effect included as

$$\frac{\partial A}{\partial z} = -\frac{\alpha_0}{2} A - \beta_1 \frac{\partial A}{\partial t} - j \frac{\beta_2}{2} \frac{\partial^2 A}{\partial t^2} - \frac{\beta_3}{6} \frac{\partial^3 A}{\partial t^3} + j\gamma |A|^2 A \qquad (2.85)$$

where $A = A(t,z)$. The proposed solution of the NLSE can be written with respect to the VSTF model of up to fifth order as

$$A(\omega, z) = H_1(\omega, z)A(\omega) + \int_{-\infty}^{\infty} \int_{-\infty}^{\infty} H_3(\omega_1, \omega_2, \omega - \omega_1 + \omega_2, z)$$

$$\times A(\omega_1)A^*(\omega_2)A(\omega - \omega_1 + \omega_2)d\omega_1 d\omega_2$$

$$+ \int_{-\infty}^{\infty} \int_{-\infty}^{\infty} \int_{-\infty}^{\infty} \int_{-\infty}^{\infty} H_5(\omega_1, \omega_2, \omega_3, \omega_4, \omega - \omega_1 + \omega_2 - \omega_3 + \omega_4, z)$$

$$\times A(\omega_1)A^*(\omega_2)A(\omega_3)A^*(\omega_4) \times A(\omega - \omega_1 + \omega_2 - \omega_3 + \omega_4)d\omega_1 d\omega_2 d\omega_3 d\omega_4$$

$$(2.86)$$

where $A(\omega) = A(\omega,0)$, that is, the amplitude envelop of the optical pulses at the input of the fiber. Taking the Fourier transform of Equation 2.3 and assuming $A(t,z)$ is of sinusoidal form, we have

$$\frac{\partial A(\omega,z)}{\partial z} = G_1(\omega)A(\omega,z)\int_{-\infty}^{\infty}\int_{-\infty}^{\infty} G_3(\omega_1,\omega_2,\omega-\omega_1+\omega_2)A(\omega_1,z)A^*(\omega_2,z)$$

$$\times A(\omega-\omega_1+\omega_2,z)d\omega_1 d\omega_2 \qquad (2.87)$$

where $G_1(\omega) = -\alpha_0/2 + j\beta_1\omega + j\beta_2/2\omega^2 - j\beta_3/6\omega_3$ and $G_3(\omega_1,\omega_2,\omega_3) = j\gamma$. ω is taking the values over the signal bandwidth and beyond in overlapping the signal spectrum of other optically modulated carriers, while $\omega_1...\omega_3$ are all also taking values over similar range as that of ω. For general expression, the limit of integration is indicted over the entire range to infinity.

Substituting Equation 2.86 into Equation 2.87 and equate both sides, the kernels can be obtained after some algebraic manipulations

$$\frac{\partial}{\partial z}\left[H_1(\omega,z)A(\omega) + \int_{-\infty}^{\infty}\int_{-\infty}^{\infty} H_3(\omega_1,\omega_2,\omega-\omega_1+\omega_2,z)\,A(\omega_1)A^*(\omega_2)\right]$$

$$\times A(\omega-\omega_1+\omega_2)d\omega_1 d\omega_2$$

$$+\int_{-\infty}^{\infty}\int_{-\infty}^{\infty}\int_{-\infty}^{\infty}\int_{-\infty}^{\infty} H_5(\omega_1,\omega_2,\omega_3,\omega_4,\omega-\omega_1+\omega_2-\omega_3+\omega_4,z)$$

$$\times A(\omega_1)A^*(\omega_2)A(\omega_3)A^*(\omega_4)A(\omega-\omega_1+\omega_2-\omega_3+\omega_4)d\omega_1 d\omega_2 d\omega_3 d\omega_4$$

$$= G_1(\omega)\left[H_1(\omega,z)A(\omega) + \int_{-\infty}^{\infty}\int_{-\infty}^{\infty i} H_3(\omega_1,\omega_2,\omega-\omega_1+\omega_2,z)\right.$$

$$\times A(\omega_1)A^*(\omega_2)A(\omega-\omega_1+\omega_2)d\omega_1 d\omega_2$$

$$+\int_{-\infty}^{\infty}\int_{-\infty}^{\infty}\int_{-\infty}^{\infty}\int_{-\infty}^{\infty} H_5(\omega_1,\omega_2,\omega_3,\omega_4,\omega-\omega_1+\omega_2-\omega_3+\omega_4,z)$$

$$\left. \times A(\omega_1)A^*(\omega_2)A(\omega_3)A^*(\omega_4)\times A(\omega-\omega_1+\omega_2-\omega_3+\omega_4)d\omega_1 d\omega_2 d\omega_3 d\omega_4 \right]$$

$$+\int_{-\infty}^{\infty}\int_{-\infty}^{\infty} G_3(\omega_1,\omega_2,\omega-\omega_1+\omega_2)$$

$$\times \left[H_1(\omega_1, z)A(\omega_1) + \int\limits_{-\infty}^{\infty} \int\limits_{-\infty}^{\infty} H_3(\omega_{11}, \omega_{12}, \omega_1 - \omega_{11} + \omega_{12}, z) \right.$$

$$\times A(\omega_{11})A^*(\omega_{12})A(\omega_1 - \omega_{11} + \omega_{12})d\omega_{11}d\omega_{12}$$

$$\times \int\limits_{-\infty}^{\infty} \int\limits_{-\infty}^{\infty} \int\limits_{-\infty}^{\infty} \int\limits_{-\infty}^{\infty} H_5(\omega_{11}, \omega_{12}, \omega_{13}, \omega_{14}, \omega_1 - \omega_{11} + \omega_{12} - \omega_{13} + \omega_{14}, z)$$

$$\left. \times A(\omega_{11})A^*(\omega_{12})A(\omega_{13})A^*(\omega_{14}) \times A(\omega_1 - \omega_{11} + \omega_{12} - \omega_{13} + \omega_{14})d\omega_{11}d\omega_{12}d\omega_{13}d\omega_{14} \right]$$

$$\times \left[H_1(\omega_1, z)A(\omega_1) + \int\limits_{-\infty}^{\infty} \int\limits_{-\infty}^{\infty} H_3(\omega_{11}, \omega_{12}, \omega_1 - \omega_{11} + \omega_{12}, z) \times A(\omega_{11})A^*(\omega_{12}) \right.$$

$$\times A(\omega_1 - \omega_{11} + \omega_{12})d\omega_{11}d\omega_{12}$$

$$+ \int\limits_{-\infty}^{\infty} \int\limits_{-\infty}^{\infty} \int\limits_{-\infty}^{\infty} \int\limits_{-\infty}^{\infty} H_5(\omega_{21}, \omega_{22}, \omega_{23}, \omega_{24}, \omega_2 - \omega_{21} + \omega_{22} - \omega_{23} + \omega_{24}, z)$$

$$\left. \times A(\omega_{21})A^*(\omega_{22})A(\omega_{23})A^*(\omega_{24}) \times A(\omega_2 - \omega_{21} + \omega_{22} - \omega_{23} + \omega_{24})d\omega_{21}d\omega_{22}d\omega_{23}d\omega_{24} \right]^*$$

$$\times \left[H_1(\omega - \omega_1 + \omega_2, z)A(\omega - \omega_1 + \omega_2) \right.$$

$$+ \int\limits_{-\infty}^{\infty} \int\limits_{-\infty}^{\infty} H_3(\omega_{31}, \omega_{32}, \omega - \omega_1 + \omega_2 - \omega_{31} + \omega_{32}, z)$$

$$\times A(\omega_{31})A^*(\omega_{32})A(\omega - \omega_1 + \omega_2 - \omega_{31} + \omega_{32})d\omega_{31}d\omega_{32}$$

$$+ \int\limits_{-\infty}^{\infty} \int\limits_{-\infty}^{\infty} \int\limits_{-\infty}^{\infty} \int\limits_{-\infty}^{\infty} H_5(\omega_{31}, \omega_{32}, \omega_{33}, \omega_{34}, \omega - \omega_1 + \omega_2 - \omega_{31} + \omega_{32} - \omega_{33} + \omega_{34}, z)$$

$$\times A(\omega_{31})A^*(\omega_{32})A(\omega_{33})A^*(\omega_{34})$$

$$\left. \times A(\omega - \omega_1 + \omega_2 - \omega_{31} + \omega_{32} - \omega_{33} + \omega_{34}) \times d\omega_{31}d\omega_{32}d\omega_{33}d\omega_{34} \right]$$

$$(2.88)$$

Equating the first-order terms on both sides we obtain

$$\frac{\partial}{\partial z} H_1(\omega, z) = G_1(\omega) H_1(\omega, z) \tag{2.89}$$

Thus the solution for the first-order transfer function (2.89) is given by

$$H_1(\omega, z) = e^{G_1(\omega)z} = e^{\left(-\frac{\alpha_0}{2} + j\beta_1\omega + j\frac{\beta_2}{2}\omega^2 - j\frac{\beta_3}{6}\omega^3\right)z} \tag{2.90}$$

This is in fact the linear transfer function of a single-mode optical fiber, with the dispersion factors β_2 and β_3 as already shown in the previous section. Similarly, for the third-order terms we have

$$\frac{\partial}{\partial z} \int\limits_{-\infty}^{\infty}\int\limits_{-\infty}^{\infty} H_3(\omega_1, \omega_2, \omega - \omega_1 + \omega_2, z) \times A(\omega_1)A^*(\omega_2)A(\omega - \omega_1 + \omega_2)d\omega_1 d\omega_2$$

$$= \int\limits_{-\infty}^{\infty}\int\limits_{-\infty}^{\infty} G_3(\omega_1, \omega_2, \omega - \omega_1 + \omega_2)H_1(\omega_1, z)A(\omega_1)H_2^*(\omega_2, z)$$

$$\times A(\omega_2)H_1(\omega - \omega_1 + \omega_2)A(\omega - \omega_1 + \omega_2)d\omega_1 d\omega_2 \tag{2.91}$$

Now letting $\omega_3 = \omega - \omega_1 + \omega_2$, then it follows

$$\frac{\partial H_3(\omega_1, \omega_2, \omega_3, z)}{\partial z} = G_1(\omega_1 - \omega_2 + \omega_3)H_3(\omega_1, \omega_2, \omega_3, z) + G_3(\omega_1, \omega_2, \omega_3)$$

$$\times H_1(\omega_1, z)H_1^*(\omega_2, z)H_1(\omega_3, z) \tag{2.92}$$

The third-kernel transfer function can be obtained as

$$H_3(\omega_1, \omega_2, \omega_3, z) = G_3(\omega_1, \omega_2, \omega_3) \times \frac{e^{(G_1(\omega_1) + G_1^*(\omega_2) + G_1(\omega_3))z} - e^{G_1(\omega_1 - \omega_2 + \omega_3)z}}{G_1(\omega_1) + G_1^*(\omega_2) + G_1(\omega_3) - G_1(\omega_1 - \omega_2 + \omega_3)} \tag{2.93}$$

The fifth-order kernel can similarly be obtained as

$$H_5(\omega_1, \omega_2, \omega_3, \omega_4, \omega_5, z)$$

$$= \frac{H_1(\omega_1, z)H_1^*(\omega_2, z)H_1(\omega_3, z)H_1^*(\omega_4, z)H_1(\omega_5, z) - H_1(\omega_1 - \omega_2 + \omega_3 - \omega_4 + \omega_5, z)}{G_1(\omega_1) + G_1^*(\omega_2) + G_1(\omega_3) + G_1^*(\omega_4) + G_1(\omega_5) - G_1(\omega_1 - \omega_2 + \omega_3 - \omega_4 + \omega_5)}$$

$$\times \left[\frac{G_3(\omega_1,\omega_2,\omega_3-\omega_4+\omega_5)G_3(\omega_3,\omega_4,\omega_5)}{G_1(\omega_3)+G_1^*(\omega_4)+G_1(\omega_5)-G_1(\omega_3-\omega_4+\omega_5)} \right.$$

$$+\frac{G_3(\omega_1,\omega_2-\omega_3+\omega_4,\omega_5)G_3^*(\omega_2,\omega_3,\omega_4)}{G_1^*(\omega_2)+G_1(\omega_3)+G_1^*(\omega_4)-G_1^*(\omega_2-\omega_3+\omega_4)}$$

$$\left. +\frac{G_3(\omega_1-\omega_2+\omega_3,\omega_4,\omega_5)G_3(\omega_1,\omega_2,\omega_3)}{G_1(\omega_1)+G_1^*(\omega_2)+G_1(\omega_3)-G_1(\omega_1-\omega_2+\omega_3)} \right]$$

$$-\frac{G_3(\omega_1,\omega_2,\omega_3-\omega_4+\omega_5)G_3(\omega_3,\omega_4,\omega_5)}{G_1(\omega_3)+G_1^*(\omega_4)+G_1(\omega_5)-G_1(\omega_3-\omega_4+\omega_5)}$$

$$\times\frac{H_1(\omega_1,z)H_1^*(\omega_2,z)H_1(\omega_1-\omega_2+\omega_3,z)-H_1(\omega_1-\omega_2+\omega_3-\omega_4+\omega_5,z)}{G_1(\omega_1)+G_1^*(\omega_2)+G_1(\omega_3-\omega_4+\omega_5)-G_1(\omega_1-\omega_2+\omega_3-\omega_4+\omega_5)}$$

$$-\frac{G_3(\omega_1,\omega_2-\omega_3+\omega_4,\omega_5)G_3^*(\omega_2,\omega_3-\omega_4)}{G_1^*(\omega_2)+G_1(\omega_3)+G_1^*(\omega_4)-G_1^*(\omega_2-\omega_3+\omega_4)}$$

$$\times\frac{H_1(\omega_1,z)H_1^*(\omega_2-\omega_3+\omega_4,z)H_1(\omega_5,z)-H_1(\omega_1-\omega_2+\omega_3-\omega_4+\omega_5,z)}{G_1(\omega_1)+G_1^*(\omega_2-\omega_3+\omega_4)+G_1(\omega_5)-G_1(\omega_1-\omega_2+\omega_3-\omega_4+\omega_5)}$$

$$-\frac{G_3(\omega_1-\omega_2+\omega_3,\omega_4,\omega_5)G_3(\omega_1,\omega_2,\omega_3)}{G_1(\omega_1)+G_1^*(\omega_2)+G_1(\omega_3)-G_1(\omega_1-\omega_2+\omega_3)}$$

$$\times\frac{H_1(\omega_1-\omega_2+\omega_3,z)H_1^*(\omega_4,z)H_1(\omega_5,z)-H_1(\omega_1-\omega_2+\omega_3-\omega_4+\omega_5,z)}{G_1(\omega_1-\omega_2+\omega_3)+G_1^*(\omega_4)+G_1(\omega_5)-G_1(\omega_1-\omega_2+\omega_3-\omega_4+\omega_5)}$$

$$(2.94)$$

Higher-order terms can be derived with ease if higher accuracy is required. However, in practice, such higher order would not exceed the 5th rank. We can understand that for a length of a uniform optical fiber the 1st to nth order frequency spectrum transfer can be evaluated, indicating the linear to nonlinear effects of the optical signals transmitting through it. Indeed, the third- and fifth-order kernel transfer functions based on the Volterra series indicate the optical field amplitude of the frequency components, which contribute to the distortion of the propagated pulses. An inverse of these higher-order functions would give the signal distortion in the time domain. Thus, the VSTFs allow us to conduct distortion analysis of optical pulses and hence an evaluation of the bit-error-rate of optical fiber communications systems.

The superiority of such Volterra transfer function expressions allow us to evaluate each effect individually, especially the nonlinear effects, so that we can design and manage the optical communications systems under linear or nonlinear operations. Currently, this linear–nonlinear boundary of operations is critical for system implementation, especially for optical systems operating at 40 Gbps where linear operation and carrier-suppressed return-to-zero format is employed. As a norm in series expansion the series need converged to a final solution. It is this convergence that would allow us to evaluate the limit of nonlinearity in a system.

2.5.3 Transmission Bit Rate and the Dispersion Factor

The effect of dispersion on the system bit rate B_r is obvious and can be estimated by using the criterion:

$$B_r \cdot \Delta \tau < 1 \tag{2.95}$$

where $\Delta \tau$ is the total pulse broadening. When the fiber length, the total dispersion $D_T = M(\lambda) + D(\lambda)$, and a source line width σ_λ, the criterion becomes

$$B_r \cdot L \cdot |D_T| \sigma_\lambda \leq 1 \tag{2.96}$$

For a total dispersion factor of 1 ps/(nm km) and a semiconductor laser of line width of 2–4 nm, the bit rate-length product cannot exceed 100 Gb/s-km. That is, if a 100 km transmission distance is used, then the bit rate cannot be higher than 1.0 Gb/s. However, with the digital signal processing algorithms (DSP algo) in coherent reception, this pulse broadening can be compensated in the electronic digital domain and the dispersive transmission distance can reach a few thousands of kms if the modulation format is QPSK and intra-dyne reception is employed (Figures 2.29 and 2.30).

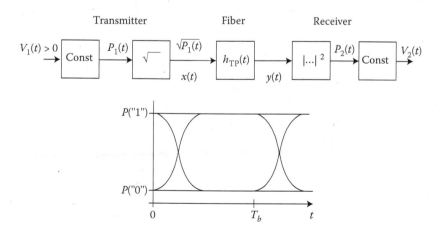

FIGURE 2.29
Schematic of an optical transmission system and its equivalent transfer functions.

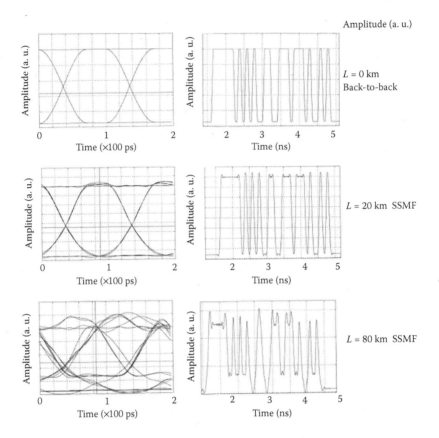

FIGURE 2.30
Eye diagram of time signals at 10 Gb/s transmission over standard SMF after 0, 20, 80 km length.

2.6 Fiber Nonlinearity Revisited

The nonlinear effects in optical fibers were described in Section 2.2. This section revisits these effects and their influence on the propagation of optical signals over long fiber lengths. The nonlinearity and linear effects in optical fibers can be classified as shown in Figure 2.31.

Fiber RI is dependent on both operating wavelengths and lightwave intensity. This intensity-dependent phenomenon is known as the Kerr effect and is the cause of fiber nonlinear effects.

2.6.1 SPM, XPM Effects

The power dependence of RI is expressed as

$$n' = n + \bar{n}_2(P/A_{eff})$$ (2.97)

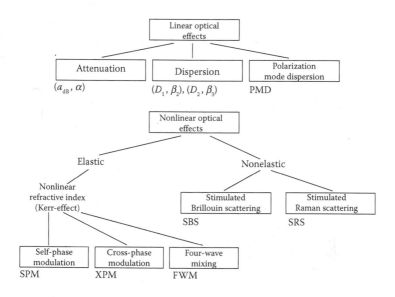

FIGURE 2.31

Linear and nonlinear fiber properties in single-mode optical fibers.

where P is the average optical power of the guided mode, $\bar{n}_2$ is the fiber non-linear coefficient, and A_{eff} is the effective area of the fiber.

Fiber nonlinear effects include intra-channel SPM, inter-channel XPM, FWM, stimulated Raman scattering (SRS), and stimulated Brillouin scattering (SBS). SRS and SBS are not the main degrading factors, as their effects are only getting noticeably large with very high optical power. On the other hand, FWM degrades severely the performance of an optical system with the generation of ghost pulses only if the phases of optical signals are matched with each other. However, with high local dispersions such as in SSMF, effects of FWM become negligible. In terms of XPM, its effects can be considered to be negligible in a DWDM system in the following scenarios: (i) highly locally dispersive system, and (ii) large-channel spacing. However, XPM should be taken into account for optical transmission systems deploying NZ-DSF fiber where local dispersion values are small. Thus, SPM is usually the dominant nonlinear effect for systems employing transmission fiber with high local dispersions, for example, SSMF and DCF. The effect of SPM is normally coupled with the nonlinear phase shift ϕ_{NL} defined as

$$\phi_{NL} = \int_0^L \gamma P(z)dz = \gamma L_{eff} P$$

$$\gamma = \omega_c \bar{n}_2 / (A_{eff} c) \tag{2.98}$$

$$L_{eff} = (1 - e^{-\alpha L}) / \alpha$$

where ω_c is the lightwave carrier, L_{eff} is the effective transmission length, and α is the fiber attenuation factor, which normally has a value of 0.17–0.2 dB/km in the 1550 nm spectral window. The temporal variation of the nonlinear phase ϕ_{NL} results in the generation of new spectral components far apart from the lightwave carrier ω_c, indicating the broadening of the signal spectrum. This spectral broadening $\delta\omega$ can be obtained from the time dependence of the nonlinear phase shift as follows:

$$\delta\omega = -\frac{\partial\phi_{NL}}{\partial T} = -\gamma\frac{\partial P}{\partial T}L_{eff} \tag{2.99}$$

Equation 2.99 indicates that $\delta\omega$ is proportional to the time derivative of the average signal power P. Additionally, the generation of new spectral components occur mainly at the rising and falling edges of optical pulses, that is, the amount of generated chirps is substantially larger for an increased steepness of the pulse edges.

The wave propagation equation can be represented as

$$\frac{\partial A(z,t)}{\partial z} + \frac{\alpha}{2}A(z,t) + \beta_1\frac{\partial A(z,t)}{\partial t} + \frac{j}{2}\beta_2\frac{\partial^2 A(z,t)}{\partial t^2} - \frac{1}{6}\beta_3\frac{\partial^3 A(z,t)}{\partial t^3}$$
$$= -j\gamma|A(z,t)|^2 A(z,t) - \frac{1}{\omega_0}\frac{\delta}{\delta t}(|A|^2 A) - T_R A\frac{\delta(|A|^2)}{\delta t} \tag{2.100}$$

in which we have ignored the pure delay factor involving β_1. The last term in the RHS represents the Raman scattering effects.

2.6.2 SPM and Modulation Instability

Nonlinear effects such as the Kerr effect where the refractive index of the fiber medium strongly depends on the intensity of the optical signal can severely impair the transmitted signals. In a nonsoliton system the Kerr effect broadens the signal optical spectrum through SPM. This broadened spectrum is mediated by fiber dispersion and causes a performance degradation. In addition, four-wave mixing (FWM) between the signal and the amplified spontaneous emission (ASE) noise generated from the inline optical amplifiers has been reported to cause performance degradations in systems using in-line optical amplifiers. This later effect is commonly referred to as modulation instability. Thus, the instability is resulted from the conversion of the phase noises to intensity noises. The gain spectrum of the modulation instability is shown in Figure 2.32 [32].

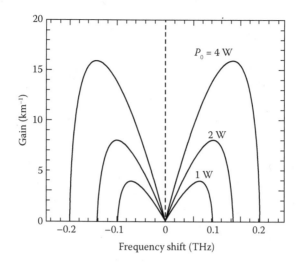

FIGURE 2.32

Spectrum of the optical gain due to modulation instability at three different average power levels in an optical fiber with $\beta_2 = 20$ ps²/km and $\gamma = 2$ W/km.

2.6.3 Effects of Mode Hopping

Up to now we have assumed that the source center emission wavelength was unaffected by the modulation. In fact, when a short current pulse is applied to a semiconductor laser, its center emission wavelength may hop from one mode to a longer neighboring wavelength. In the case where a multi-longitudinal mode laser is used, this hopping effect is negligible; however, it is very significant for a single longitudinal laser.

2.6.4 SPM and Intra-Channel Nonlinear Effects

Under considerations only the SPM of all the nonlinear effects on the optical signals transmitting through dispersive transmission link we can drop the cross coupling terms but $\Delta\Omega$, the nonlinear effect is thus also contributed the additional intra-channel effects with ω_1, ω_2 take the values with the spectra of the optical signal and not cross over the spectra of other adjacent channels. By substituting of the fundamental order transfer function we arrive at

$$H_3(\omega_1, \omega_2, \omega)_{\text{SPM,inter}} = -\frac{j\gamma_L}{4\pi^2}\left(e^{-j\omega^2 N_s \beta_2 L_s/2}\right)$$

$$\times (1 - e^{-(\alpha + j\beta_2\Delta\Omega)L_s})\left(L_s\sqrt{\alpha^2 + \beta_2^2\Delta\Omega^2}\, e^{j\tan^{-1}\frac{\alpha}{\beta_2\Delta\Omega}}\right)\sum_{k=0}^{N_s-1}e^{-jk\beta_2 L_s\Delta\Omega}$$

$$(2.101)$$

The nonlinear distortion noises contributed to the signals when operating under the two regimes of large and negligible dispersion are given in Ref. [33], depending on the dispersion factor of the fiber spans. The nonlinear transfer function H_3 indicates the power penalty due to nonlinear distortion, and can be approximated as [34]

$$H_3^i(\omega_1, \omega_2, \omega) \approx \begin{cases} j\gamma_L e^{-[\alpha/2 - j\beta_2(\omega_1 - \omega_2)(\omega - \omega_2)]} \\ \times [L_s^{eff} - j\beta_2(\omega_1 - \omega_2)(\omega - \omega_2)]\left(\dfrac{L_s - L_s^{eff}}{\alpha} - L_s L_s^{eff}\right) \end{cases} \tag{2.102}$$

Thus if the ASE noise of the inline optical amplifier is weak compared with signal power, then we can obtain the nonlinear distortion noises for *highly dispersive* fiber spans (e.g., G.652 standard single-mode fiber SSMF) via

$$K_{N,\beta}(\omega_0) = N_s \left[Q(\omega_0) + 2\left(\frac{\gamma_L}{2\pi}\right)^2 \frac{\Omega^2}{\alpha^2} \frac{P^3}{\Delta\omega_c^3} \partial\left(\frac{\omega_0}{\Delta\omega_c}, \frac{\beta\Omega^2}{\alpha}\right)\right] \tag{2.103}$$

and for *mildly dispersive* fiber spans (e.g., G.655 fiber spans):

$$K_{N,\beta \ll 20}(\omega_0) \approx N_s \left[Q(\omega_0) + 2\left(\frac{\gamma_L}{2\pi}\right)^2 \frac{\Omega^2}{\alpha^2} \frac{P^3}{\Delta\omega_c^3} \partial\left(\frac{\omega_0}{\Delta\omega_c}, 0\right)\right] \tag{2.104}$$

with

$$\partial_\chi(x, \xi) = \int_{-\infty}^{\infty} dx_1 \int_{-\infty}^{\infty} dx_2 * \left(\frac{(1 + e^{-\alpha L}) - 2e^{-\alpha L}\cos[\alpha L \xi(x - x_1)(x_1 - x_2)]}{1 + \xi^2(x - x_1)^2(x_1 - x_2)^2} \\ * \eta(x_1)\eta(x_2)\eta(x - x_1 + x_2)\right) \tag{2.105}$$

and

$$\eta(x) = \begin{cases} 1 & \text{for } x = [-1/2, 1/2] \\ 0 & \text{elsewhere} \end{cases} \tag{2.106}$$

The nonlinear power penalty thus consists of the linear optical amplifier noises; the second is the SPM noises from the input signal and nonlinear interference between the input and the optical amplifier noises, which may be ignored when the ASE is weak. Equations 2.103 and 2.104 show the variation of the penalty, hence channel capacity of dispersive fibers, of transmission systems operating under the influence of nonlinear effects with optically amplified multi-span transmission lines whose fiber dispersion parameter varying from 0 to −20 ps²/km. Under the scenario of nondispersive fiber, the spectral efficiency has been limited to about 3–4 b/s/Hz and 9–6 b/s/Hz, with 4 and 32 spans, respectively, for a dispersion factor of −20 ps²/km with 100 DWDM channels of 50 GHz spacing between the channels with the optical spectral noise density of 1 µW/GHz. The fiber length of each span is 80 km.

By definition, the nonlinear threshold is determined at a 1 dB penalty deviation level from the linear OSNR, the contribution of the nonlinear noise term, from Equation 2.103, we can obtain the maximal launched power at which there is an onset of the degradation of the channel capacity as

$$\max\_P = \sqrt{0.1\frac{\omega_c^3}{2N_s(\gamma_L/2\pi)^2(\Omega^2/\alpha^2)}} \qquad (2.107)$$

An example of the estimation of the maximum level of power per channel to be launched to the fiber before reaching the nonlinear threshold 1-dB penalty level: for 100 overall channels of 150 GHz spacing, $\Omega_T \approx 200$ nm, then $P_{th} \approx 58$ µW/GHz. Thus for 25 GHz bandwidth, we have the threshold power level at $P_{th\,\beta low} = 0.15$mW per channel. For highly dispersive and 8 wavelength channels we have $P_{th\,\beta high} \rightarrow 7\text{–}10 P_{th\,\beta low}$, or the threshold level may reach 1.5 mW/channel. The estimations given here, as an example, are consistent with the analytical expression obtained in Equation 2.107. Thus this shows clearly that (i) dispersive multi-span long-distance transmission under coherent ideal receiver would lead to better channel capacity than low dispersive transmission line. (ii) If a combination of low and high dispersive fiber spans is used, then we expect that, from our analytical Volterra approach, the penalty would reach the same level of threshold power so that a 1-dB penalty on the OSNR is reached. Note that this approach relies on the average level of optical power of the lightwave modulated sequence. This may not be easy to estimate if the simulation model is employed; and (iii) however under simulation, the estimation of average power cannot be done without costing extremely high time, thus commonly, the instantaneous power is estimated at the sampled time interval of a symbol. This sampled amplitude and hence the instantaneous power can be deduced. Therefore, the nonlinear phase is estimated and superimposed on the sampled complex envelope for further propagation along the fiber length. The sequence high-low dispersive spans would offer slightly better performance than low high combination. This can be due to the fact that for low dispersive fiber the output optical pulse would be higher in amplitude, which is to be launched into the high dispersive fibers, thus this would suffer higher nonlinear effects due to the fact that the instantaneous power launched into the fiber would be different—even the average power would be the same for both cases.

The argument in step (iii) can be further strengthened by representing a fiber span by the Volterra series transfer functions (VSTF) as shown in

Figure 2.33. Any swapping of the sequence of low and high dispersion fiber spans would offer the same power penalty due to nonlinear phase distortion, except the accumulated noises contributed from the ASE noises of the in-line optical amplifiers of all spans. Thus we could see that the noise figure (NF) of both configurations can be approximated as the same. This is contrasting to the simulated results reported in Ref. [35]. We believe that the difference in the power penalty in different order of arrangement of low and high dispersion fiber spans reported in Ref. [15] is due to numerical error, as the split-step Fourier method (SSFM) was possibly employed and the instantaneous amplitude of the complex envelope was commonly used. This does not indicate the total average signal power of all channels. Therefore we can conclude that the simulated nonlinear threshold power level would be suffering additional artificial OSNR, penalty due to the instantaneous power of the sampled complex amplitude of the propagating amplitude.

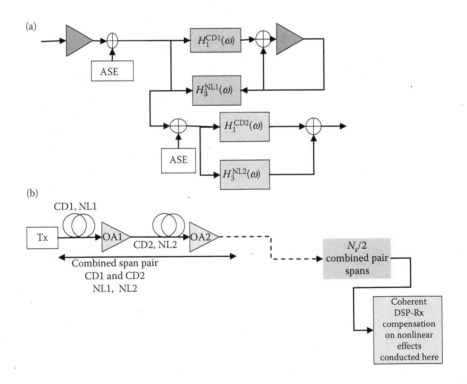

FIGURE 2.33
(a) System of concatenation of fiber spans consisting of pairs of different CDs and NL, model for simulation; (b) Optically amplified N_s-span fiber link without DCF, a DSP-based optical receiver with compensation of nonlinear effects by BP or Volterra series transfer function conducted by digital signal processing. This model represents real-time processing in practical optical transmission systems.

Ref. [36] reported the variation of the channel capacity against the input power/channel with dispersion as a parameter $-2 \rightarrow -20 \text{ ps}^2/\text{Km}$ and a noise power spectral density of 10 µW/GHz over 4 spans, and that for 4 and 32 spans of dispersive fibers of $0 \text{ µ}–20 \text{ ps}^2/\text{km}$ with a channel spacing of 50 GHz and 100 channels, the noise spectral density is 10 µW/GHz. The deviation of the capacity is observed at the onset of the power per channel of 0.1, 2, and 5 mW. Further observations can be made here. The noise responses indicate that the nonlinear frequency transfer function of a highly dispersive fiber link is related directly to the fundamental linear transfer function of the fiber link. When the transmission is highly dispersive the linear transfer function acts as a low-pass filter and thus all the energy concentrates in the passband of this filter, which may be lower than that of the signal at the transmitting end. This may thus lower the nonlinear effects as given in Equation 2.101. While for lower dispersive fiber this transfer function would represent a low-pass filter with 3 dB roll off frequency much higher than that of a dispersive fiber. For example, the G.655 would have a dispersion factor of about three times lower than that of the G.652 fiber. This wideband low-pass filter will allow the nonlinear effects of intra-channels and inter-channel interactions. The dispersive accumulation term $\sum_{k=0}^{N_s-1} e^{-jk\beta_2 L_s \Delta\Omega}$ dominates when the number of spans is high.

In the simulation results given in Ref. [37], the Volterra series transfer functions were applied for dispersive fiber spans. It is expected from Equation 2.101 that the arrangement of alternating position between G.655 and G.652 would not exert any penalty. The simulation reported in Ref. [38] indicates 1.5 dB difference at 10×2 spans (SSMF + TWC) and no difference at 20×2 spans. The contribution by the ASE noises of the optical amplifiers at the end of each span would influence the phase noises and hence the effects on the error vector magnitude (EVM) of the sampled signal detected constellation.

From the transfer functions, including both linear and nonlinear kernels of the dispersive fibers, we could see that if the noises are the same then the nonlinear effects would not be different regardless of whether high or low dispersive fiber spans are placed at the front or back. However, if the nonlinear noises are accounted for, and especially the intra-channel effects, we could see that if less dispersive fibers are placed in the front then higher noises are expected, and thus a lower nonlinear threshold (at which 1-dB penalty is reached on the OSNR). This is the opposite of the simulation results presented in Ref. [39]. However, these accumulated noises are much smaller than the average signal power. Under simulation, depending on the numerical approach to solve the NLSE, the estimation of signal power at the sampled instant is normally obtained from the sampled amplitude at this instant, and thus different with the average launched power into the fiber span. This creates discrepancies in the order of the high- or low-dispersion fiber spans. Thus, there are possibilities that the amplitude of the very dispersive pulse sequence at some instants along the propagation path can be superimposed. This amplitude may reach a level much higher than the nonlinear threshold and thus create different distortion penalty due to nonlinear effects.

Volterra transfer function (VSTF) offers better accuracy and covers a number of SPM and parametric scattering, but suffers costs of computing resources due to two-dimensional FFT for the SPM and XPM. This model should be employed when such extra nonlinear phase noises are required, such as in the case of superchannel transmission (Figure 2.33).

2.6.5 Nonlinear Phase Noises

Gordon and Mollenauer [40] showed that when optical amplifiers are used to compensate for fiber loss, the interaction of amplifier noise and the Kerr effect causes phase noise, even in systems using constant-intensity modulation. This nonlinear phase noise, often called the Gordon–Mollenauer effect or, more precisely, SPM-induced nonlinear phase noise, or simply nonlinear phase noise (NLPN), corrupts the received phase and limits transmission distance in systems using M-ary QAM. The NLPN in turn would create random variation in the intensity, thus a transfer or conversion of the NLPN into intensity noise, or modulation intensity.

Under the cascade of optically amplified spans to form a multi-span long-haul link without using dispersion compensating fiber (DCF) has emerged as the most modern optical link structure with coherent detection and digital signal processing at the receivers. Ho and Kahn [41] have studied and derived the variances and co-variances of DWDM optical transmission systems. For an electric field E_0 of the optical waves launched at the input of the first span, the field at the input of the k_{th} span would be the launched field superimposed by the noises accumulated over k spans as $E_k = E_0 + n_1 + n_2 + \cdots + n_{k-1}$, then the variance $\sigma_{\phi_{NL}}$ of the nonlinear phase shift is given as

$$\sigma_{NLPN}(\alpha_{NL}) = (\gamma L_{eff})^2 \left[\sigma_{NL}^2 (N-1) + (\alpha_{NL}-1)^2 f(N\sigma_1^2) - 2(\alpha_{NL}-1) \sum_{k=1}^{N} f(N\sigma_1^2) \right]$$

(2.108)

where α is the scaling factor, $f(N\sigma^2)$ is the expected value of the optical electric field between two consecutive spans, and σ is the variance of the field under superposition with the noises. N is the total number of optically amplified spans, the optimal factor can be found by differentiating Equation 2.108 with respect to this factor to give:

$$\alpha_{NL} \approx -\gamma L_{eff} \frac{N+1}{2}$$

(2.109)

At high OSNR $\gg 1$, this variance can be found to be

$$\sigma_{NLNP}^2 \approx \frac{4}{3} N^3 (\gamma L_{eff} \sigma |E_0|^2)^2 \tag{2.110}$$

with $\sigma \equiv \sigma_{|E_0 + n_1 + n_2 + \ldots n_k|}$ as the variance of the field superimposed by noises after kth span of k cascade spans. The expected value of the nonlinear phase shift can be approximated as

$$\langle \phi_{NL} \rangle \simeq N \gamma L_{eff} |E_0|^2 \tag{2.111}$$

Then the NLPN variance of N cascaded spans can then be given as

$$\sigma_{NLNP}^2 \approx \frac{4}{3} N^3 (\gamma L_{eff} \sigma |E_0|^2) = \frac{4}{3} N \frac{\langle \phi_{NL} \rangle^2}{OSNR_L^2} \tag{2.112}$$

where $OSNR_L$ is the optical signal-to-noise ratio in linear scale and the mean phase noise is given by Equation 2.111. Thus we can see that the nonlinear phase rotation due to SPM in N-cascade-span link is the total phase rotation accumulated over the spans.

The variances of the residual NLPN is also given as

$$\sigma_{NLNP,res}^2 \approx \frac{1}{6} \frac{\langle \phi_{NL} \rangle^2}{OSNR_L^2} \tag{2.113}$$

The NLPN power variance is also proportional to the square of the accumulated phase rotation. This allows a compensation algorithm for nonlinear impairments by rotating the phase of the digital sampled of the in phase and quadrature phase components at the end of each span. This is indeed a linear operation based on the derived and observed rotation of the constellation due to SPM. This linear-phase rotation simplifies the numerical and hence the computing resources of the DSP. The phase to intensity conversion, the instability problem, will create some degradation of the OSNR due to this increase of the noise intensity over the 0.4 nm band commonly measured of the noises in practice. Thus, we expect a logrithmic reduction of the OSNR with respect to the number of cascaded optically amplified fiber spans to the SPM-induced and modulation instability.

2.7 Special Dispersion Optical Fibers

At the beginning of the 1980s, there was great interest to reduce the total dispersion $[M(\lambda) + D(\lambda)]$ of single-mode optical fiber at 1550 nm where the loss is lowest for silica fiber. There were two significant trends; one is to reduce the line-width and to stabilize the laser center wavelength, and the other is to reduce the dispersion at this wavelength. The fibers designed for long-haul transmission systems usually exhibit a near-zero dispersion at a certain spectral window. These are called dispersion-shifted fibers, within which the total dispersion approaches zero, $[M(\lambda) + D(\lambda)] \sim 0$. The material dispersion factor $M(\lambda)$ is natural and slightly affected by vPulse response from near fieldariation of doping material and concentration. The waveguide dispersion factor, $D(\lambda)$ can be tailored by designing appropriate refractive index profiles and geometrical structure, to balance the material dispersion effects so that the total dispersion factor reaches a null value at a specific wavelength or spectral window. Note that the dispersion factors due to material and waveguide take algebraic values, thus they can be designed to take opposite values to cancel each other.

The problems facing dispersion-shifted fibers are that the four-wave mixing (FWM) can occur easily due to the phase matching condition that can be satisfied at the zero dispersion from the equally spaced wavelength channels. Thus, usually the zero dispersion wavelength is shifted to outside the C-band to avoid FWM. These types of fibers are called nonzero dispersion-shifted fibers (NZ-DSF), within which the zero-dispersion wavelength commonly placed around 1510 nm so that only some small dispersion amount occur in the C-band to avoid the FWM problems.

Advanced optical fiber design technique can offer the design of dispersion-flattening fibers where the dispersion factor is flat over the wavelength range from 1300 to 1600 nm by tailoring the refractive index profile of the core of optical fibers in such distribution as the W-profile, the segmented profile and multilayer core structure, and so on.

Another type of optical fiber which would be required for compensating the dispersion effect on the optical signal after transmitted over a length of optical fiber is the dispersion compensated fiber whose dispersion factor is many times larger than that of the standard communication fiber with an opposite sign. This can be designed by setting the total dispersion to the required compensated dispersion and thus the waveguide dispersion can be found over the required operating range. Therefore, optical fiber structures can be designed to obtain the core radius, and the refractive index profile with optimum mode spot size.

2.8 SMF Transfer Function: Simplified Linear and Nonlinear Operating Region

In this section, a closed expression of the frequency transfer function of dispersive and nonlinear single-mode optical fibers for broadband operation can be derived, similar for the case under microwave photonics. The expression takes into account both chromatic dispersion and SPM effects, and is valid for optical double-sideband modulation, optical single-sideband (SSB) modulation, and chirped optical transmitters.

The evolution along the propagation path z of the "small-signal" intensity modulation (IM), or complex power $\bar{p}(\omega, z)$ and phase rotation (PM) $\bar{\phi}(\omega, z)$ during the propagation of the guided mode through the single-mode optical fiber (SMF), taking into account both the chromatic dispersion and the nonlinearity (SPM) effects, is governed by the following set of differential equations [42–45]:

$$\frac{\delta \bar{p}(\omega, z)}{\delta z} = \beta_2 \omega^2 \bar{P}_0 \bar{\phi}(\omega, z); \quad \bar{A}(\omega, z) = \sqrt{\bar{p}(\omega, z)} \tag{2.114}$$

$$\frac{\delta \bar{\phi}(\omega, z)}{\delta z} = -\left[\frac{\beta_2 \omega^2}{4 \bar{P}_0} + \gamma e^{-\alpha z} \right] \bar{p}(\omega, z) \tag{2.115}$$

where $\bar{A}(\omega, z)$ and $\bar{\phi}(\omega, z)$ are defined as the normalized complex amplitude and phase, respectively, of the optical field in the Fourier domain, ω is the radial frequency of the RF or broadband signal, z is the distance along the propagation axis of the fiber, α is the attenuation coefficient in the linear scale of SMF, and β_2 is the first order dispersion coefficient, that is, the group delay factor as a function of the optical wavelength given by

$$\beta_2 = -\frac{\lambda^2 D(\lambda)}{2\pi c} \tag{2.116}$$

whereby c is the velocity of light in vacuum, and $D(\lambda)$ is the dispersion factor of the fiber, typically taking value of 17 ps/nm/km for silica SMF at the operating wavelength $\lambda = 1550$ nm.

γ is the nonlinear SPM coefficient defined by

$$\gamma = \frac{2\pi n_2}{\lambda A_{eff}}; \quad \text{with } A_{eff} = \pi r_0^2 \tag{2.117}$$

with n_2 as the nonlinear index coefficient of the fiber, typically $n_2 = 1.3 \times 10^{-23}$ m^2/W for SMF-28, and $A_{eff} = \pi r_0^2$ is the effective area of the fiber, which is the area of the Gaussian mode spot of the guided mode in a single-mode optical fiber under the weakly guiding condition [46].

These equations are derived from the observer positioned on the moving frame of the phase velocity of the waves, which is normally expressed by the nonlinear Schrodinger equation (NLSE):

$$\frac{\partial A(t, z)}{\partial z} = -\left[\alpha A(t, z) + \frac{j}{2} \beta_2 \frac{\partial A^2(t, z)}{\partial t^2} \right] + j\gamma |A(t, z)|^2 A(t, z) \tag{2.118}$$

Present coherent optical receiver incorporating ADC and digital signal processing (DSP), under the quadrature amplitude modulation (QAM), the amplitudes of both the inphase and quadrature phase components are recovered which is proportional to the power of the optical signals incident at the front end of the optical hybrid coupler followed by balanced photodetector pairs. A schematic of the transmission is shown in Figure 2.34, in which the transmitter can generate an optical sequence or near single frequency sinusoidal waves at a frequency reaching 30 GHz using a Fujitsu DAC sampling rate of 65 GSa/s. The optical modulator is a typical Fujitsu IQ modulator modulated by electrical signals output from the DAC and phase shifted in RF domain by an electrical phase shifter PS. The RF phase can be set such that when they are $\pi/2$ shifted with respect to each other, the suppression of one of the single sidebands can be achieved at the output spectrum. The main carrier can be suppressed by biasing the "children" Mach–Zehnder intensity modulators (MZIM) at the minimum transmission point (Figure 2.35).

By differentiating Equation 2.114 and substituting into Equation 2.38 we obtain

$$\frac{\delta^2 \overline{p}(\omega, z)}{\delta z^2} = -\left[\frac{\beta_2^2 \omega^4}{4} - \beta_2 \omega^2 + \gamma \overline{P}_0 e^{-\alpha z} \right] \overline{p}(\omega, z) \qquad (2.119)$$

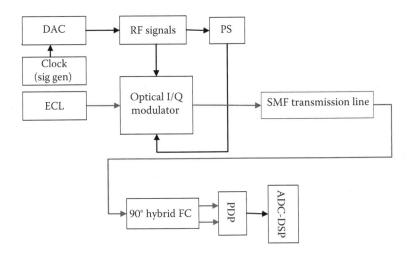

FIGURE 2.34
Digital-based optical transmitter and coherent reception with real-time sampling and digital signal processing. DAC = digital to analog converter; ADC = analog-to-digital converter; DSP = digital signal processing; PDP = photodetector pair; FC = fiber coupler; I/Q = inphase/quadrature phase.

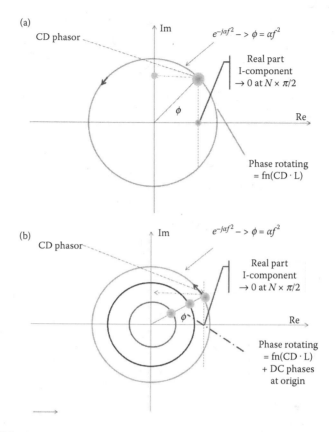

FIGURE 2.35
Phasor of the amplitude evolution along the propagation path, assuming the amplitude is not affected by attenuation. (a) Phasor; (b) phase or the phase constellation of M-QAM ($M = 16$ square QAM), three amplitude level.

Subject to the initial conditions of

$$\bar{p}(\omega,0) = \bar{p}_{in}(\omega) \quad \text{and} \quad \frac{\delta \bar{p}(\omega,0)}{\delta z} = \beta_2 \omega^2 \bar{P}_0 e^{-\alpha z} \phi(\omega,0) = \beta_2 \omega^2 \bar{P}_0 e^{-\alpha z} \phi_{in}(\omega) \quad (2.120)$$

where the subscript "*in*" indicates the input location, which is the starting of the propagation of the modulated optical waves through the SMF.

Now by changing some variables with the setting of

$$x = 2\sqrt{B}e^{-\alpha z/2}; \quad B = -\frac{\beta_2 \omega^2 \gamma P_0}{\alpha^2} \quad (2.121)$$

then Equation 2.119 can be rewritten as

$$\left[x\frac{\partial^2}{\partial x^2} + x\frac{\partial}{\partial x} - (x^2 - v^2) \right]\overline{p}(\omega, z) = 0 \qquad (2.122)$$

The solution of this equation is a combination of purely imaginary Bessel's functions L and K and subject to the initial conditions of Equation 2.120. Thus, the evolution of the complex amplitude of the intensity modulated (IM) optical waves along the propagation path is given by [47,48]

$$\overline{p}(\omega, z) = \frac{2\sinh(\pi v)}{\pi}\left\{ \begin{matrix} \sqrt{B}\overline{p}_{in}(\omega)\left[\dfrac{\partial L_{iv}}{\partial x}(2\sqrt{B})K_{iv}(x) - \dfrac{\partial K_{iv}}{\partial x}(2\sqrt{B})I_{iv}(x) \right] \\ + \dfrac{\alpha B}{\gamma}\phi_{in}(\omega)\left[K_{iv}(2\sqrt{B})L_{iv}(x) - L_{iv}(2\sqrt{B})K_{iv}(x) \right] \end{matrix} \right\}$$

$$(2.123)$$

with $v = -(\beta_2\omega^2/\alpha)$. The first term on the RHS of Equation 2.123 is the magnitude part and the second is the phase part, that is the inphase and quadrature components of the QAM signal. Thus the in phase and quadrature parts of the complex magnitude can be expressed as

$$\overline{p}_I(\omega, z) = \frac{2\sinh(\pi v)}{\pi}\left\{ \sqrt{B}\overline{p}_{in}(\omega)\left[\frac{\partial L_{iv}}{\partial x}(2\sqrt{B})K_{iv}(x) - \frac{\partial K_{iv}}{\partial x}(2\sqrt{B})I_{iv}(x) \right] \right\} \qquad (2.124)$$

$$\overline{p}_Q(\omega, z) = \frac{\alpha B}{\gamma}\phi_{in}(\omega)\frac{2\sinh(\pi v)}{\pi}[K_{iv}(2\sqrt{B})L_{iv}(x) - L_{iv}(2\sqrt{B})K_{iv}(x)] \qquad (2.125)$$

The variations of the inphase (real part) and quadrature phase (imaginary part) of the complex power can be estimated by referring to Figures 2.36 and 2.37, respectively. The inphase and quadrature phase components move along the horizontal and vertical axis within the normalized ±1 limits, meaning that as the phases rotate around the unit circle, these components oscillate in a manner such that when the phase is $(2M + 1)$ $(M = 0,1,2...)$ or an odd number of $\pi/2$, then the inphase component becomes nullified and so on. Likewise, the quadrature phase are zero at $N\pi$ $(N = 0,1,2...)$. In the case of QAM scheme, for example, 16QAM, there would be three amplitude levels of the phase constellation and these levels are rotating as shown in Figure 2.37, the initial phase is set by the initial position of the constellation point of square 16 QAM, but the oscillation and nullified locations would be very much similar to that of Figure 2.36.

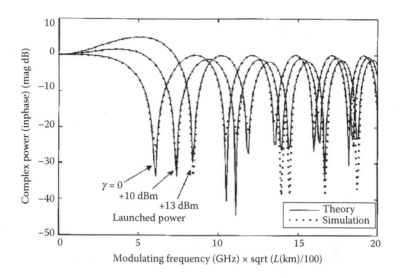

FIGURE 2.36
Variation of the magnitude of the optical field intensity with frequency, the intensity frequency response of standard SMF of $L = 100$ km (at $z = 100$ km) under coherent detection with normalized amplitude of the intensity of the optical waves under linear ($\gamma = 0$) and nonlinear operating conditions. SSMF parameters: $\alpha = 0.2$ dB/km, $D = 17$ ps^2/(nm km); $n_2 = 3.2 \times 10^{-23}$ m^2/W. $L = 100$ km.

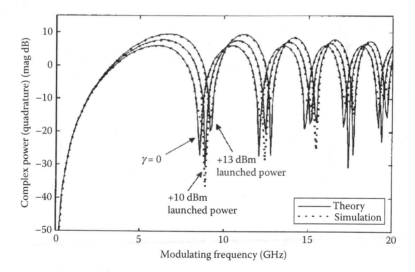

FIGURE 2.37
Variation of the magnitude of phase component with frequency (PM-IM conversion) of standard SMF of $L = 100$ km (at $z = 100$ km) under coherent detection with normalized amplitude of the intensity of the optical waves under linear ($\gamma = 0$) and nonlinear operating conditions. SSMF parameters: $\alpha = 0.2$ dB/km, $D = 17$ ps^2/(nm km); $n_2 = 3.2 \times 10^{-23}$ m^2/W. $L = 100$ km. (Extracted from F. Ramos and J. Marti, *IEEE Photonic Tech Lett*, 12(5), 549–551, May 2000.)

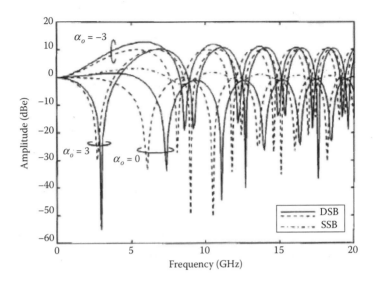

FIGURE 2.38
Frequency response of 100 km SSMF for DSB and SSB spectral signals under linear and non-linear (+10 dBm). – – – = linear ; continuous line nonlinear DSB and —·—·— SSB with chirp parameters α of a directly modulated laser diode as a parameter. Note the flat response of SSB signals.

Under linear operating regime $\gamma = 0$, we can obtain the expressions for the complex power amplitude and phase and the overall fiber transfer function under a modulation transfer as:

$$\bar{p}_I(\omega, z) = \cos \frac{\beta_2 \omega^2 z}{2}$$

$$p_\phi(\omega, z) = 2\sin \frac{\beta_2 \omega^2 z}{2}$$

$$H_F(\omega, z) = \bar{p}_I(\omega, z) + \frac{H_{PM}(\omega, z)}{2} p_Q(\omega, z)$$

$$H_{PM}(\omega, z) = \text{frequency response modulated signals} \tag{2.126}$$

For single-sideband (SSB) modulated signals $H_{PM}(\omega, z) = j$, which are purely complex, then we can obtain the transfer characteristics as shown in Figure 2.38. Note that for SSB signals the frequency response is flat over a very wide band and the notches of the linear and nonlinear responses are significantly reduced. This is due to the single side band signals that are at least half the band of that of the DSB, and thus the nonlinear effects are also reduced. The nonlinearity is estimated with the average power launched into the 100 km fiber at 10 dBm.

2.9 Numerical Solution: Split-Step Fourier Method

In practice, with the extremely high-speed operation of the transmission systems, it is very costly to simulate by experiment, especially when the fiber transmission line is very long, for example, a few thousands of kms. Then it is preferred to conduct computer simulations to guide the experimental set-up. In such simulation the propagation of modulated lightwave channels play a crucial role so as to achieve transmission performance of the systems closed to practical ones. The main challenge in the simulation of the propagation of lightwave channels employing the nonlinear Schroedinger equation (NLSE), which can be derived from Maxwell equations, is whether the signal presented in the time domain can be propagating though the fiber and its equivalent in the frequency domain when the nonlinearity is effective. The propagation techniques for such modulated channels are described in the subsection of this part.

2.9.1 Symmetrical Split-Step Fourier Method

The evolution of slow varying complex envelopes $A(z,t)$ of optical pulses along a single-mode optical fiber is governed by the nonlinear Schroedinger equation (NLSE):

$$\frac{\partial A(z,t)}{\partial z} + \frac{\alpha}{2} A(z,t) + \beta_1 \frac{\partial A(z,t)}{\partial t} + \frac{j}{2} \beta_2 \frac{\partial^2 A(z,t)}{\partial t^2} - \frac{1}{6} \beta_3 \frac{\partial^3 A(z,t)}{\partial t^3}$$

$$= -j\gamma |A(z,t)|^2 A(z,t) \tag{2.127}$$

where z is the spatial longitudinal coordinate, α accounts for fiber attenuation, β_1 indicates DGD, β_2 and β_3 represent second- and third-order dispersion factors of fiber CD, and γ is the nonlinear coefficient as also defined above. In a single-channel transmission, (2.127) includes the following effects: fiber attenuation, fiber CD and PMD, dispersion slope, and SPM nonlinearity. Fluctuation of optical intensity caused by Gordon–Mollenauer effect is also included in this equation. We can observe that the term involves β_2, and β_3 relates to the phase evolution of optical carriers under the pulse envelop. Respectively, the term β_1 relates to the delay of the pulse when propagating through a length of the fiber. So if the observer is situated on the top of the pulse envelop then this delay term can be eliminated.

The solution of NLSE and hence the modeling of pulse propagation along a single-mode optical fiber is solved numerically by using the SSFM so as to facilitate the solution of a nonlinear equation when nonlinearity is involved. In SSFM, fiber length is divided into a large number of small segments δz. In practice, fiber dispersion and nonlinearity are mutually interactive at any distance along the fiber. However, these mutual effects are small within δz and, thus, effects of fiber dispersion and fiber nonlinearity over δz are assumed to be statistically independent of each other. As a result, SSFM can separately define two operators: (i) the linear operator that involves fiber attenuation and fiber dispersion effects and, (ii) the nonlinearity operator that takes into account fiber nonlinearities. These linear and nonlinear operators are formulated as follows:

$$\hat{D} = -\frac{j\beta_2}{2}\frac{\partial^2}{\partial T^2} + \frac{\beta_3}{6}\frac{\partial^3}{\partial T^3} - \frac{\alpha}{2} \tag{2.128}$$

$$\hat{N} = j\gamma\,|A|^2$$

where $j = \sqrt{-1}$, A replaces $A(z,t)$ for simpler notation, and $T = t - z/v_g$ is the reference time frame moving at the group velocity, meaning that the observer is situated on top of the pulse envelop. Equation 2.128 can be rewritten in a shorter form, given by

$$\frac{\partial A}{\partial z} = (\hat{D} + \hat{N})A \tag{2.129}$$

and the complex amplitudes of optical pulses propagating from z to $z + \delta z$ are calculated using the following approximation:

$$A(z + h, T) \approx \exp(h\hat{D})\exp(h\hat{N})A(z, T) \tag{2.130}$$

Equation 2.130 is accurate to the second order of the step size δz. The accuracy of SSFM can be improved by including the effect of fiber nonlinearity in the middle of the segment rather than at the segment boundary. This modified SSFM is known as the symmetric SSFM (Figure 2.39).

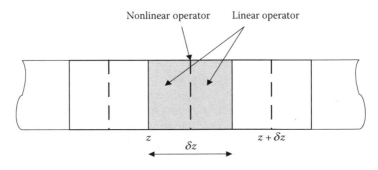

FIGURE 2.39
Schematic illustration of symmetric SSFM.

Equation 2.130 can now be modified as

$$A(z + \delta z, T) \approx \exp\left(\frac{\delta z}{2}\hat{D}\right)\exp\left(\int\limits_{z}^{z+\delta z}\hat{N}(z')dz'\right)\exp\left(\frac{\delta z}{2}\hat{D}\right)A(z,T) \qquad (2.131)$$

This method is accurate to the third order of the step size δz. In symmetric SSFM, the optical pulse propagates along a fiber segment δz in two stages. First, the optical pulse propagates through the linear operator that has a step of $\delta z/2$ in which the fiber attenuation and dispersion effects are taken into account. An FFT is used here in the propagation step so that the output of this half-size step is in the frequency domain. Note that the carrier is removed here and the phase of the carrier is represented by the complex part, the phase evolution. Hence, the term complex amplitude is coined.

Then, the fiber nonlinearity is superimposed to the frequency domain pulse spectrum at the middle of the segment. After that, the pulse propagates through the second half of the linear operator via an inverse FFT to get the pulse envelop back in the time domain.

The process continues repetitively over consecutive segments of size δz until the end of the fiber length. It should be again noted that the linear operator is computed in the time domain while the nonlinear operator is calculated in the frequency domain.

2.9.1.1 Modeling of Polarization Mode Dispersion

As described above, polarization mode dispersion (PMD) is a result of the delay difference between the propagation of each polarized mode of the LP modes LP_{01}^{H} and LP_{01}^{V} of the horizontal and vertical directions, respectively, as illustrated in Figure 2.4. The parameter differential group delay (DGD), determines the first-order PMD, which can be implemented by modeling the optical fiber as two separate paths representing the propagation of two polarization states. The symmetrical SSFM can be implemented in each step on each polarized transmission path and then their outputs are superimposed to form the output optical field of the propagated signals. The transfer function to represent the first-order PMD is given by

$$H(f) = H^{+}(f) + H^{-}(f) \qquad (2.132)$$

where

$$H^{+}(f) = \sqrt{k}\exp\left[j2\pi f\left(-\frac{\Delta\tau}{2}\right)\right] \qquad (2.133)$$

and

$$H^-(f) = \sqrt{k} \exp\left[j2\pi f\left(-\frac{\Delta\tau}{2} \right) \right] \qquad (2.134)$$

in which k is the power splitting ratio, $k = 0.5$ when a 3-dB or 50:50 optical coupler/splitter is used, and $\Delta\tau$ is the instantaneous DGD value in which the average value of a statistical distribution follows a Maxwell distribution (refer to Equation 2.78) [50,51]. This randomness is due to the random variations of the core geometry, the fiber stress and hence anisotropy due to the drawing process, variation of temperatures, and so on in installed fibers.

2.9.1.2 Optimization of Symmetrical SSFM

2.9.1.2.1 Optimization of Computational Time

A huge amount of time can be spent for the symmetric SSFM via the uses of FFT and IFFT operations, in particular when fiber nonlinear effects are involved. In practice, when optical pulses propagate toward the end of a fiber span, the pulse intensity has been greatly attenuated due to the fiber attenuation. As a result, fiber nonlinear effects are getting negligible for the rest of that fiber span and, hence, the transmission is operating in a linear domain in this range. In this research, a technique to configure symmetric SSFM is proposed in order to reduce the computational time. If the peak power of an optical pulse is lower than the nonlinear threshold of the transmission fiber, for example around −4 dBm, symmetrical SSFM is switched to a linear mode operation. This linear mode involves only fiber dispersions and fiber attenuation and its low-pass equivalent transfer function for the optical fiber is

$$H(\varpi) = \exp\left\{ -j[(1/2)\beta_2\varpi^2 + (1/6)\beta_3\varpi^3] \right\} \qquad (2.135)$$

If β_3 is not considered in this fiber transfer function, which is normally the case due to its negligible effects on 40 Gb/s and lower bit rate transmission systems, the above transfer function has a parabolic phase profile [52,53].

2.9.1.2.2 Mitigation of Windowing Effect and Waveform Discontinuity

In symmetric SSFM, mathematical operations of FFT and IFFT play very significant roles. However, due to a finite window length required for FFT and IFFT operations, these operations normally introduce overshooting at two boundary regions of the FFT window, commonly known as the windowing effect of FFT. In addition, since the FFT operation is a block-based process, there exists the issue of waveform discontinuity, that is, the right-most sample of the current output block does not start at the same position of the

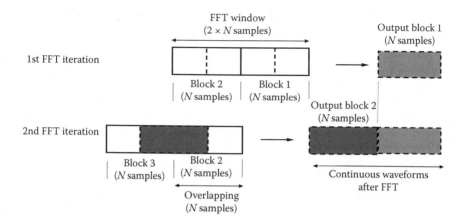

FIGURE 2.40
Proposed technique for mitigating windowing effect and waveform discontinuity caused by FFT/IFFT operations.

left-most sample of the previous output block. The windowing effect and the waveform discontinuity problems are resolved with the following technique, also seen in Figure 2.40.

The actual window length for FFT/IFFT operations consists of two blocks of samples, hence $2N$ sample length. The output, however, is a truncated version with the length of one block (N samples) and output samples are taken in the middle of the two input blocks. The next FFT window overlaps the previous one by one block of N samples.

2.10 Nonlinear Fiber Transfer Functions and Compensations in Digital Signal Processing

Nonlinear effects have been considered in the previous section in which the SPM effects play the major role in the distortion of modulated signals, besides the linear chromatic dispersion effects. We have also seen that the NLSE has been used extensively in the modeling of the modulated lightwave signals propagating through optical fiber links in which both linear and nonlinear effects are included.

In practice we have seen many optical components such as the fiber Bragg gratings, dispersion compensating fibers (DCF), or optical fiber filter structures [54] to compensate for chromatic dispersion effects in the optical domain as described in Sections 2.2 through 2.4. Nonlinear dispersion compensation can also be compensated in the optical domain by phase conjugators [55,56], but these are required to be placed exactly at the midway of optical fiber links that would be hard to be determined. However, under current high-speed optical communication technology, DSP-based receiver, the signals are processed in the electronic domain after the electronic pre-amplifier and the ADC. Thus it is possible to compensate for both the linear and nonlinear dispersion if algorithms can be found to do the reverse dispersion processes in the electronic domain to minimize the signal distortion. These algorithms would be developed if such transfer functions of the fibers operating in linear and nonlinear regions can be simplified so as to cost the least number of processing steps for processors working at ultra-high speed [57,58]. The schematic of the optical coherent receiver in the long-haul optical fiber communication system is shown in Figure 2.41. Both transmitters and receivers can integrate digital signal processors before and after the optical amplifier multi-span optical fiber transmission link. The fiber link can be represented by a canonical form of transfer functions. It is noted here that the sampler must operate at a very high rate, normally at about 56 or 64 GSa/s [59], thus the DSP would have minimum memory banks and processing speed must be high enough so that real-time processing of signals can be achieved. Hence, algorithms must be very efficient and take minimal time.

This section thus is dedicated to some recent development in representing the transfer functions of optical fibers for signal propagation and compensations, with applications especially in the electronic domain.

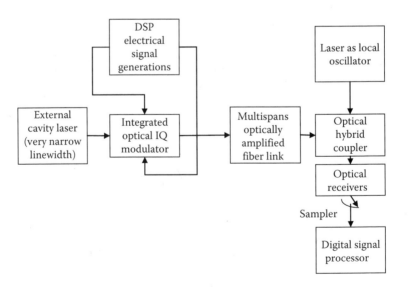

FIGURE 2.41

Schematic form of an optical receiver based on a digital signal processor using coherent detection in a modern optically amplified fiber link transmission system.

2.10.1 Cascades of Linear and Nonlinear Transfer Functions in Time and Frequency Domains

To reduce computational requirements at the receiver and assuming that the nonlinear phase rotation on the optical carrier can be separable from the linear phase effects, one can represent the transfer functions of the propagation of the optical pulse sequence over a length L by a cascade of linear and nonlinear phase rotation as

$$E_n(t, z + h) = E(t, z)e^{j\gamma h|E(t,z)|^2} \tag{2.136}$$

and

$$\tilde{E}(\omega, z + h) = \tilde{E}_n(\omega, z + h)e^{-j\left(\frac{\alpha}{2} + \frac{\beta_2}{2}\omega^2\right)h} \tag{2.137}$$

where the nonlinear phase is multiplied by the signals envelope at the input of a fiber length. This nonlinear phase is estimated under a number of considerations so that it is valid under certain constraints. h is the step size, as we have assumed in the previous section, but can also take a much larger distance, thus reduction of computational resources. $\tilde{E}(\omega, z + h)$ and $\tilde{E}_n(\omega, z + h)$ are the approximated fields at the input and output of the fiber over a step of order n. Clearly from Equation 2.136 we can observe that the phase accumulated over the distance step h is contributed to the rotation of the phase of the carrier, while Equation 2.137 represents the rotation of the phase of the carrier after propagating through h by the linear GVD effect evaluated in spectral domain. Thus the transfer function of the linear dispersion effect given can be employed together with the nonlinear phase contribution as shown in Figure 2.42 over the whole transmission link of N spans or cascades of span by span over the whole link.

The assumptions and observation through experiments of the nonlinear phase effects on transmission of signals are as follows:

- Amplitude-dependent phase rotation to improve system performance has been demonstrated in Refs. [60] and [61] over short fiber spans with nearly perfect CD compensation per span.

- The received signal has a spiral-shaped constellation as reported in Ref. [62]. It is possible to exploit the correlation between the received amplitude and nonlinear phase shift to reduce nonlinear phase noise variance under simulation using the SSFM. This spiral rotation leads to conclusions that under possibly weak nonlinear effects the phase can be superimposed on the modulated signals as an additional phase of components. Thus the cascade of nonlinear-phase superposition and linear transfer function.

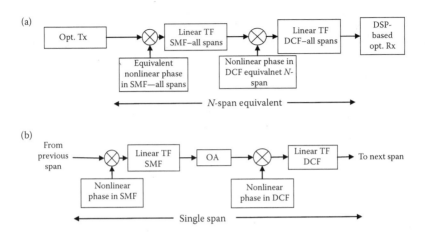

FIGURE 2.42
Representation of optical signals propagating through optical fibers with separable nonlinear and linear transfer function (a) equivalent all spans; (b) equivalent for each span.

The effective length of each step must be evaluated so as not to over compensate. This effective length can be estimated as given in Ref. [63], in which SSMF is typically about 22 km, with a nonlinear coefficient of 2.1e–20 m/W (Figure 2.43).

Once the nonlinear phase noises can be represented as a phase superposition on the signals, under coherent detection the optical field would be detected and presented as an electronic current or voltages at the output of an optical receiver whose signals are then sampled by a high-speed sampler to convert to digital domain and processed by digital signal processor (DSP). The compensation of nonlinear-phase noises is then conducted in the digital domain and thus a BP algorithm is required. This algorithm can be

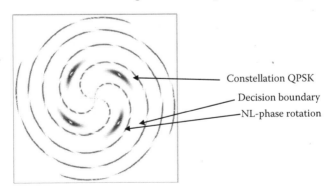

FIGURE 2.43
Received signal constellation of QPSK under coherent detection over 5000 km of SSMF under nonlinear effects and linear dispersion with decision boundaries (spiral lines) for detection. (Extracted from A.P.T. Lau and J.M. Kahn, *IEEE J. Lightw Tech.*, 25(10), 3008, Oct. 2007. With permission.)

implemented by forward propagation with a nonlinear coefficient of sign opposite to that of the transmission fiber. The numerical implementation of such transfer function and phase superposition given in Equations 2.136 and 2.137 is quite straight forward and numerically effective, as the phase over the propagation step can be over one span or sections of spans or even the whole transmission link [65]. However, the compensation may be too much and thus distortion does also happen. In this case, there must be an adaptive technique to monitor the compensation process so that when the nonlinear phase distortion is just completely compensated, then the process is ceased.

2.10.2 Volterra Nonlinear Transfer Function and Electronic Compensation

As described in Section 2.5, the wave propagation inside an SMF can be governed by a simplified version of the NLSE, in which only SPM are effected. It is noted that $A = A(t,z)$ is the electric field envelop of the optical signal, β_2 is the second-order dispersion parameter, α is the fiber attenuation coefficient, and γ_L is the nonlinear coefficient of the fiber. The solution of the NLSE can be written with Volterra series transfer functions (VSTFs) of kernels of the fundamental order and $(2N + 1)$th order as described in Ref. [66]. However, up to the third order is sufficient to represent the weak nonlinear effects in the slowly varying amplitude of the guide wave propagating in a single-mode weakly guiding fiber. The frequency domain of the amplitude along the transmission line is given as [67]

$$A(\omega,z) = \begin{cases} H_1(\omega,z)A(\omega) + \displaystyle\iint_{-\infty}^{\infty} H_3(\omega_1,\omega_2,\omega,z) \\ \times A(\omega_1)A^*(\omega_2)A(\omega - \omega_1 + \omega_2)d\omega_1 d\omega_2 \end{cases} \tag{2.138}$$

$$H_1(\omega,z) = e^{-\alpha z/2}e^{-j\omega^2 \beta_2 z/2} \tag{2.139}$$

$$H_3(\omega_1,\omega_2,\omega,z) = \begin{cases} -\dfrac{j\gamma}{4\pi^2} H_1(\omega,z) \\ \times \dfrac{1 - e^{-(\alpha + j\beta_2(\omega_1 - \omega)(\omega_1 - \omega_2))z}}{\alpha + j\beta_2(\omega_1 - \omega)(\omega_1 - \omega_2)} \end{cases} \tag{2.140}$$

where $A(\omega) = A(\omega, z = 0)$ represents the optical pulse at the input of the fiber in the frequency domain, and ω_1, ω_2 and ω are the dummy variables acting as parameters indicating the cross interactions of the light waves at different frequencies, that is, intra- or inter- channels, especially the inter-channel interaction effects. The range of these spectral variables changes

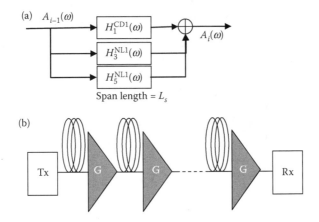

FIGURE 2.44
(a) Representation of a fiber span by VSTFs of first and higher order (up to fifth order), (b) cascade of optically amplified N_s-span fiber link without DCF.

from $(-\infty, +\infty)$. Thus, we can observe that (ω_1, ω_2) form a plane of the angular frequency components and the angular frequency ω can be scanned across all regions to see the interactions of the nonlinear effects. We can distinguish the regions on this plane, and different nonlinear effects after the propagation of the lightwaves in the nonlinear regime.

Thus there are regions where there are cross terms indicating the interaction of different and nonidentical frequency components of the signal spectra. Thus these cross terms are the inter-modulation terms, that is, due to XPM as commonly known. The term $j\beta_2(\omega_1 - \omega)(\omega_1 - \omega_2)$ accounts for the waveform distortion within a single span. Higher-order kernels, for example the fifth-order kernel $H_5(\omega_1, \omega_2, \omega_3, \omega_4, \omega)$, can be used if higher accuracy is required. These nonlinear transfer functions indicate the nonlinear distortion effects on the linear transfer part, thus they are the power penalty or distortion noise that degrades the channel capacity (Figure 2.44).

2.10.3 Inverse of Volterra Expansion and Nonlinearity Compensation in Electronic Domain

In this section, we describe compensation and equalization of nonlinear effects of optical signals transmitted over linear CD and nonlinear single-mode optical fiber. The mathematical representation of the equalization scheme is based on the inverse of the nonlinear transfer function represented by the Volterra series. The implementation of such nonlinear equalization schemes is in the electronic domain. That is, at the stage where the optical signals have been received and converted into electrical domain as the voltage output of the electronic preamplifier, then digitized by an analog-to-digital converter (ADC) and processed in a digital signal processor.

The electronic nonlinearity compensation scheme based on the inverse of Volterra expansion can be implemented in the electronic domain in a DSP, where 1.2 dB in Q improvement can be achieved with 256 Gb/s PDM-16QAM, and simultaneously reduce the compensation complexity by the reduction of the processing rate.

To meet the ever increasing demands of the data traffic, improvement in spectral efficiency is desired. Data signals modulate light waves via optical modulators using advanced modulation formats and multiplexing using sub-carriers or polarization. Multi-level modulation formats such as 16QAM or 64QAM with higher spectrum efficiency are considered for realization of a future target rate of 400 Gb/s or 1 Tb/s per channel. However, the multi-level modulation formats require that the received signal has a higher level of optical signal to noise ratio (OSNR) that significantly reduces the possible transmission distance. To achieve a higher-received OSNR, suppression of the nonlinear penalty is inevitable to keep sufficient optical power for launch into the transmission fiber. There are several approaches to suppress the non-linearity, such as dispersion management, employing new fibers with larger core diameter and electronic nonlinear compensation in digital domain. A typical structure of a digital coherent receiver for QPSK and polarization demultiplexing of optical channels is shown in Figure 2.45. First the optical channels are fed into a 90° hybrid coupler to mix with a local oscillator. Their polarizations are split so that they can be aligned for maximum efficiency. The opto-electronic device, a balanced pair of photodiodes, converts the optical into electronic currents and then amplifies them by an electronic wideband pre-amplifier. At this stage the signals are in the electronic analog domain. The signals are then conditioned in their analog form, for example, by an automatic gain control, converted into digital quantized levels, and then processed by the DSP unit.

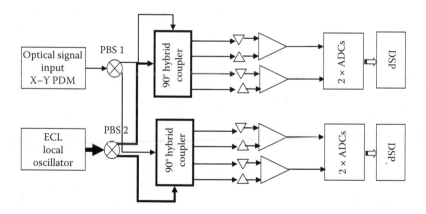

FIGURE 2.45
Typical structure of a digital coherent receiver incorporating ADCs and digital processing unit (DSP). One ADC is assigned per detected channel of the inphase and quadrature quantities.

DSP techniques make possible the compensation of large amounts of accumulated chromatic dispersion at the receiver. As a result, we can achieve the benefit of suppressing inter-channel nonlinearities in the WDM system by removing in-line optical dispersion compensation, hence reduction of inline optical amplifiers and thus increasing or extending the transmission distance. Under this scenario, intra-channel nonlinearity becomes the dominant impairment [68]. Fortunately, due to its deterministic nature, intra-channel nonlinearity can be compensated. Several approaches have been proposed to compensate the intra-channel nonlinearity, such as the digital BP algorithm [69], the adaptive nonlinear equalization [70], and the maximum likelihood sequence estimation (MLSE) [71,72]. All of the proposed methods suffer from the difficulty that the implementation complexity is too high, especially its demand for ultra-fast memory storage. In this chapter we propose a new electronic nonlinearity compensation scheme based on the inverse of Volterra expansion. We show that 1.2 dB in Q improvement can be achieved with 256 Gb/s PDM-16QAM transmission over a fiber link of 1000 km without inline dispersion compensation at 3 dBm launch power. We also simplify the implementation complexity by reducing the nonlinear processing rate. Negligible performance degradation with the same *baud* rate of the modulated data sequence for the nonlinearity compensation can also be achieved.

2.10.3.1 Inverse of Volterra Transfer Function

The Manakov-PMD equations that describe the evolution of optical electromagnetic fields enveloped in an optical fiber operating in the nonlinear SPM (self-phase modulation) region can be expressed as [73]

$$\frac{\partial A_x}{\partial z} - j\frac{\beta_2}{2}\frac{\partial^2 A_x}{\partial t^2} + \frac{\alpha}{2}A_x = -j\frac{8}{9}\gamma(|A_x|^2 + |A_y|^2)A_x \qquad (2.141)$$

$$\frac{\partial A_y}{\partial z} - j\frac{\beta_2}{2}\frac{\partial^2 A_y}{\partial t^2} + \frac{\alpha}{2}A_y = -j\frac{8}{9}\gamma(|A_x|^2 + |A_y|^2)A_y \qquad (2.142)$$

where A_x and A_y are the electric field envelop of the optical signals measured relative to the axes of the linear polarized mode of the fiber, β_2 is the second-order dispersion parameter related to the group velocity dispersion (GVD) of the single-mode optical fiber, and α is the fiber attenuation coefficient. Both linear polarization-mode dispersion (PMD) and nonlinear polarization dispersion are not included for simplicity. The first- and third-order Volterra series transfer function for the solution of Equations 2.141 and 2.142 can be written as [74]

$$H_1(\omega) = e^{-(\alpha + j\beta_2\omega^2)L/2} \qquad (2.143)$$

$$H_3(\omega_1,\omega_2,\omega) = -\frac{8}{9}\frac{j\gamma}{4\pi^2}H_1(\omega) \times \frac{1 - e^{-(\alpha+j\beta_2(\omega_1-\omega)(\omega_1-\omega_2))L}}{\alpha + j\beta_2(\omega_1 - \omega)(\omega_1 - \omega_2)} \qquad (2.144)$$

where $j\beta_2(\omega_1 - \omega)(\omega_1 - \omega_2)$ accounts for the impact of the waveform distortion due to the linear dispersion effects within a span. The complex term indicates the evolution of the phase of the carrier under the pulse envelope. The third-order kernel as a function of the nonlinear effects contains the main frequency component and the cross coupling between different frequency components. Higher transfer functions h5 can be used as well, but consuming a large chunk of memory may not be possible for ultra-high-speed ADC and DSP chips. Furthermore, the accuracy of the contribution of this order function would not be sufficiently high to warrant its inclusion. When the linear dispersion is extracted, the third-order transfer function can be further simplified as

$$H_3(\omega_1,\omega_2,\omega) \approx -\frac{8}{9}\frac{j\gamma}{4\pi^2} \times \frac{1 - e^{-\alpha L}}{\alpha}H_1(\omega) \qquad (2.145)$$

For optically amplified N spans fiber links without using dispersion compensating module (DCM), the whole fiber transfer functions are given by [75]

$$H_1^{(N)}(\omega) = e^{-j\omega^2 N\beta_2 L/2} \qquad (2.146)$$

$$H_3^{(N)}(\omega_1,\omega_2,\omega_3) = -\frac{8}{9}\frac{j\gamma}{4\pi^2}H_1^{(N)}(\omega) \times \frac{1 - e^{-(\alpha+j\beta_2(\omega_1-\omega)(\omega_1-\omega_2))L}}{\alpha + j\beta_2(\omega_1 - \omega)(\omega_1 - \omega_2)}$$

$$\times \sum_{k=0}^{N-1} e^{-jk\beta_2 L(\omega_1-\omega)(\omega_1-\omega_2)} \qquad (2.147)$$

where $\sum_{k=0}^{N-1} e^{-jk\beta_2 L(\omega_1-\omega)(\omega_1-\omega_2)}$ accounts for the waveform distortion at the input of each span. The third-order inverse kernel of this nonlinear system [76] can be obtained as

$$K_1^{(N)}(\omega) = e^{j\omega^2 N\beta_2 L/2} \qquad (2.148)$$

$$K_3^{(N)}(\omega_1,\omega_2,\omega_3) = \frac{8}{9}\frac{j\gamma}{4\pi^2}K_1^{(N)}(\omega) \times \frac{1 - e^{-(\alpha+j\beta_2(\omega_1-\omega)(\omega_1-\omega_2))L}}{\alpha + j\beta_2(\omega_1 - \omega)(\omega_1 - \omega_2)}$$

$$\times \sum_{k=0}^{N-1} e^{j(N-k)\beta_2 L(\omega_1-\omega)(\omega_1-\omega_2)} \qquad (2.149)$$

If taking the waveform distortion at the input of each span into consideration, Equation 2.149 can be approximated by

$$K_3^{(N)}(\omega_1,\omega_2,\omega_3) \approx \frac{8}{9}\frac{j\gamma}{4\pi^2} \times \frac{1-e^{-\alpha L}}{\alpha} K_1^{(N)}(\omega) \sum_{k=1}^{N} e^{jk\beta_2 L(\omega_1-\omega)(\omega_1-\omega_2)} \quad (2.150)$$

2.10.3.2 Electronic Compensation Structure

Equations 2.148 and 2.150 can be realized by the scheme shown in Figure 2.46. This structure or algorithm can be implemented in the digital domain in the DSP after the electronic preamplifier and the ADC as described in Figure 2.45. Here $c = (8\gamma/9) \times ((1 - e^{-\alpha} L)/\alpha)$ is a constant, $H_{CD} = e^{j\omega^2\beta_2 L/2}$ and compensates the residue dispersion of each span. Figure 2.46a shows the general structure to realize Equations 2.148 and 2.150. The compensation can be separated into linear compensation part and nonlinear compensation part. The linear compensation part is simply CD compensation. The nonlinear compensation part can be divided into N stages, where N is the span number. The detailed realization of each nonlinear inverse compensating stage is shown in Figure 2.46b, which is a realization of

$$K_{3,k}^{(N)}(\omega_1,\omega_2,\omega_3) \approx \frac{8}{9}\frac{j\gamma}{4\pi^2} \times \frac{1-e^{-\alpha L}}{\alpha} K_1^{(N)}(\omega) e^{jk\beta_2 L(\omega_1-\omega)(\omega_1-\omega_2)} \quad (2.151)$$

$$S_{x,k}(\omega) = jc(H_{CD})^N \int\!\!\!\int_{-\infty}^{\infty} e^{jk\beta_2 L(\omega_1-\omega)(\omega_1-\omega_2)} \left[A_x(\omega_1)A_x^*(\omega_2) + A_y(\omega_1)A_y^*(\omega_2) \right]$$

$$\times A_x(\omega - \omega_1 + \omega_2)d\omega_1 d\omega_2 \quad (2.152)$$

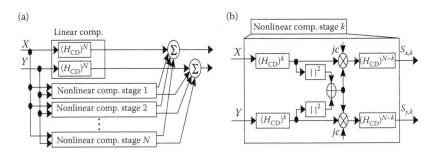

FIGURE 2.46
Electronic nonlinear compensation based on third-order inverse of Volterra expansion to be implemented in electronic DSP. (a) Block diagram of the proposed compensation scheme. (b) Detailed realization of nonlinear compensation stage k. (Adapted from L.N. Binh, L. Liu, L.C. Li, in: *Nonlinear Optical Systems*, CRC Press, Boston, 2012.)

$$S_{y,k}(\omega) = jc(H_{CD})^N \iint\limits_{-\infty}^{\infty} e^{jk\beta_2 L(\omega_1-\omega)(\omega_1-\omega_2)} \left[A_x(\omega_1)A_x^*(\omega_2) + A_y(\omega_1)A_y^*(\omega_2) \right]$$

$$\times A_y(\omega - \omega_1 + \omega_2)d\omega_1 d\omega_2 \qquad\qquad (2.153)$$

$S_{x,k}(\omega)$ and $S_{y,k}(\omega)$ are derived by first passing the received signal of X and Y though $(H_{CD})^k$, and then implementing the nonlinear compensation of $jc(|\ |_X^2 + |\ |_Y^2 \cdot (\ )_X$ and $jc(|\ |_X^2 + |\ |_Y^2 \cdot (\ )_Y$ [31]. Finally, the residual dispersion can be compensated by passing through the linear inverse function $(H_{CD})^{N-k}$.

For linear compensation, a processing rate equal to doubling the baud rate is common, but further reduction of the sampling rate can also be possible [77,78]. For the nonlinear compensation, it is possible by simulation that a single *baud* rate results comparable to that of the doubling *baud* rate, hence reduction of the implementation complexity. It is very critical, as the DSP at ultra-high sampling rate is very limited in the number of numerical operation.

The simulation platform is shown in Figure 2.47 [79]. The parameters of the transmission are also indicated in Figure 2.47. The transmission scheme 256 Gb/s NRZ PDM-16QAM with periodically inserted pilots is generated at the transmitter and then transmitted through 13 spans of fiber link. Each span consists of a standard single-mode fiber (SSMF) with a CD coefficient of 16.8 ps/(nm km), a Kerr nonlinearity coefficient of 0.0014 $m^{-1}W^{-1}$, a loss coefficient of 0.2 dB/km, and an Erbium-doped fiber amplifier (EDFA) with a noise figure of 5.5 dB. The span length is 80 km, and we assume 6 dB connector losses at the fiber span output, so the gain of each EDFA is 22 dB. There is also a pre- and a post-amplifier at transmitter and receiver to ensure a specified launch power and receiving power to satisfy the operational condition of the fiber nonlinear effects and the sensitivity of the coherent

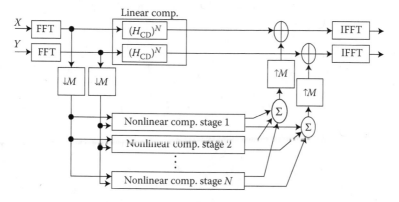

FIGURE 2.47
Simulation system setup and electronic nonlinearity compensation scheme at the digital coherent receiver with a nonlinear dispersion scheme in an electronic domain employing the inverse Volterra series algorithm.

receiver. Random polarization rotation is also considered along the transmission line, but no PMD is assumed. In order to have sufficient statistics, the same symbol sequence is transmitted 16 times with different ASE noise realization in the link and a total of 262,144 symbols are used for the estimation of the bit-error rate (BER). The transmitted signal is received with a polarization diversity coherent detector, sampled at the rate of twice the baud rate, and then processed in the digital electronic domain including the nonlinear inverse Volterra algorithm given herewith. The sampled signal first goes through the nonlinear compensator shown in Figure 2.47c, where both chromatic dispersion (CD) and intra-channel nonlinearity are compensated, and then quadruple butterfly-structured finite difference recursive filters for polarization de-multiplexing and residual distortion compensation. Feed-forward carrier phase recovery is carried out before making a decision. With the help of periodically sent pilots, Gray mapping without differential coding can be used to minimize the complexity of the coder. The proposed nonlinear compensation scheme is shown in Figure 2.47c. Both linear and nonlinear compensation sections share the same FFT and IFFT to reduce the computing complexity. For the linear compensation, the processing rate of twice the *baud rate*. For the nonlinear compensation part, a reduced processing rate is possible to reduce the demand on the processing power of the DSP at extremely high speed.

Figure 2.48 shows the performance of the proposed electronic nonlinear compensator. Figure 2.48 shows that the proposed nonlinearity compensator improves Q by 1.2 dB at 3 dBm launch power. This improvement is very significant in terms of sensitivity of the receiver. It was found that there is only negligible performance degradation with nonlinear processing rate of baud rate.

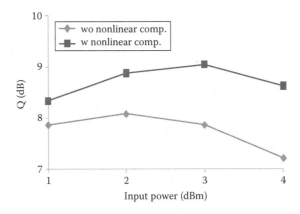

FIGURE 2.48

Performance of the proposed nonlinearity compensator. Quality factor Q versus input launched power with and without nonlinearity compensator. (Extracted from J. Tang, *J. Lightw Tech.*, 20(7), July 2002, 1095–1101.)

2.10.3.3 Remarks

In this section, based on the nonlinear transfer function of the single-mode optical fiber via the representation of the Volterra kernels, electronic digital processing technique is proposed with the algorithm based on the analytical version of the inverse Volterra series for compensation of the nonlinear. For example, optical modulated signals under polarization multiplexed 16-QAM scheme can be transmitted over noncompensating fiber links and then received coherently by an optical receiver to produce sampled pulse values in the digital electronic domain, for example, sampled by a real-time oscilloscope and then stored in memory for further off-line or real-time processing.

2.10.4 Back-Propagation Techniques for Compensation of Nonlinear Distortion

Recently, back propagation (BP) was proposed as a universal technique for jointly compensating linear and nonlinear impairments [80–82]. BP involves solving an inverse nonlinear Schrödinger equation (NLSE) through the fiber to estimate the transmitted signal. BP has been shown to enable the compensation of nonlinear effects of signals propagated through the fiber transmission line within which the power surpassed that of the nonlinear threshold. The main drawbacks of BP are its excessive computational requirement, and the difficulty in applying it in the presence of PMD. A computationally simpler algorithm for solving the NLSE based on a noniterative asymmetric split-step Fourier method (SSFM) can be developed to overcome these difficulties.

The BP technique developed by Lau and Kahn offers the compensation of nonlinear effects by linear superposition of the nonlinear phase (NLP) on the temporal-domain signals and, even more important, the propagation steps can be over the entire effective length of the span. Comparing with BP, the third order Volterra TF involves two dimensional FFT (ω, ω_1, ω_2). This BP is superior due to its low cost in computational resources that may be most preferred in the processing power of realtime DSP-coherent transmission systems. However, attention should be paid to the accuracy of the linear BP. Thus, adaptive BP should also be done at the receiver DSP. The phase rotation due to NLPN as observed by simulation and experiment is shown in Figure 2.49. We can see that the constellation of the modulation format QPSK rotates circularly due to the linear CD, and the blur of the constellation points is due to the nonlinear phase rotation. Having identified this feature of the nonlinear phase effects, the BP can be developed to compensate these effects in the DSP-based coherent receiver.

The model of the digital compensator by BP in the receiver is shown in Figure 2.50. The optical signals are received by the coherent photoreception, sampled by an ADC, then processed by the BP to compensate both the linear and nonlinear dispersion effects. A typical compensated constellation for QPSK is shown in Figure 2.51. Clearly the bit error rate (BER) of the

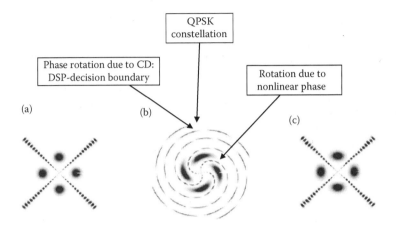

FIGURE 2.49
QPSK constellation and phase rotation due to linear and nonlinear effects: (a) no CD effect;
(b) nonlinear and linear CD rotation; (c) compensated NL and linear CD.

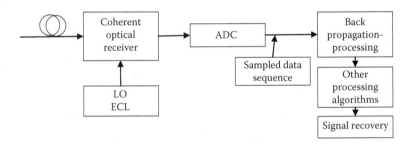

FIGURE 2.50
Receiver model incorporating BP to compensate nonlinear effects.

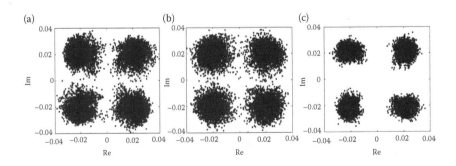

FIGURE 2.51
Equalized constellations for 21.4 Gb/s 50% RZ-QPSK transmitted over 25 × 2 80-km-SSMF spans
with 10% CD under compensation per span at 0 dBm launched power. The algorithms used are:
(a) linear equalization only; (b) NLPN compensation + linear equalization; and (c) BP-1S.

compensated constellation (b) and (c) would be much better than that of the non-compensated (a) by the nonlinear rotation and BP-1S.

The field of the guided wave can be represented at each step size h in the time with a superposition of the nonlinear phase SPM effects and with the superposition of the linear CD effects in the spectral domains by (Figure 2.52):

$$E_n(t, z + h) = E(t, z)e^{j\gamma h|E(t,z)|^2} \tag{2.154}$$

$$\tilde{E}_n(\omega, z + h) = \tilde{E}_n(\omega, z + h)e^{j\left[\frac{\alpha}{2} + \frac{\beta_2\omega^2}{2}\right]h} \tag{2.155}$$

The schematic of the BP algorithm can be represented in Figure 2.50. It is noted that the nonlinear superposition on the field in the time domain can be considered a phase rotation.

Thus the BP can be implemented either at the receiver or at the transmission, with the name post compensation or pre-equalization respectively. The BP at the receiver can be applied by propagating the received sampled channels back toward the transmitter with the nonlinear coefficient in opposite sign of that of the forward propagation fiber in the time domain, and hence using the fiber transfer function described in Section 2.8 for the linear domain to propagate to compensate for the CD effects.

A further simplification of the BP has been reported by Mateo and Huang [83] for dispersion-managed links and including Raman optical amplification. The low-complexity SPM compensation is experimentally demonstrated in WDM transmission over dispersion-managed links with and

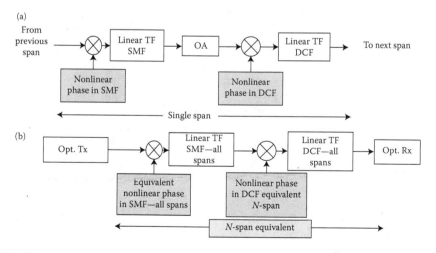

FIGURE 2.52
Schematic of multi-span equivalent transmission system for application of the back propagation by superposition of nonlinear and linear CD effects in (a) single span and (b) multispan.

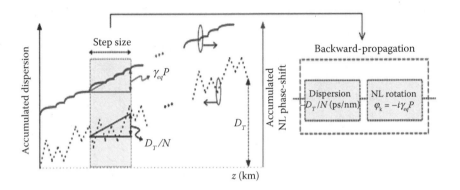

FIGURE 2.53
Schematic of the BP with low complexity in nonuniform dispersion-managed transmission systems with dispersion compensation steps and then nonlinear phase rotation.

without Raman amplification. Significant improvement of about 1–2 dB in optical launched power can be achieved with the longest propagation step size of 530 km reported by backward-propagation. The schematic diagram of the principles of operation of such BP is shown in Figure 2.53.

In this algorithm, the transmission line is divided into N steps. Each step composes of one dispersion compensation stage within which GVD can be found by diving the total residual dispersion by the number of steps, D_T/N. The total residual dispersion D_T is defined as the dispersion left over or over compensated by the dispersion-compensating fiber or nondispersion-compensating length in all spans, and the opposite sign value of the dispersion factor is taken, thus effective backward propagation.

The nonlinear phase rotation is found by applying a reversed nonlinear phase shift accumulated over the whole line divided by the number of steps. For dispersion managed transmission link, the nonlinear coefficients for the uncompensated transmission and compensating span sections are different and thus the application of the BP must be altered accordingly for each section.

The application of BP for polarization multiplexed channels remains challenged due to the cross-coupling and polarization state rotation. The system transmission and BP compensating for nonlinear phase rotation effects will be treated in later chapters of this book.

2.11 Concluding Remarks

Although the treatment of optical and in particular single-mode fibers has been extensively presented by several research papers and textbook, currently the single-mode and few-mode fibers have been extensively revisited for the design and considerations for compensation of linear and nonlinear distortion effects in the digital domain of coherent reception end and pre-equalization at the transmitting end. This chapter has thus presents the fiber characteristics in static parameters and dynamic characteristics in term of dispersion properties and nonlinear phase rotation effects.

A brief introduction of the compensation of nonlinear effects via the use of nonlinear transfer functions and BP techniques is given. Further treatment of these digital processing techniques will be given in later chapters of this book.

The applications of digital signal processing for spatial multiplexed channels warrantees that the revisiting of the few mode fibers and the guiding as well as splitting and combining of these modes and their polarized partners is appropriate at this time of the development of high-capacity, extremely long-haul optical fiber transmission systems. However, the availability of inline optical amplification for all these modal channels remains the important topic for system engineering.

References

1. C. Xia, N. Bai, I. Ozdur, X. Zhou, and G. Li, Supermodes for optical transmission, *Optics Express*, 19(17), 16653, August 15, 2011.
2. A. Safaai-Jazi and J.C. McKeeman, Synthesis of intensity patterns in few-mode optical fibers, *IEEE J. Lightw. Tech.*, 9(9), 1047, September 1991.
3. M. Salsi, C. Koebele, D. Sperti, P. Tran, P. Brindel, H. Mardoyan, S. Bigo et al., Transmission at 2×100 Gb/s, over two modes of 40 km-long prototype few-mode fiber, using LCOS based mode multiplexer and demultiplexer, *Optical Fiber Conference, OFC 2011, and National Fiber Optic Engineers Conference (NFOEC)*, Los Angeles, California, March 6, 2011, Postdeadline Session B (PDPB).
4. S. Randel, R. Ryf, A. Sierra, P. J. Winzer, A.H. Gnauck, C.A. Bolle, R.-J. Essiambre, D.W. Peckham, A. McCurdy, and R. Lingle, Jr., 6×56-Gb/s mode-division multiplexed transmission over 33-km few-mode fiber enabled by 6×6 MIMO equalization, *Optics Express*, 19(17), 16697, August 2011.
5. C.D. Poole, J.M. Wiesenfeld, D.J. DiGiovanni, and A.M. Vengsarkar, Optical fiber-based dispersion compensation using higher order modes near cutoff, *IEEE J. Lightw. Tech.*, 12(10), 1746–1758, October 1994.
6. L.N. Binh, L. Ling, and L.C Li, Volterra series transfer function in optical transmission and nonlinear compensation, Chapter 10, in: *Nonlinear Optical Systems*, Ed. L.N. Binh and D.V. Liet, CRC Press, Boca Raton, 2012.
7. A. Mecozzi, C.B. Clausen, and M. Shtaif, Analysis of intrachannel nonlinear effects in highly dispersed optical pulse transmission, *IEEE Photon. Technol. Lett.* 12, 392–394, 2000.
8. K. Kikuchi, Coherent detection of phase-shift keying signals using digital carrier-phase estimation, in *Proceedings of IEEE Conference on Optical Fiber Communications* (Institute of Electrical and Electronics Engineers, Anaheim, 2006), Paper OTuI4.

9. G.P. Agrawal, *Fiber Optic Communications Systems*, 4th Ed., J. Wiley, NY, 2010.
10. R.H. Stolen, E.P. Ippen, and A.R. Tynes, Raman oscillation in glass optical waveguide, *Appl. Phys. Let.*, 20, 62–64, 1972.
11. R.H. Stolen and E.P. Ippen, Raman gain in glass optical waveguides, *Appl. Phys. Lett.*, 22, 276–278, 1973.
12. R.G. Smith, Optical power handling capacity of low loss optical fibers as determined by stimulated Raman and Brillouin scattering, *Appl. Optics*, 11, 1972, 2489.
13. G.P. Agrawal, *Fiber Optic Communications Systems*, 3rd Ed., J. Wiley, NY, 2002, p. 60.
14. G.P. Agrawal, *Fiber Optic Communications Systems*, 3rd Ed., J. Wiley, NY, 2002, p. 67.
15. J.P. Gordon and H. Kogelnik, PMD fundamentals: Polarization mode dispersion in optical fibers, *PNAS*, 97(9), 4541–4550, April 2000.
16. Corning. Inc, An introduction to the fundamentals of PMD in fibers, White Paper July 2006.
17. A. Galtarossa and L. Palmieri, Relationship between pulse broadening due to polarisation mode dispersion and differential group delay in long singlemode fiber, *Electronics Letters*, 34(5), 492–493, March 1998.
18. J.M. Fini and H.A. Haus, Accumulation of polarization-mode dispersion in cascades of compensated optical fibers, *IEEE Photonics Technology Letters*, 13(2), 124–126, February 2001.
19. A. Carena, V. Curri, R. Gaudino, P. Poggiolini, and S. Benedetto, A time-domain optical transmission system simulation package accounting for nonlinear and polarization-related effects in fiber, *IEEE Journal on Selected Areas in Communications*, 15(4), 751–765, 1997.
20. S.A. Jacobs, J.J. Refi, and R.E. Fangmann, Statistical estimation of PMD coefficients for system design, *Electronics Letters*, 33(7), 619–621, March 1997.
21. G.P. Agrawal, *Fiber Optic Communication Systems*, Academic Press, NY, 2002.
22. A.F Elrefaie, R.E Wagner, D.A. Atlas, and D.G. Daut Chromatic dispersion limitations in coherent lightwave transmission systems, *IEEE J. Lightw. Tech.*, 6(6), 704–709, 1998.
23. J. Tang, The channel capacity of a multispan DWDM system employing dispersive nonlinear optical fibers and an ideal coherent optical receiver, *IEEE J. Lightw. Tech.*, 20(7), 1095–1101, 2002.
24. B. Xu and M. Brandt-Pearce, Comparison of FWM- and XPM-induced crosstalk using the Volterra series transfer function method, *IEEE J. Lightwave Tech.*, 21(1), 40–54, 2003.
25. J. Tang, The Shannon channel capacity of dispersion-Free nonlinear optical fiber transmission, *IEEE J. Lightwave Tech.*, 19(8), 1104–1109, 2001.
26. J. Tang, A comparison study of the Shannon channel capacity of various nonlinear optical fibers, *IEEE J. Lightwave Tech*, 24(5), 2070–2075, 2006.
27. J.G. Proakis, *Digital Communications*, 4th ed., McGraw-Hill, New York, 2001, pp. 185–213.
28. L.N. Binh, *Digital Optical Communications*, CRC Press, Boca Raton, FL, 2009.
29. L. N. Binh and N. Nguyen, Generation of high-order multi-bound-solitons and propagation in optical fibers, *Optics Communications*, 282, 2394–2406, 2009.
30. K.V. Peddanarappagari and M. Brandt-Pearce, Volterra series transfer function of single-mode fibers, *J. Lightw. Technol.*, 15(12), 2232–2241, December 1997.

31. L.N. Binh, L. Liu, and L.C. Li, Volterra series transfer function in optical transmission and nonlinear compensation Chapter 10, in: *Nonlinear Optical Systems*, Ed. L.N. Binh and D.V. Liet, CRC Press, Boston, 2012.

32. G.P. Agrawal, *Nonlinear Fiber Optics*, 3rd Ed., Academic Press, San Diego, CA, 2001.

33. J. Tang, The channel capacity of a multi-span DWDM system employing dispersive nonlinear optical fibers and ideal coherent receiver, *IEEE J. Lightw Tech.*, 20(7), July 2002, 1095–1101.

34. K.V. Peddanarappagari and M. Brandt-Pearce, Volterra series approach for optimizing fiber-optic communications systems designs, *IEEE J. Lightw. Tech.*, 16(11), 2046–2055, 1998.

35. J. Pina, C. Xia, A.G. Strieger, and D.V.D. Borne, *Nonlinear Tolerance of Polarization-Multiplexed QPSK Transmission Over Fiber Links*, ECOC2011, Geneva, 2011.

36. J. Tang, Channel capacity of a multi-span DWDM system employing nonlinear optical fibers and ideal coherent receiver, *IEEE J. Lightw. Tech*, 20(7), 1095–1101, 2002.

37. L.N. Binh, Linear and nonlinear transfer functions of single-mode fiber for optical transmission systems, *JOSA A*, 26(7), 1564–1575, 2009.

38. L.N. Binh, L. Liu, and L.C. Li, Volterra series transfer function in optical transmission and nonlinear compensation, Chapter 10 in: *Nonlinear Optical Systems*, Ed. L.N. Binh and D.V. Liet, January 2012, CRC Press, Taylors and Francis Pub., Boca Raton, FL.

39. J. Tang, A comparison study of the Shannon channel capacity of various nonlinear optical fibers, *IEEE J. Lightwave Tech*, 24(5), 2070–2075, 2006.

40. J.P. Gordon and L.F. Mollenauer, Phase noise in photonic communications systems using linear amplifiers, *Opt. Lett.*, 15, 1351–1353, December 1990.

41. K.-P. Ho and J.M. Kahn, Electronic compensation technique to mitigate nonlinear phase noise, *IEEE J. Lightw Tech.*, 22(3), March 2004, 779.

42. A.V.T. Cartaxo, B. Wedding, and W. Idler, New measurement technique of nonlinearity coefficient of optical fibre using fibre transfer function, in *ProcECOC98*, Madrid, Spain, 1998, pp. 169–170.

43. A.V.T. Cartaxo, B. Wedding, and W. Idler, Influence of fiber nonlinearity on the phase noise to intensity noise conversion in fiber transmission: Theoretical and experimental analysis, *IEEE J. Lightwave Tech.*, 16(7), 1187, July 1998.

44. F. Ramos and J. Martí, Frequency transfer function of dispersive and nonlinear single-mode optical fibers in microwave optical systems, *IEEE Photonics Tech. Lett.*, 12(5), 549, May 2000.

45. G. Agrawal, *Nonlinear Fiber Optics*, 2nd Edn, Academic Press, San Diego, 1995.

46. L.N. Binh, *Guided Wave Photonics*, CRC Press, Taylor & Francis, Boston, 2012.

47. F. Ramos and J. Martí, Frequency transfer function of dispersive and nonlinear single-mode optical fibers in microwave optical systems, *IEEE Photonics Tech. Lett.*, 12(5), 549, May 2000.

48. T.M. Dunster, Bessel functions of purely imaginary order, with an application to second-order linear differential equations having a large parameter, *SIAM J. Math. Anal.*, 21(4), 995–1018, July 1990.

49. F. Ramos and J. Martí, Frequency transfer function of dispersive and nonlinear single-mode optical fibers in microwave optical systems, *IEEE Photonic Tech Lett*, 12(5), 549–551, May 2000.

50. A.F. Elrefaie and R.E. Wagner, Chromatic dispersion limitations for FSK and DPSK systems with direct detection receivers, *IEEE Photonics Technology Letters*, 3(1), 71–73, 1991.
51. A.F. Elrefaie, R.E. Wagner, D.A. Atlas, and A.D. Daut, Chromatic dispersion limitation in coherent lightwave systems, *IEEE Journal of Lightwave Technology*, 6(5), 704–710, 1988.
52. A.F. Elrefaie and R.E. Wagner, Chromatic dispersion limitations for FSK and DPSK systems with direct detection receivers, *IEEE Photonics Technology Letters*, 3(1), 71–73, 1991.
53. A.F. Elrefaie, R.E. Wagner, D.A. Atlas, and A.D. Daut, Chromatic dispersion limitation in coherent lightwave systems, *IEEE Journal of Lightwave Technology*, 6(5), 704–710, 1988.
54. L.N. Binh, *Photonic Signal Processing*, CRC Press, Taylor & Francis, Boca Raton, FL, 2007.
55. D.M. Pepper, Applications of optical phase conjugation, *Scientific American*, 254, 74–83, January 1986.
56. P. Minzioni, V. Pusino, I. Cristiani, L. Marazzi, M. Martinelli, C. Langrock, M.M. Fejer, and V. Degiorgio, Optical phase conjugation in phase-modulated transmission systems: Experimental comparison of different nonlinearity-compensation methods*Optics Express*, 16 August 2010, 18(17), 18119.
57. X. Liu and D.A. Fishman, A fast and reliable algorithm for electronic pre-equalization of SPM and chromatic dispersion, *Optical Fiber Communications, Conference, OFC2007*, CA. 2007.
58. E. Ip and J.M. Kahn, Compensation of dispersion and nonlinear impairments using digital back propagation, *IEEE J. Lightw. Tech.*, 26(20), 3416, October 15, 2008.
59. Fujitsu Inc., Reality Check: Challenges of mixed-signal VLSI design for high-speed optical communications, ECOC2011, Geneva, 2011, http://www.fujitsu.com/downloads/MICRO/fme/dataconverters/ECOC-2009.pdf.
60. K. Kikuchi, M. Fukase, and S. Kim, Electronic post-compensation for nonlinear phase noise in a 1000-km 20-Gb/s optical QPSK transmission system using the homodyne receiver with digital signal processing, *Proc. Opt. Fiber Comm. Conf. (OFC' 07)*, Los Angeles, CA, 2007, Paper OTuA2.
61. G. Charlet, N. Maaref, J. Renaudier, H. Mardoyan, P. Tran, and S. Bigo, Transmission of 40 Gb/s QPSK with coherent detection over ultralong distance improved by nonlinearity mitigation, *Proc. European Conference on Optical Communication (ECOC 2006)*, Cannes, France, 2006, Paper Th4.3.4.
62. A.P.T. Lau and J.M. Kahn, Signal design and detection in presence of nonlinear phase noise, *J. Lightw. Technol.*, 25(10), 3008–3016, October 2007.
63. G.P. Agrawal, *Nonlinear Fiber Optics*, 5th Ed., Academic Press, Oxford, 2013.
64. A.P.T. Lau and J.M. Kahn, Signal design and detection in presence of nonlinear phase noise, *IEEE J. Lightw Tech.*, 25(10), 3008, October 2007.
65. A. Dochhan, R. Rath, C. Hebebrand, J. Leibrich, and W. Rosenkranz, *Evaluation of Digital Back-propagation Performance Dependent on Stepsize and ADC Sampling Rate for Coherent NRZ- and RZ-DQPSK Experimental Data*, ECOC 2011, Geneva, 2011.
66. K.V. Peddanarappagari and M. Brandt-Pearce, Volterra series transfer function of single-mode fibers, *J. Lightw. Technol.*, 15(12), 2232–2241, 1997.

67. L.N. Binh, L. Liu, and L.C. Li, Volterra series transfer function in optical transmission and nonlinear compensation, Chapter 10 in *Nonlinear Optical Systems*, L.N. Binh and dang Van Liet (eds), CRC Press, 2012, Baco Raton, FL, USA.

68. E. Yamazaki et al., *Mitigation of Nonlinearities in Optical Transmission Systems*, OFC2011, OThF1, Los Angeles, CA.

69. E. Ip and J.M. Kahn, Compensation of dispersion and nonlinear impairments using digital backpropagation, *J. Lightw. Technol.*, 26(20), 3416–3425, Oct. 2008.

70. Y. Gao et al., Experimental demonstration of nonlinear electrical equalizer to mitigate intra-channel nonlinearities in coherent QPSK systems, ECOC2009, Paper 9.4.7.

71. N. Stojanovic et al., MLSE-based nonlinearity mitigation for WDM 112 Gbit/s PDM-QPSK transmission with digital coherent receiver, OFC2011, paper OWW6, 2011.

72. L.N. Binh, T.L. Huynh, K.K. Pang, and T. Sivahumara, MLSE equalizers for frequency discrimination receiver of MSK optical transmission systems, *IEEE Journal of Lightwave Technology*, 26(12), 1586–1595, June 15, 2008.

73. P.K.A. Wai et al., Analysis of nonlinear polarization-mode dispersion in optical fibers with randomly varying birefringence, OFC'97 , paper ThF4, 1997.

74. K.V. Peddanarappagari and M. Brandt-Pearce, Volterra series transfer function of single-mode fibers, *J. Lightw. Technol.*, 15(12), 2232–2241, December 1997.

75. J.K. Fischer et al., Equivalent single-span model for dispersion-managed fiber-optic transmission systems, *J. Lightw. Technol.*, 27(16), 3425–3432, August 2009.

76. M. Schetzen, Theory of pth-order inverse of nonlinear systems, *IEEE Transactions on Circuits and Systems*, cas-23(5), May 1976.

77. E. Ip and J.M. Kahn, Digital equalization of chromatic dispersion and polarization mode dispersion, *J. Lightw. Technol.*, 25(8), 2033–2043, August 2007.

78. T. Pfau et al., Hardware-efficient coherent digital receiver concept with feed-forward carrier recovery for M-QAM constellations, *J. Lightw. Technol.*, 27(8), 989–999, April 2009.

79. C. Behrens, Mitigation of nonlinear impairments for advanced optical modulation formats, Doctoral Disertation, University College, London, August 2012.

80. X. Li, X. Chen, G. Goldfarb, E. Mateo, I. Kim, F. Yaman, and G. Li, Electronic post-compensation of WDM transmission impairments using coherent detection and digital signal processing, *Opt. Expr.*, 16(2), 881–888, January 2008.

81. W. Shieh, H. Bao, and Y. Tang, Coherent optical OFDM: Theory and design, *Opt. Expr.*, 16(2), 841–859, January 2008.

82. E. Ip and J.M. Kahn, Compensation of dispersion and nonlinear impairments using digital back propagation, *IEEE J. Lightwave Technology*, 26(20), October 15, 2008.

83. E. Mateo, M.-F. Huang, F. Yaman, T. Wang, Y. Aono, and T. Tajima, Nonlinearity compensation using very-low complexity backward propagation in dispersion managed links, *Proc. Optical Fiber Communication Conference (OFC)* Los Angeles, California, March 4, 2012, OFC2012, Section Nonlinearity Mitigation (OTh3C).

3

External Modulators for Coherent Transmission and Reception

Under coherent transmission it is essential that the linewidth of the laser source remain as stable and narrow as possible so that the mixing with a local oscillator laser would give the baseband signals. Thus the modulation of the lightwaves generated by the light source must be done outside the lasing cavity, hence external modulators. This chapter deals with the modulation formats employed to manipulate the phase and amplitude of the electromagnetic fields of lightwave sources by electro-optic modulators.

The external modulation is implemented by using optical modulators within which the input is coupled with the output of a CW laser source. The laser is thus turned on at all times and the generated optical continuous waves (CW) are then modulated—frequency, phase, or amplitude—through an external optical modulator. The uses of these transmitters in optical communication transmission systems are given, especially those for long-haul transmission at very high bit rate. We focus here on the operational principles of the photonic modulation section and discuss electrical sections only when essential.

3.1 Introduction

A photonic transmitter would consist of a single or multiple lightwave sources that can be either modulated directly by manipulating the driving current of the laser diode or externally via an integrated optical modulator. These are called direct and external modulation techniques.

This chapter presents the techniques for generation of lightwaves and modulation techniques of lightwaves, either directly or externally. Direct modulation is the technique that directly manipulates the stimulated emission from inside the laser cavity, via the use of electro-optic effects. In external modulation, the laser is turned on at all times then the generated lightwaves are coupled with an integrated optic modulator, through which the electro-optic effect is used with the electrical traveling waves that incorporate the coded information signals, and the amplitude and/or phases of the lightwaves are modulated. Advanced modulation formats have recently attracted much attention for enhancement of the transmission efficiency since the mid-1980s for coherent optical communications [1–6]. Hence, the preservation of the narrow linewidth of the laser source is critical for operation bit rates in the range of several tens of Gb/s. Thus, external modulation is essential.

This chapter describes the modulation techniques for optical communication systems. This includes modulation of the phase and amplitude of lightwaves and pulse-shaping NRZ or RZ, and schemes such as amplitude shift keying (ASK), differential phase shift keying (DPSK), minimum shift keying (MSK), frequency shift keying (FSK), multi-level amplitude and phase modulation such as quadrature amplitude modulation (QAM), and multi-carrier such as orthogonal frequency division modulation (OFDM). Appropriate Simulink® models are given for different modulation schemes.

3.2 External Modulation and Advanced Modulation Formats

The modulation of lightwaves via an external optical modulator can be classified into three types depending on the special effects that alter the lightwave's property, especially the intensity or the phase of the lightwave carrier. In an external modulator the intensity is normally manipulated by manipulating the phase of the carrier lightwaves guided in one path of an interferometer. Mach–Zehnder interferometric structure is the most common type [7–9].

3.2.1 Electro-Absorption Modulators

Because the electric field in the active region not only modulates the absorption characteristics, but also the refractive index, the EAM produces some chirp. However, this chirp usually is much less than that of a directly modulated laser. A small on-state (bias) voltage of around 0–1 V often is applied to minimize the modulator chirp.

The EA modulator (EAM) employs the Franz and Keldysh effect, which is observed as lengthening the wavelength of the absorption edge of a semiconductor medium under the influence of an electric field [10]. In quantum structure such as the multi-quantum well (MQW) structure this effect is called the Stark effect, or the electro-absorption (EA) effect. The EAM can be integrated with a laser structure on the same integrated circuit chip. For a LiNbO$_3$ modulator, the device is externally connected to a laser source via an optical fiber.

The total insertion loss of semiconductor intensity modulator is about 8–10 dB including fiber-waveguide coupling loss, which is rather high. However, this loss can be compensated by a semiconductor optical amplifier (SOA) that can be integrated on the same circuit. Compared to LiNbO$_3$ its total insertion loss is about 3–4 dB, which can be affordable as erbium-doped fiber amplifier (EDFA—see Chapter 5) is now readily available.

The driving voltage for an EAM is usually lower than that required for LiNbO$_3$. However, the extension ratio is not as high as that of the LiNbO$_3$ type, which is about 25 dB as compared to 10 dB for EAM. This feature contrasts the operating characteristics of the LiNbO$_3$ and EAMs. Although the driving voltage for EAM is about 3–4 V and 5–7 V for LiNbO$_3$, the former type would be preferred for intensity or phase modulation formats due to their high extinction ratio, which offers a much lower "zero" noise level and hence a high-quality factor.

EAM is small and can be integrated with the laser on the same substrate as shown in Figure 3.1. An EAM combined with a CW laser source is known as an electroabsorption modulated laser (EML).

An EML consists of a CW DFB laser followed by an EAM, as shown above. Both devices can be integrated monolithically on the same InP substrate, leading to a compact design and low coupling losses between the two devices. The EAM consists of an active semiconductor region sandwiched between a p- and n-doped layer, forming a p–n junction. As mentioned above, the EAM works on the principle known as Franz–Keldysh effect, according to which the effective bandgap of a semiconductor decreases with increasing electric field.

Without bias voltage across the p–n junction, the bandgap of the active region is just wide enough to be transparent at the wavelength of the laser light. However, when a sufficiently large reverse bias is applied across the p–n junction, the effective bandgap is reduced to the point where the active region begins to absorb the laser light and thus becomes opaque.

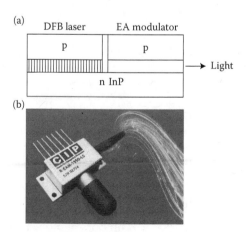

FIGURE 3.1
(a) Typical structure of EAM integrated with a distributed feedback laser on the same InP substrate; (b) package EAM of Huawei Center for Integrated Photonics of England.

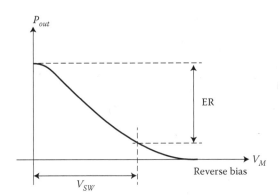

FIGURE 3.2
Power versus applied voltage of EAM.

In practical EAMs, the active region usually is structured as an MQW, providing a stronger field-dependent absorption effect known as the quantum-confined Stark effect.

The relationship between the optical output power P_{out} and the applied reverse voltage V_m of an EAM is described by the so-called switching curve as shown in Figure 3.2, which illustrates such a curve together with the achievable ER for a given switching voltage V_{SW}.

The voltage for switching the modulator from the on state to the off state, the switching voltage V_{SW} is typically in the range of 1.5–4 V, and the dynamic ER is usually in the range of 11–13 dB.

Compared with the electro-opticmodulator (EOM) described in detail below, EAM can operate with much lower voltages—a few volts instead of 10 V or more. It can also be operated at very high speed; a modulation bandwidth of tens of GHz can be achieved, which makes these devices useful for optical fiber communication. A convenient feature is that an EAM can be integrated with a distributed-feedback laser diode on a single chip to form a data transmitter in the form of a photonic-integrated circuit. Compared with direct modulation of the laser diode, a higher bandwidth and reduced chirp can be obtained. One major drawback of EAM is that the extinction ratio is not very high as compared with EOM, but its linear relationship between the applied voltage and output power is high.

In simple terms, the operation of an EAM can be considered that of an optical attenuator at very high speed depending on the level of reverse bias.

3.2.2 Electro-Optic Modulators

Integrated electro-optic modulator is an optical modulation device in which the lightwaves are confined to optical waveguides. The propagation speed of the guided lightwaves are sped up or slowed down with a reduction or increase of the refractive index, hence their phases retarded or advanced, respectively, by the change of the refractive index of the guided region with the applied electric field, that is, applied voltage via the electrodes. The change of the refractive index is through the electro-optic effects of the medium, usually a crystal substrate. Thus, in this section the phase modulation is introduced.

3.2.2.1 Phase Modulators

The phase modulator is a device that manipulates the phase of optical carrier signals under the influence of an electric field created by an applied voltage. When voltage is not applied to the RF-electrode, the number of periods of the lightwaves, n, exists in a certain path length. When voltage is applied to the RF-electrode, one or a fraction of one period of the wave is added, which now means $(n + 1)$ waves exist in the same length. In this case, the phase has been changed by 2π and the half-voltage of this is called the driving voltage. In case of long-distance optical transmission, the waveform is susceptible to degradation due to nonlinear effect such as self-phase modulation, and so on. A phase modulator can be used to alter the phase of the carrier to compensate for this degradation. The magnitude of the change of the phase depends on the change of the refractive index created via the electro-optic effect, which in turn depends on the orientation of the crystal axis with respect to the direction of the established electric field by the applied signal voltage.

An integrated optic phase modulator operates in a similar manner except that the lightwave carrier is guided via an optical waveguide, for which a diffused or ion-exchanged confined regions for LiNbO$_3$, and rib-waveguide structures for semiconductor type. Two electrodes are deposited so an electric field can be established across the cross section of the waveguide so that a change of the refractive index via the electro-optic effect results, as shown in Figure 3.3. For ultra-fast operation, one of the electrodes is a traveling-wave or hot-electrode type, and the other is a ground electrode. The traveling wave electrode must be terminated with matching impedance at the end so as to avoid wave reflection. Usually a quarter wavelength impedance is used to match the impedance of the traveling wave electrode to that of the 50 Ω transmission line.

A phasor representation of a phase-modulated lightwave can be by the circular rotation at a radial speed of ω_c. Thus, the vector with an angle ϕ represents the magnitude and phase of the lightwave.

3.2.2.2 Intensity Modulators

Basic structured LN modulator comprises of (i) two waveguides, (ii) two Y-junctions, and (iii) RF/DC travelling wave electrodes (Figure 3.4). Optical signals coming from the lightwave source are launched into the LN modulator through the polarization-maintaining fiber; it is then equally split into two branches at the first Y-junction on the substrate. When no voltage is

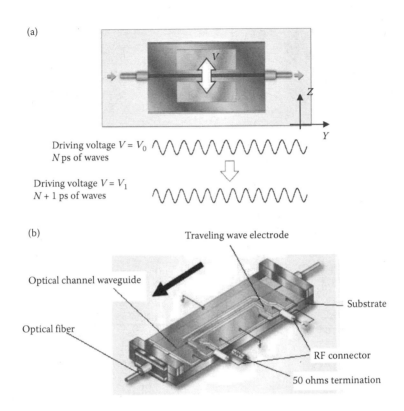

FIGURE 3.3
Electro-optic phase modulation in an integrated modulator using LiNbO$_3$. Electrode impedance matching is not shown. (a) Schematic diagram; (b) integrated optic structure.

applied to the RF electrodes, the two signals are re-combined constructively
at the second Y-junction and coupled into a single output. In this case, output signals from the LN modulator are recognized as "ONE." When voltage
is applied to the RF electrode, due to the electro-optic effects of LN crystal
substrate, the waveguide refractive index is changed, and hence the carrier

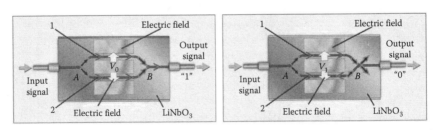

FIGURE 3.4
Intensity modulation using interferometric principles in guided-wave structures in LiNbO$_3$
(a) constructive interference mode—ON; (b) destructive interference mode—OFF. Optical
guided wave paths 1 and 2. Electric field is established across the optical waveguide.

phase in one arm is advanced though retarded in the other arm. Thence the two signals are re-combined destructively at the second Y-junction. They are transformed into a higher-order mode and radiated at the junction. If the phase retarding is in multiple odd factor of π, the two signals are completely out of phase, the combined signals are radiated into the substrate, and the output signal from the LN modulator is recognized as a "ZERO." The voltage difference that induces this "0" and "1" is called the driving voltage of the modulator, and is one important parameter of the figure of merit of the modulator.

3.2.2.3 Phasor Representation and Transfer Characteristics

Consider an interferometric intensity modulator, which consists of an input waveguide then split into two branches and then recombines to a single output waveguide. If the two electrodes are initially biased with voltages V_{b1} and V_{b2} then the initial phases exerted on the lightwaves would be $\phi_1 = \pi V_{b1}/V_\pi = -\phi_2$, which are indicated by the bias vectors shown in Figure 3.5. From these positions the phasors are swinging according to the magnitude and sign of the pulse voltages applied to the electrodes. They can be switched to the two positions that can be constructive or destructive. The output field of the lightwave carrier can be represented by

$$E_0 = \frac{1}{2} E_{iRMS} e^{j\omega_c t} \left(e^{j\phi_1(t)} + e^{j\phi_2(t)} \right) \tag{3.1}$$

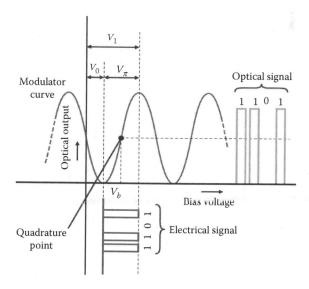

FIGURE 3.5
Electrical to optical transfer curve of an interferometric intensity modulator.

where ω_c is the carrier radial frequency, E_{iRMS} is the root mean square value of the magnitude of the carrier, and $\phi_1(t)$ and $\phi_2(t)$ are the temporal phases generated by the two time-dependent pulse sequences applied to the two electrodes. With the voltage levels varying according to the magnitude of the pulse sequence, one can obtain the transfer curve as shown in Figure 3.5. This phasor representation can be used to determine exactly the biasing conditions and magnitude of the RF or digital signals required for driving the optical modulators to achieve 50%, 33%, or 67% bit period pulse shapes.

The power transfer function of Mach–Zehnder modulator is expressed as[*]

$$P_0(t) = \alpha_M P_i \cos^2 \frac{\pi V(t)}{V_\pi} \tag{3.2}$$

where $P_0(t)$ is the output-transmitted power, α_M is the modulator total insertion loss, P_i is the input power (usually from the laser diode), $V(t)$ is the time-dependent signal-applied voltage, and V_π is the driving voltage so that a π phase shift is exerted on the lightwave carrier. It is necessary to set the static bias on the transmission curve through bias electrode. It is common practice to set the bias point at the half-intensity transmission point, or a $\pi/2$ phase difference between the two optical waveguide branches, the quadrature bias point. As shown in Figure 3.5 the electrical digital signals are transformed into optical digital signal by switching voltage to both ends of quadrature points on the positive and negative.

3.2.2.4 Bias Control

One factor that affects the modulator performance is the drift of the bias voltage. For the Mach–Zehnder interferometric modulator (MZIM) type it is very critical that it bias at the quadrature point or at minimum or maximum locations on the transfer curve. DC drift is the phenomena occurred in $LiNbO_3$ due to the build-up of charges on the surface of the crystal substrate. Under this drift the transmission curve gradually shifts in the long term [30,31]. In the case of the $LiNbO_3$ modulator, the bias-point control is vital, as the bias point will shift long term. To compensate for the drift, it is necessary to monitor the output signals and feed them back into the bias control circuits to adjust the DC voltage so that operating points stay at the same point, as shown in Figure 3.6, for example, the quadrature point. It is the manufacturer's responsibility to reduce DC drift so that DC voltage is not beyond the limit over the lifetime of device.

---

[*] Note this equation represents single drive MZIM—it is the same for dual drive MZIM provided that the bias voltages applied to the two electrodes are equal and opposite in signs. The transfer curve of the field representation would have half the periodic frequency of the transmission curve shown in Figure 3.5.

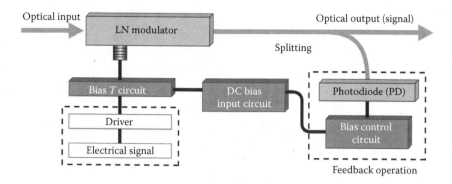

FIGURE 3.6
Arrangement of bias control of integrated optical modulators.

3.2.2.5 Chirp-Free Optical Modulators

Owing to the symmetry of the crystal refractive index of the uniaxial anisotropy of the class 3 m LiNbO$_3$, the crystal cut and the propagation direction of the electric field affect both modulator efficiency, denoted as driving voltage and modulator chirp. As shown in Figure 3.7, in the case of the Z-cut structure, as a hot electrode is placed on top of the waveguide, RF field flux is more concentrated, and this results in the improvement of overlap between RF and optical field. However, overlap between RF in ground electrode and

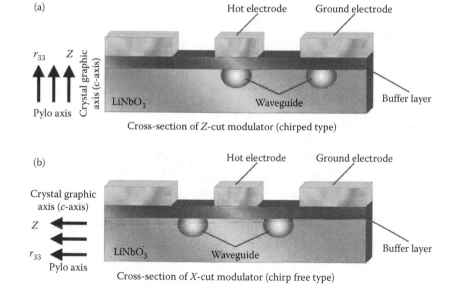

FIGURE 3.7
Different crystal cuts of LiNbO$_3$ integrated structures. (a) Integration of electrodes and optical waveguides in Z-cut and (b) X-cut.

waveguide is reduced in the Z-cut structure so that overall improvement of driving voltage for Z-cut structure compared to X-cut is approximately 20%. The different overlapping area for the Z-cut structure results in a chirp parameter of 0.7, whereas X-cut and Z-propagation has almost zero chirp due to its symmetric structure. A number of commonly arranged electrode and waveguide structures are shown in Figure 3.7 to maximize the interaction between the traveling electric field and the optical guided waves. Furthermore, a buffer layer, normally SiO_2, is used to match the velocities between these waves so as to optimize to optical modulation bandwidth.

3.2.2.6 Structures of Photonic Modulators

Figures 3.8a and 3.8b show the structure of an MZ intensity modulator using single- and dual-electrode configurations, respectively. The thin line electrode is called the "hot" electrode, or traveling wave electrode. RF connectors are required for launching the RF data signals to establish the electric field required for electro-optic effects. Impedance termination is also required. Optical fiber pig tails are also attached to the end faces of the diffused waveguide. The mode spot size of the diffused waveguide is not symmetric and

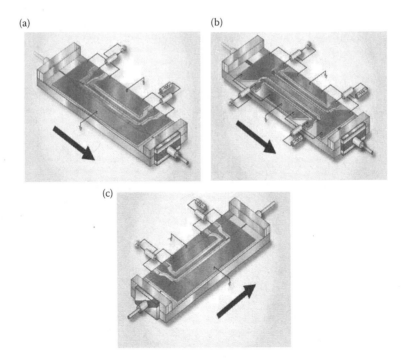

FIGURE 3.8
Intensity modulators using $LiNbO_3$. (a) Single drive electrode; (b) dual electrode structure; (c) electro-optic polarization scrambler using $LiNbO_3$.

hence some diffusion parameters are controlled so that maximizing the coupling between the fiber and the diffused or rib waveguide can be achieved. Due to this mismatching between the mode spot sizes of the circular and diffused optical waveguides there occurs coupling loss. Furthermore, the difference between the refractive indices of fiber and LiNbO$_3$ is quite substantial, and thus Fresnel reflection loss would also occur.

Figure 3.8c shows the structure of a polarization modulator, which is essential for multiplexing of two polarized data sequences so as to double the transmission capacity, For example, 40 G to 80 Gb/s. Furthermore, this type of polarization modulator can be used as a polarization rotator in a polarization dispersion-compensating sub-system.

3.2.2.7 Typical Operational Parameters

See Table 3.1.

3.2.3 ASK Modulation Formats and Pulse Shaping

3.2.3.1 Return-to-Zero Optical Pulses

Figure 3.9 shows the conventional structure of a RZ-ASK transmitter in which two external LiNbO$_3$ MZIMs can be used. The MZIM shown in this transmitter can be either a single- or dual-drive (push–pull) type. Operational principles of the MZIM were presented in Section 3.2 of the previous chapter. The optical OOK transmitter would normally consist of a narrow linewidth laser source to generate lightwaves in which wavelength satisfies the International Telecommunication Union (ITU) grid standard.

The first MZIM, commonly known as the pulse carver, is used to generate the periodic pulse trains with a required return-to-zero (RZ) format. The suppression of the lightwave carrier can also be carried out at this stage, if necessary, which is commonly known as the carrier-suppressed RZ (CSRZ). Compared to other RZ types, CSRZ pulse shape is found to have attractive

TABLE 3.1

Typical Operational Parameters of Optical Intensity Modulators

Parameters	Typical Values	Definition/Comments
Modulation speed	10 Gb/s	Capability to transmit digital signals
Insertion loss	Max 5 dB	Defined as the optical power loss within the modulator
Driving voltage	Max 4 V	The RF voltage required to have a full modulation
Optical bandwidth	Min 8 GHz	3 dB roll-off in efficiency at the highest frequency in the modulated signal spectrum
ON/OFF extinction ratio	Min 20 dB	The ratio of maximum optical power (ON) and minimum optical power (OFF)
Polarization extinction ratio	Min 20 dB	The ratio of two polarization states (TM- and TE-guided modes) at the output

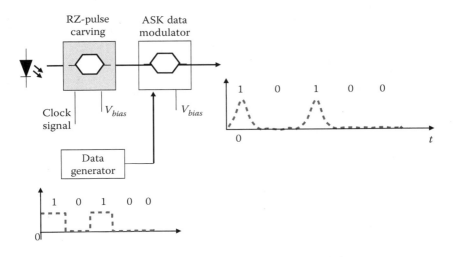

FIGURE 3.9
Conventional structure of an OOK optical transmitter utilizing two MZIMs.

attributes for long-haul WDM transmissions, including the π phase difference of adjacent modulated bits, suppression of the optical carrier component in optical spectrum, and narrower spectral width.

Different types of RZ pulses can be generated depending on the driving amplitude of the RF voltage and the biasing schemes of the MZIM. The equations governing the RZ pulse electric field waveforms are

$$
E(t) = \begin{cases} \sqrt{\dfrac{E_b}{T}}\,\sin\left[\dfrac{\pi}{2}\cos\left(\dfrac{\pi t}{T}\right)\right] & \text{67\% duty-ratio RZ pulses or CSRZ} \\[4mm] \sqrt{\dfrac{E_b}{T}}\,\sin\left[\dfrac{\pi}{2}\left(1+\sin\left(\dfrac{\pi t}{T}\right)\right)\right] & \text{33\% duty-ratio RZ pulses or RZ33} \end{cases}
$$

$$(3.3)$$

where E_b is the pulse energy per transmitted bit and T is one bit period.

The 33% duty-ratio RZ pulse is denoted as RZ33 pulse, whereas the 67% duty cycle RZ pulse is known as the CSRZ type. The art in generating these two RZ pulse types stays at the difference of the biasing point on the transfer curve of an MZIM.

The bias voltage conditions and the pulse shape of these two RZ types, the carrier suppression and nonsuppression of maximum carrier, can be implemented with the biasing points at the minimum and maximum transmission point of the transmittance characteristics of the MZIM, respectively. The peak-to-peak amplitude of the RF driving voltage is $2V_\pi$, where V_π is the required driving voltage to obtain a π phase shift of the lightwave carrier. Another important point is that the RF signal is operating at only half the transmission bit rate. Hence, pulse carving is actually implementing the frequency doubling. The generations of RZ33 and CSRZ pulse train are demonstrated in Figures 3.10a and 3.10b.

The pulse carver can also utilize a dual drive MZIM, which is driven by two complementary sinusoidal RF signals. This pulse carver is biased at $-V_{\pi/2}$ and $+V_{\pi/2}$, with the peak-to-peak amplitude of $V_{\pi/2}$. Thus, a π phase shift is created between the state "1" and "0" of the pulse sequence, and hence the RZ with alternating phase 0 and π. If the carrier suppression is required then the two electrodes are applied with voltages V_{π} and swing voltage amplitude of V_{π}.

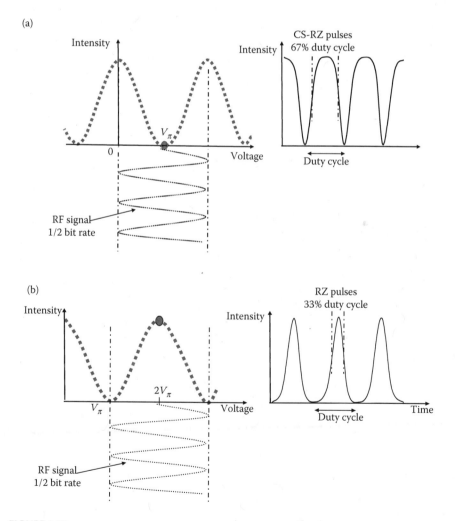

FIGURE 3.10
Bias point and RF driving signals for generation of (a) CSRZ and (b) RZ33 pulses.

Although RZ modulation offers improved performance, RZ optical systems usually require more complex transmitters than those in the NRZ ones. Compared to only one stage for modulating data on the NRZ optical signals, two modulation stages are required for generation of RZ optical pulses.

3.2.3.2 Phasor Representation

Recalling Equation 3.1 we have

$$E_0 = \frac{E_i}{2}\left[e^{j\phi_1(t)} + e^{j\phi_2(t)}\right] = \frac{E_i}{2}\left[e^{j\pi v_1(t)/V_\pi} + e^{j\pi v_2(t)/V_\pi}\right] \tag{3.4}$$

It can be seen that the modulating process for generation of RZ pulses can be represented by a phasor diagram as shown in Figure 3.11. This technique gives a clear understanding of the superposition of the fields at the coupling output of two arms of the MZIM. Here, a dual-drive MZIM is used, that is, the data driving signals $[V_1(t)]$ and inverse data ($\overline{\text{data}}$: $V_2(t) = -V_1(t)$) are applied to each arm of the MZIM, respectively, and the RF voltages swing in inverse directions. Applying the phasor representation, vector addition, and simple trigonometric calculus, the process of generating RZ33 and CSRZ is explained in detail and verified.

The width of these pulses are commonly measured at the position of full-width half-maximum (FWHM). It is noted that the measured pulses are intensity pulses whereas we are considering the addition of the fields in the MZIM. Thus, the normalized E_0 field vector has the value of $\pm 1\sqrt{2}$ at the FWHM intensity pulse positions and the time interval between these points gives the FWHM values.

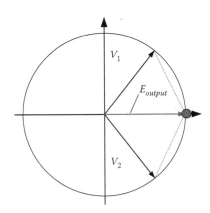

FIGURE 3.11
Phasor representation for generation of output field in dual-drive MZIM.

3.2.3.3 Phasor Representation of CSRZ Pulses

Key parameters including the V_{bias}, the amplitude of the RF driving signal, are shown in Figure 3.12a. Accordingly, its initialized phasor representation is demonstrated in Figure 3.12b.

The values of the key parameters are outlined as follows: (i) V_{bias} is $\pm V_{\pi}/2$; (ii) swing voltage of driving RF signal on each arm has the amplitude of $V_{\pi}/2$ (i.e., $V_{p-p} = V_{\pi}$); (iii) RF signal operates at half of bit rate ($B_R/2$); (iv) at the FWHM position of the optical pulse, the $E_{out} = \pm 1/\sqrt{2}$ and the component vectors V_1 and V_2 form with the vertical axis a phase of $\pi/4$, as shown in Figure 3.13.

Considering the scenario for generation of 40 Gb/s CSRZ optical signal, the modulating frequency is f_m (f_m = 20 GHz = $B_R/2$). At the FWHM positions of the optical pulse, the phase is given by the following expressions:

$$\frac{\pi}{2}\sin(2\pi f_m) = \frac{\pi}{4} \Rightarrow \sin 2\pi f_m = \frac{1}{2} \Rightarrow 2\pi f_m = \left(\frac{\pi}{6}, \frac{5\pi}{6}\right) + 2n\pi \quad (3.5)$$

Thus, the calculation of TFWHM can be carried out and, hence, the duty cycle of the RZ optical pulse can be obtained as given in the following expression:

$$T_{FWHM} = \left(\frac{5\pi}{6} - \frac{\pi}{6}\right)\frac{1}{R2\pi} = \frac{1}{3}\pi \times \frac{1}{R} \Rightarrow \frac{T_{FWHM}}{T_{BIT}} = \frac{1.66 \times 10^{-4}}{2.5 \times 10^{-11}} = 66.67\% \quad (3.6)$$

The result obtained in Equation 3.6 clearly verifies the generation of CSRZ optical pulses from the phasor representation.

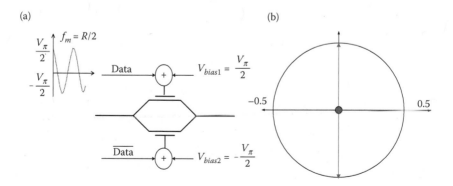

(a)　　　　　　　　　　　　　　　　　　　(b)

FIGURE 3.12
Initialized stage for generation of CSRZ pulse: (a) RF driving signal and the bias voltages; (b) initial phasor representation.

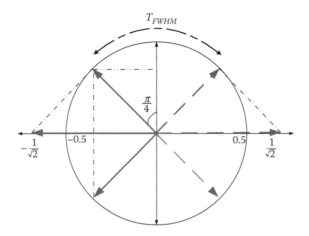

FIGURE 3.13
Phasor representation of CSRZ pulse generation using dual-drive MZIM.

3.2.3.4 Phasor Representation of RZ33 Pulses

Key parameters including the V_{bias}, the amplitude of driving voltage, and its correspondent initialized phasor representation are shown in Figures 3.14a and 3.14b, respectively.

The values of the key parameters are (i) V_{bias} is V_π for both arms; (ii) swing voltage of driving RF signal on each arm has the amplitude of $V_\pi/2$ (i.e., $V_{p-p} = V_\pi$); (iii) RF signal operates at half of bit rate ($B_R/2$).

At the FWHM position of the optical pulse, the $E_{output} = \pm 1/\sqrt{2}$ and the component vectors V_1 and V_2 form with the horizontal axis a phase of $\pi/4$, as shown in Figure 3.15.

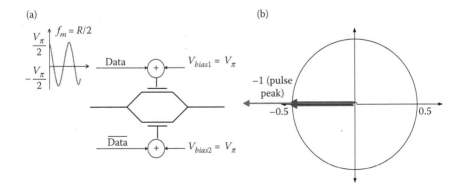

FIGURE 3.14
Initialized stage for generation of an RZ33 pulse: (a) RF driving signal and the bias voltage; (b) initial phasor representation.

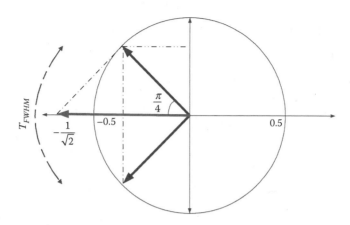

FIGURE 3.15
Phasor representation of RZ33 pulse generation using dual-drive MZIM.

Considering the scenario for generation of 40 Gb/s CSRZ optical signal, the modulating frequency is f_m ($f_m = 20$ GHz $= B_R/2$). At the FWHM positions of the optical pulse, the phase is given by the following expressions:

$$\frac{\pi}{2}\cos(2\pi f_m t) = \frac{\pi}{4} \Rightarrow t_1 = \frac{1}{6f_m} \tag{3.7}$$

$$\frac{\pi}{2}\cos(2\pi f_m t) = -\frac{\pi}{4} \Rightarrow t_2 = \frac{1}{3f_m} \tag{3.8}$$

$$T_{FWHM} = \frac{1}{3f_m} - \frac{1}{6f_m} = \frac{1}{6f_m} \therefore \frac{T_{FWHM}}{T_b} = \frac{1/6f_m}{1/2f_m} = 33\% \tag{3.9}$$

Thus, the calculation of TFWHM can be carried out and the duty cycle of the RZ optical pulse can be obtained. The result obtained in Equation 3.9 clearly verifies the generation of RZ33 optical pulses from the phasor representation.

3.2.4 Differential Phase Shift Keying

3.2.4.1 Background

Digital encoding of data information by modulating the phase of the lightwave carrier is referred to as optical phase shift keying (OPSK). In the early days, optical PSK was studied extensively for coherent photonic transmission systems. This technique requires the manipulation of the absolute phase of the lightwave carrier. Thus, precise alignment of the transmitter and demodulator center frequencies for the coherent detection is required. These coherent optical PSK systems face severe obstacles such as broad linewidth and chirping problems of the laser source. Meanwhile, the DPSK scheme

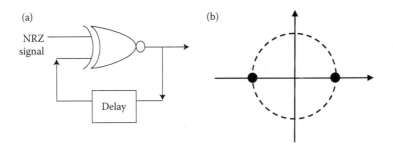

FIGURE 3.16
(a) DPSK pre-coder; (b) signal constellation diagram of DPSK. (Adapted from R.C. Alferness, *Science*, 14 November 1986:234(4778), 825–829.)

overcomes those problems, since the DPSK optically modulated signals can be detected incoherently. This technique only requires the coherence of the lightwave carriers over one bit period for the comparison of the differentially coded phases of the consecutive optical pulses.

A binary "1" is encoded if the present input bit and the past encoded bit are of opposite logic, whereas a binary 0 is encoded if the logics are similar. This operation is equivalent to an XOR logic operation. Hence, an XOR gate is employed as a differential encoder. NOR can also be used to replace XOR operation in differential encoding, as shown in Figure 3.16a. In DPSK, the electrical data "1" indicates a π phase change between the consecutive data bits in the optical carrier, while the binary "0" is encoded if there is no phase change between the consecutive data bits. Hence, this encoding scheme gives rise to two points located exactly at π phase difference with respect to each other in signal constellation diagram. For continuous PSK such as the MSK, the phase evolves continuously over a quarter of the section, thus a phase change of $\pi/2$ between one phase state to the other. This is indicated by the inner bold circle as shown in Figure 3.16b.

3.2.4.2 Optical DPSK Transmitter

Figure 3.17 shows the structure of a 40 Gb/s DPSK transmitter in which two external LiNbO$_3$ MZIMs are used. Operational principles of a MZIM were presented above. The MZIMs shown in Figure 3.17 can be either of single- or dual-drive type. The optical DPSK transmitter also consists of a narrow linewidth laser to generate a lightwave wherein the wavelength conforms to the ITU grid.

The RZ optical pulses are then fed into the second MZIM, through which the RZ pulses are modulated by the pre-coded binary data to generate RZ-DPSK optical signals. Electrical data pulses are differentially pre-coded in a precoder using the XOR coding scheme. Without a pulse carver, the structure shown in Figure 3.18 is an optical NRZ-DPSK (non return-to-zero differential phase-shift keying) transmitter. In data modulation for DPSK format, the second MZIM is biased at the minimum transmission point.

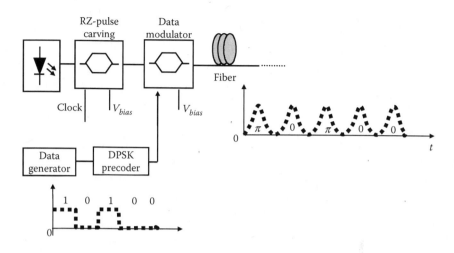

FIGURE 3.17
DPSK optical transmitter with RZ pulse carver.

The pre-coded electrical data has a peak-to-peak amplitude equal to $2V_\pi$ and operates at the transmission bit rate. The modulation principles for generation of optical DPSK signals are demonstrated in Figure 3.18.

The electro-optic phase modulator might also be used for generation of DPSK signals instead of MZIM. Using an optical phase modulator, the transmitted optical signal is chirped, whereas using MZIM, especially the X-cut type with Z-propagation, chirp-free signals can be produced. However, in practice, a small amount of chirp might be useful for transmission [23].

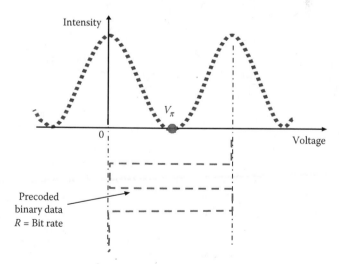

FIGURE 3.18
Bias point and RF driving signals for generation of the optical DPSK format.

3.3 Generation of Modulation Formats

Modulation is the process facilitating the transfer of information over a medium; for example, a wireless or optical environment. Three basic types of modulation techniques are based on the manipulation of a parameter of the optical carrier to represent the information digital data. These are ASK, phase shift keying (PSK), and FSK. In addition to the manipulation of the carrier, the occupation of the data pulse over a single period would also determine the amount of energy concentrates and the speed of the system required for transmission. The pulse can remain constant over a bit period or return to zero level within a portion of the period. These formats would be named nonreturn to zero (NRZ) or return to zero (RZ). They are combined with the modulation of the carrier to form the various modulation formats that are presented in this section.

3.3.1 Amplitude Modulation ASK-NRZ and ASK-RZ

Figure 3.19 shows the base band signals of the NRZ and RZ formats and their corresponding block diagrams of a photonic transmitter. There are a number of advanced formats used in advanced optical communications; based on the intensity of the pulse they may include NRZ, RZ, and duobinary. These ASK formats can also be integrated with the phase modulation to generate discrete or continuous phase NRZ or RZ formats. Currently the majority of 10 Gb/s installed optical communications systems have been developed with NRZ due to its simple transmitter design and bandwidth-efficient characteristic. However, RZ format has higher robustness to fiber nonlinearity and polarization mode dispersion (PMD). In this section, the RZ pulse is generated by an MZIM commonly known as a *pulse carver*, as arranged in. A number of variations in RZ format based on the biasing point in the transmission curve can be employed to generate various types, such as carrier-suppressed return to zero (CS-RZ) [11].

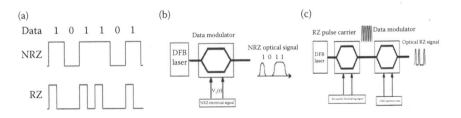

FIGURE 3.19
(a) Baseband NRZ and RZ line coding for 101101 data sequence; (b) block diagram of NRZ photonics transmitter; and (c) RZ photonics transmitter incorporating a pulse carver.

CSRZ has been found to have more attractive attributes in long-haul WDM transmissions compared to the conventional RZ format due to the possibility of reducing the upper level of the power contained in the carrier that serves no purpose in the transmission but only increases the total energy level, and therefore approaching the nonlinear threshold level faster. CSRZ pulse has optical phase difference of π in adjacent bits, removing the optical carrier component in optical spectrum and reducing the spectral width. This offers an advantage in compact WDM channel spacing.

3.3.2 Amplitude Modulation Carrier-Suppressed RZ Formats

The suppression of the carrier can be implemented by biasing the MZ interferometer in such a way so that there is a pi phase shift between the two arms of the interferometer. The magnitude of the sinusoidal signals applied to one or both arms would determine the width of the optical output pulse sequence.

3.3.3 Discrete Phase Modulation NRZ Formats

The term discrete phase modulation refers to the differential phase shift keying, whether DPSK or quadrature DPSK (DQPSK), to indicate the states of the phases of the lightwave carrier are switching from one distinct location on the phasor diagram to the other state; for example, from 0 to π or $-\pi/2$ to $-\pi/2$ for binary PSK (BPSK), or even more evenly spaced PSK levels as in the case of M-ary PSK.

3.3.3.1 Differential Phase Shift Keying

Information encoded in the phase of an optical carrier is commonly referred to as OPSK. In early days, PSK required precise alignment of the transmitter and demodulator center frequencies. Hence, PSK system is not widely deployed. With DPSK scheme introduced, coherent detection is not critical since DPSK detection only requires source coherence over one bit period by comparison of two consecutive pulses.

A binary "1" is encoded if the present input bit and the past encoded bit are of opposite logic and a binary 0 is encoded if the logic is similar. This operation is equivalent to XOR logic operation. Hence, an XOR gate is usually employed in a differential encoder. NOR can also be used to replace XOR operation in differential encoding, as shown in Figure 3.20.

In optical application, electrical data "1" is represented by a π phase change between the consecutive data bits in the optical carrier, while state "0" is encoded with no phase change between the consecutive data bits. Hence, this encoding scheme gives rise to two points located exactly at π phase difference with respect to each other in the signal constellation diagram, as indicated in Figure 3.20b.

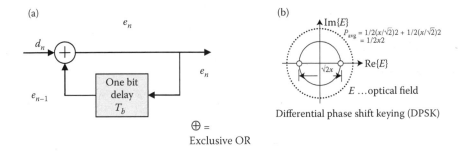

FIGURE 3.20
(a) The encoded differential data are generated by $e_n = d_n \oplus e_{n-1}$; (b) signal constellation diagram of DPSK.

An RZ-DPSK transmitter consists of an optical source, pulse carver, data modulator, differential data encoder, and a channel coupler (Figure 3.21). The channel coupler model is not developed in simulation by assuming no coupling losses when optical RZ-DPSK modulated signal is launched into the optical fiber. This modulation scheme has combined the functionality of dual drive MZIM modulator of pulse carving and phase modulation.

The pulse carver, usually an MZ interferometric intensity modulator, is driven by a sinusoidal RF signal for single-drive MZIM and two complementary electrical RF signals for dual-drive MZIM, to carve pulses out from optical carrier signal-forming RZ pulses. These optical RZ pulses are fed into the second MZ intensity modulator where RZ pulses are modulated by differential NRZ electrical data to generate RZ-DPSK. This data phase modulation can be performed using a straight-line phase modulator, but the MZ waveguide structure has several advantages over the phase modulator due to its chirpless property. Electrical data pulses are differentially precoded in a differential precoder as shown in Figure 3.20a. Without a pulse carver and sinusoidal RF signal, the output pulse sequence follows the NRZ-DPSK format, that is, the pulse would occupy 100% of the pulse period and there is no transition between the consecutive "1s."

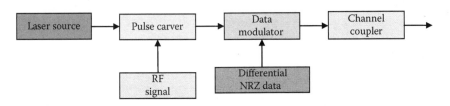

FIGURE 3.21
Block diagrams of RZ-DPSK transmitter.

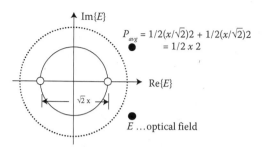

Differential phase shift keying (DPSK)

FIGURE 3.22
Signal constellation diagram of DQPSK. Two bold dots are orthogonal to the DPSK constellation.

3.3.3.2 Differential Quadrature Phase Shift Keying

This differential coding is similar to DPSK except that each symbol consists of two bits represented by the two orthogonal axial discrete phases at $(0, \pi)$ and $(-\pi/2, +\pi/2)$, as shown in Figure 3.22, or two additional orthogonal phase positions located on the imaginary axis of Figure 3.20b.

Alternatively, the QPSK modulation format can be generated by using two PSKs or DPSKs placed in two light paths of an interferometer, where one of the paths is optically phase shifted by $\pi/2$ to create a quadrature phase difference. Thus, amplitude modulation of the two MZIMs can be done to generate QPSK. Further details of this type of optical modulator will be described in a later section of this chapter, the IQ modulator.

3.3.3.3 Non Return-to-Zero Differential Phase Shift Keying

Figure 3.23 shows the block diagram of a typical NRZ-DPSK transmitter. The differential precoder of electrical data is implemented using the logic explained in the previous section. In phase modulating of an optical carrier, the MZ modulator known as data phase modulator is biased at the minimum point and driven by a data swing of $2V_\pi$. The modulator showed an excellent behavior that the phase of the optical carrier will be altered by π exactly when the signal transiting the minimum point of the transfer characteristic.

3.3.3.4 Return-to-Zero Differential Phase Shift Keying

The arrangement of RZ-DPSK transmitter is essentially similar to RZ-ASK, as shown in Figure 3.24, with the data intensity modulator replaced with the data phase modulator. The difference between them is the biasing point and the electrical voltage swing. Different RZ formats can also be generated.

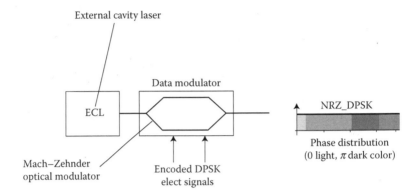

FIGURE 3.23
Block diagram of NRZ-DPSK photonics transmitter.

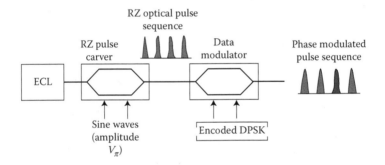

FIGURE 3.24
Block diagram of RZ-DPSK photonics transmitter including a DFB laser, two cascaded MZIMs incorporating applied electrodes and signal voltages.

3.3.3.5 Generation of M-Ary Amplitude Differential Phase Shift Keying (M-Ary ADPSK) Using One MZIM

As an example, a 16-ary MADPSK signal can be represented by a constellation shown in Figure 3.25. It is, indeed, a combination of a 4-ary ASK and a DQPSK scheme. At the transmitting end, each group of four bits $[D_3D_2D_1D_0]$ of user data are encoded into a symbol. Among them, the two least significant bits $[D_1D_0]$ can be encoded into four phase states, and the other two most significant bits, $[D_3D_2]$, are encoded into four amplitude levels. At the receiving end, as MZ delay interferometers (MZDI) or coherently mixing with a local oscillator laser can be used for phase comparison and detection, a diagonal arrangement of the signal constellation shown in Figure 3.25a is preferred. This simplifies the design of transmitter and receiver, and minimizes the number of phase detectors, hence leading to high receiver sensitivity.

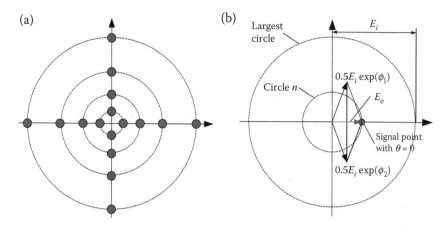

FIGURE 3.25
Signal constellation of 4-ary ADPSK format and phasor representation of a point on the constellation point for driving voltages applied to dual dive MZIM.

In order to balance the bit error rate (BER) between ASK and DQPSK components, the signal levels corresponding to four circles of the signal space should be adjusted to a reasonable ratio, which depends on the noise power at the receiver. As an example, if this ratio is set to $[I_0/I_1/I_2/I_3] = [1/1.4/2/2.5]$, where I_0, I_1, I_2, and I_3 are the intensity of the optical signals corresponding to circle 0, circle 1, circle 2, and circle 3, respectively, then by selecting E_i equal to the amplitude of the circle 3 and V_π equal to 1, the driving voltages should have the values given in Table 3.2. Inversely speaking, one can set the outermost level such that its peak value is below the nonlinear SPM threshold, the voltage level of the outermost circle would be determined. The innermost circle is limited to the condition that the largest signal to noise ratio (SNR) should be achieved. This is related to the optical SNR (OSNR) required for a certain BER. Thus, from the largest and smallest amplitude levels we can then design the other points of the constellation.

Furthermore, to minimize the effect of inter-symbol interference, 66% RZ and 50% RZ pulse formats are also used as alternatives to the traditional

TABLE 3.2

Driving Voltages for 16-Ary MADPSK Signal Constellation

	Circle 0			Circle 1			Circle 2			Circle 3	
Phase	$V_t(t)$ (V)	$V_2(t)$ (V)	Phase	$V_t(t)$ (V)	$V_2(t)$ (V)	Phase	$V_t(t)$ (V)	$V_2(t)$ (V)	Phase	$V_t(t)$ (V)	$V_2(t)$ (V)
0	0.38	−0.38	0	0.30	−0.30	0	0.21	−0.21	0	0.0	0.0
$\pi/2$	0.88	0.12	$\pi/2$	0.80	0.20	$\pi/2$	0.71	0.29	$\pi/2$	0.5	0.5
π	−0.62	0.62	π	−0.7	0.70	π	−0.79	0.79	π	−1.0	1.0
$3\pi/2$	−0.12	−0.88	$3\pi/2$	−0.20	−0.8	$3\pi/2$	−0.29	−0.71	$3\pi/2$	−0.5	−0.5

NRZ counterpart. These RZ pulse formats can be created by a pulse carver that precedes or follows the dual-drive MZIM modulator. Mathematically, waveforms of NRZ and RZ pulses can be represented by the following equations, where E_{on}, $n = 0, 1, 2, 3$ are peak amplitude of the signals in circles 0, 1, 2, and 3 of the constellation, respectively:

$$p(t) = \begin{cases} E_{on} & \text{for NRZ} \\ E_{on} \cos\left(\dfrac{\pi}{2} \cos^2\left(\dfrac{1.5\pi t}{T_s}\right)\right) & \text{for 66\% RZ} \\ E_{on} \cos\left(\dfrac{\pi}{2} \cos^2\left(\dfrac{2\pi t}{T_s}\right)\right) & \text{for 50\% RZ} \end{cases} \tag{3.10}$$

3.3.3.6 Continuous Phase Modulation PM-NRZ Formats

In the previous section the optical transmitters for discrete PSK modulation formats have been described. Obviously, the phase of the carrier has been used to indicate the digital states of the bits or symbols. These phases are allocated in a noncontinuous manner around a circle corresponding to the magnitude of the wave. Alternatively, the phase of the carrier can be continuously modulated and the total phase changes at the transition instants, usually at the end of the bit period, would be the same as those for discrete cases. Since the phase of the carrier continuously varies during the bit period, this can be considered an FSK modulation technique, except that the transition of the phase at the end of one bit and the beginning of next bit would be continuous. The continuity of the carrier phase at these transitions would reduce the signal bandwidth and hence be more tolerable to dispersion effects and higher energy concentration for effective transmission over the optical guided medium. One of the examples of the reduction of the phase at the transition is the offset DQPSK, which is a minor but important variation on the QPSK or DQPSK. In OQPSK the Q-channel is shifted by half a symbol period so that the transition instants of I and Q channel signals do not happen at the same time. The result of this simple change is that the phase shifts at any one time are limited, and hence the offset QPSK is a more constant envelope than the normal QPSK.

The enhancement of the efficiency of the bandwidth of the signals can be further improved if the phase changes at these transitions are continuous. In this case, the change of the phase during the symbol period is continuously changed by using half-cycle sinusoidal driving signals with the total phase transition over a symbol period a fraction of π, depending on the levels of this PSK modulation. If the change is $\pi/2$ then we have an MSK scheme. The orthogonality of the I- and Q-channels will also reduce further the bandwidth of the carrier-modulated signals.

In this section we describe the basic principles of optical MSK and the photonic transmitters for these modulation formats. Ideally, the driving signal to the phase modulator should be a triangular wave so that a linear phase variation of the carrier in the symbol period is linear. However, when a sinusoidal function is used there is some nonlinear variation; we thus term this type of MSK a nonlinear MSK format. This nonlinearity contributes to some penalty in the OSNR, which will be explained in a later chapter. Furthermore, the MSK as a special form of ODQPSK is also described for optical systems.

3.3.3.7 Linear and Nonlinear MSK

3.3.3.7.1 Signals and Precoding

MSK is a special form of continuous phase FSK (CPFSK) signal in which the two frequencies are spaced in such way that they are orthogonal and hence have a minimal spacing between them, defined by

$$s(t) = \sqrt{\frac{2E_b}{T_b}} \cos[2\pi f_1 t + \theta(0)] \quad \text{for symbol 1} \tag{3.11}$$

$$s(t) = \sqrt{\frac{2E_b}{T_b}} \cos[2\pi f_2 t + \theta(0)] \quad \text{for symbol 0} \tag{3.12}$$

As shown by the equations above, the signal frequency change corresponds to higher frequency for data-1 and lower frequency for data-0. Both frequencies, f_1 and f_2, are defined by

$$f_1 = f_c + \frac{1}{4T_b} \quad \text{and } f_2 = f_c - \frac{1}{4T_b} \tag{3.13}$$

Depending on the binary data, the phase of signal changes, data-1 increases the phase by $\pi/2$, while data-0 decreases the phase by $\pi/2$. The variation of phase follows paths of sequence of straight lines in phase trellis (Figure 3.26), in which the slopes represent frequency changes. The change in carrier frequency from data-0 to data-1, or vice versa, is equal to half the bit rate of incoming data [5]. This is the minimum frequency spacing that allows the two FSK signals representing symbols 1 and 0 to be coherently orthogonal in the sense that they do not interfere with one another in the process of detection.

An MSK signal consists of both I and Q components, which can be written as

$$s(t) = \sqrt{\frac{2E_b}{T_b}} \cos[\theta(t)] \cos[2\pi f_c t] - \sqrt{\frac{2E_b}{T_b}} \sin[\theta(t)] \sin[2\pi f_c t] \tag{3.14}$$

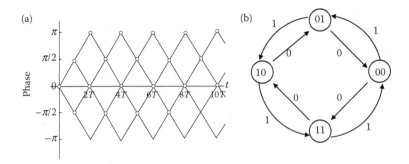

FIGURE 3.26
(a) Phase trellis for MSK; (b) state diagram for MSK. (L.N. Binh, *Optics Communications*, 281(17), 4245–4253, Sept. 2008.)

The in-phase component consists of half-cycle cosine pulse defined by

$$s_I(t) = \pm\sqrt{\frac{2E_b}{T_b}}\cos\left(\frac{\pi t}{2T_b}\right), \quad -T_b \leq t \leq T_b \tag{3.15}$$

While the quadrature component would take the form

$$s_Q(t) = \pm\sqrt{\frac{2E_b}{T_b}}\sin\left(\frac{\pi t}{2T_b}\right), \quad 0 \leq t \leq 2T_b \tag{3.16}$$

During even bit interval, the I-component consists of positive cosine waveform for phase of 0, while negative cosine waveform for phase of π; during odd-bit interval, the Q-component consists of positive sine waveform for phase of $\pi/2$, while negative sine waveform for phase of $-\pi/2$ (see Figure 3.26). Any of four states can arise: 0, $\pi/2$, $-\pi/2$, π. However, only state 0 or π can occur during any even bit interval, furthermore only $\pi/2$ or $-\pi/2$ can occur during any odd-bit interval. The transmitted signal is the sum of I and Q components and its phase is continuous with time.

Two important characteristics of MSK are that each data bit is held for a two-bit period, meaning the symbol period is equal to a two-bit period ($h = 1/2$), and the I- and Q-components are interleaved. I- and Q- components are delayed by one-bit period with respect to each other. Therefore, the I-component or Q-component can only change one at a time (when one is at zero-crossing, the other is at maximum peak). The precoder can be a combinational logic as shown in Figure 3.26.

A truth table can be constructed based on the logic state diagram and combinational logic diagram above. For positive half-cycle cosine wave and positive half-cycle sine wave, the output is 1; for negative half-cycle cosine wave and negative half-cycle sine wave, the output is 0. Then, a K-map can

TABLE 3.3

Truth Table Based on MSK State Diagram

$b_n S_0' S_1'$	$S_0' S_1'$	Output
100	01	1
001	00	1
100	01	1
101	10	0
010	01	1
101	10	0
110	11	0
111	00	1
000	11	0
011	10	0

be constructed to derive the logic gates of the pre-coder, based on the truth table (see Table 3.3). The following three pre-coding logic equations are derived:

$$S_0 = \overline{b_n}\,\overline{S_0'}\,\overline{S_1'} + b_n\,\overline{S_0'}S_1' + \overline{b_n}S_0'S_1' + b_n S_0'\overline{S_1'} \tag{3.17}$$

$$S_1 = \overline{S_1'} = \overline{b_n}\,\overline{S_1'} + b_n\,\overline{S_1'} \tag{3.18}$$

$$\text{Output} = \overline{S_0} \tag{3.19}$$

The resultant logic gates construction for the precoder is as shown in Figure 3.27.

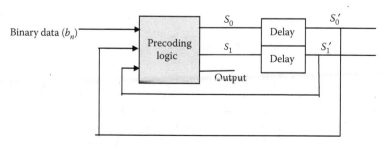

FIGURE 3.27
Combinational logic, the basis of the logic for constructing the precoder.

3.3.3.7.2 MSK as a Special Case of Continuous Phase FSK (CPFSK)

CPFSK signals are modulated in the upper and lower sides of band frequency carriers f_1 and f_2, which can be expressed as

$$s(t) = \sqrt{\frac{2E_b}{T_b}} \cos[2\pi f_1 t + \theta(0)] \quad \text{symbol 1} \tag{3.20}$$

$$s(t) = \sqrt{\frac{2E_b}{T_b}} \cos[2\pi f_2 t + \theta(0)] \quad \text{symbol 0} \tag{3.21}$$

where $f_1 = f_c + (1/4T_b)$ and $f_2 = f_c - (1/4T_b)$ with T_b is the bit period.

The phase slope of lightwave carrier changes linearly or nonlinearly with the modulating binary data. In the case of linear MSK, the carrier phase linearly changes by $\pi/2$ at the end of the bit slot with data "1," while it linearly decreases by $\pi/2$ with data "0." The variation of phase follows paths of well-defined phase trellis in which the slopes represent frequency changes. The change in carrier frequency from data-0 to data-1, or vice versa, equals half the bit rate of incoming data [12]. This is the minimum frequency spacing that allows the two FSK signals representing symbols 1 and 0 to be coherently orthogonal in the sense that they do not interfere with one another in the process of detection. MSK carrier phase is always continuous at bit transitions. The MSK signal in Equations 3.20 and 3.21 can be simplified as

$$s(t) = \sqrt{\frac{2E_b}{T_b}} \cos[2\pi f_c t + d_k \frac{\pi t}{2T_b} + \Phi_k], \quad kT_b \le t \le (k+1)T_b \tag{3.22}$$

and the base-band equivalent optical MSK signal is represented as

$$\tilde{s}(t) = \sqrt{\frac{2E_b}{T_b}} \exp\left\{ j\left[d_k \frac{\pi t}{2T} + \Phi(t, k) \right] \right\}, \quad kT \le t \le (k+1)T$$

$$= \sqrt{\frac{2E_b}{T_b}} \exp\left\{ j\left[d_k 2\pi f_d t + \Phi(t, k) \right] \right\} \tag{3.23}$$

where $d_k = \pm 1$ are the logic levels, f_d is the frequency deviation from the optical carrier frequency, and $h = 2f_d T$ is defined as the frequency modulation index. In case of optical MSK, $h = 1/2$ or $f_d = 1/(4T_b)$.

3.3.3.7.3 MSK as Offset Differential Quadrature Phase Shift Keying

Equation 3.16 can be rewritten to express MSK signals in the form of I–Q components as

$$s(t) = \pm\sqrt{\frac{2E_b}{T_b}}\cos\left(\frac{\pi t}{2T_b}\right)\cos[2\pi f_c t] \pm \sqrt{\frac{2E_b}{T_b}}\sin\left(\frac{\pi t}{2T_b}\right)\sin[2\pi f_c t] \quad (3.24)$$

The I- and Q-components consist of half-cycle sine and cosine pulses defined by

$$s_I(t) = \pm\sqrt{\frac{2E_b}{T_b}}\cos\left(\frac{\pi t}{2T_b}\right) \quad -T_b < t < T_b \quad\quad\quad (3.25)$$

$$s_Q(t) = \pm\sqrt{\frac{2E_b}{T_b}}\sin\left(\frac{\pi t}{2T_b}\right) \quad 0 < t < 2T_b \quad\quad\quad (3.26)$$

During even-bit intervals, the in-phase component consists of positive cosine waveform for phase of 0, while negative cosine waveform for phase of π; during odd-bit interval, the Q-component consists of positive sine waveform for a phase of $\pi/2$, while negative sine waveform for a phase of $-\pi/2$. Any of four states can arise: 0, $\pi/2$, $-\pi/2$, π. However, only state 0 or π can occur during any even bit interval and only $\pi/2$ or $-\pi/2$ can occur during any odd-bit interval. The transmitted signal is the sum of I- and Q-components and its phase is continuous with time.

Two important characteristics of MSK are that each data bit is held for a two-bit period, meaning the symbol period is equal to a two-bit period ($h = 1/2$) and the I- and Q-components are interleaved. I- and Q-components are delayed by one-bit period with respect to each other. Therefore, only one I-component or Q-component can change at a time (when one is at zero-crossing, the other is at maximum peak).

3.4 Photonic MSK Transmitter Using Two Cascaded Electro-Optic Phase Modulators

Electro-optic phase modulators and interferometers operating at high frequency using resonant-type electrodes have been studied and proposed in Refs. [13–15]. In addition, high-speed electronic driving circuits evolved with the ASIC (advanced specific integrated circuit) technology using 0.1 µm GaAs P-HEMT or InP HEMTs, enabling the feasibility in realization to the optical MSK transmitter structure. The base-band equivalent optical MSK signal is represented in Equations 3.23 and 3.24.

The first electro-optic phase modulator (E-OPM) enables the frequency modulation of data logics into upper sidebands (USB) and lower sidebands (LSB) of the optical carrier, with a frequency deviation of f_d. Differential

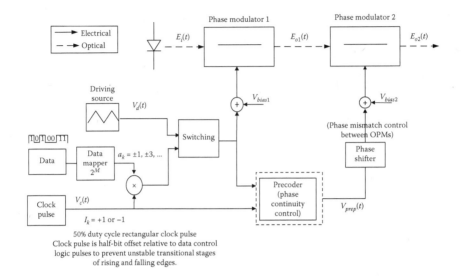

FIGURE 3.28
Block diagram of optical MSK transmitter employing two cascaded optical phase modulators.

phase precoding is not necessary in this configuration due to the nature of the continuity of the differential phase trellis. By alternating the driving sources $V_d(t)$ to sinusoidal waveforms for simple implementation or combination of sinusoidal and periodic ramp signals, which was first proposed by Amoroso in 1976 [16], different schemes of linear and nonlinear phase-shaping MSK-transmitted sequences can be generated, for which spectra are shown in Figure 3.28.

The second E-OPM enforces the phase continuity of the light wave carrier at every bit transition. The delay control between the E-OPMs is usually implemented by the phase shifter shown in Figure 3.28. The driving voltage of the second E-OPM is pre-coded to fully compensate the transitional phase jump at the output $E_{o1}(t)$ of the first E-OPM. Phase continuity characteristic of the optical MSK signals is determined by

$$\Phi(t,k) = \frac{\pi}{2}\left(\sum_{j=0}^{k-1} a_j - a_k I_k \sum_{j=0}^{k-1} I_j \right) \tag{3.27}$$

where $a_k = \pm 1$ are the logic levels, $I_k = \pm 1$ is a clock pulse wherein the duty cycle is equal to the period of the driving signal, $V_d(t)$ f_d is the frequency deviation from the optical carrier frequency, and $h = 2f_d T$ is previously defined as the frequency modulation index. In a case of optical MSK, $h = 1/2$ or $f_d = 1/(4T)$. The phase evolution of the continuous-phase optical MSK signals is explained in Figure 3.29. To mitigate the effects of unstable stages

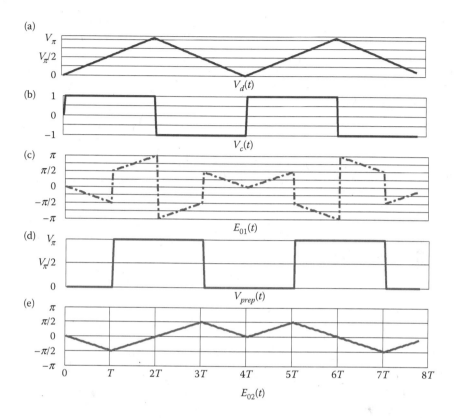

FIGURE 3.29

Evolution of time-domain phase trellis of optical MSK signal sequence [–1 1 1 –1 1 –1 1 1] as inputs and outputs at different stages of the optical MSK transmitter. The notation is denoted in Figure 3.28 accordingly: (a) $V_d(t)$: periodic triangular driving signal for optical MSK signals with duty cycle of 4-bit period, (b) $V_c(t)$: the clock pulse with duty cycle of $4T$, (c) $E_{01}(t)$: phase output of OPM1, (d) $V_{prep}(t)$: pre-computed phase compensation driving voltage of OPM2, and (e) $E_{02}(t)$: phase trellis of a transmitted optical MSK sequences at output of OPM2.

of rising and falling edges of the electronic circuits, the clock pulse $V_c(t)$ is offset with the driving voltages $V_d(t)$.

3.4.1 Configuration of Optical MSK Transmitter Using Mach–Zehnder Intensity Modulators: I–Q Approach

The conceptual block diagram of an optical MSK transmitter is shown in Figure 3.30. The transmitter consists of two dual-drive electro-optic MZM modulators generating chirpless I and Q components of MSK-modulated signals, which is considered a special case of staggered or offset QPSK. The binary logic data are pre-coded and de-interleaved into even and odd-bit streams, which are interleaved with each other by one-bit duration offset.

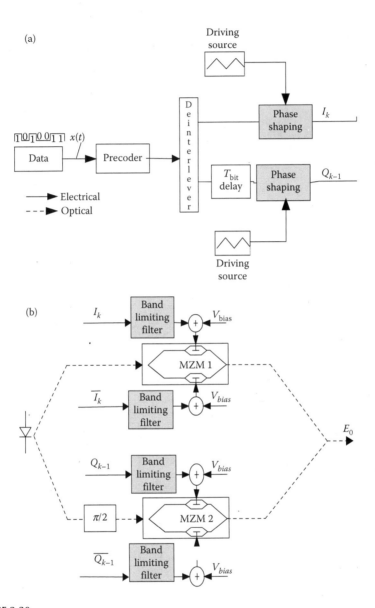

FIGURE 3.30
Block diagram configurations of band-limited phase-shaped optical MSK. (a) Encoding circuit and pulse shaping; (b) parallel MZIM optical modulators.

Two arms of the dual-drive MZM modulator are biased at $V_\pi/2$ and $-V_\pi/2$ and driven with *data* and $\overline{data}$. Phase-shaping driving sources can be a periodic triangular voltage source in the case of linear MSK generation or simply a sinusoidal source for generating a nonlinear MSK-like signal, which also obtain linear phase trellis property but with small ripples introduced

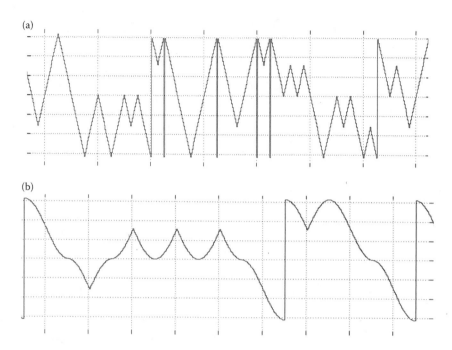

FIGURE 3.31
Phase trellis of linear and nonlinear MSK transmitted signals: (a) triangular and (b) sinusoidal.

in the magnitude. The magnitude fluctuation level depends on the magnitude of the phase-shaping driving source. High spectral efficiency can be achieved with tight filtering of the driving signals before modulating the electro-optic MZMs. Three types of pulse-shaping filters are investigated, including Gaussian, raised cosine, and square-root raised cosine filters. The optical carrier phase trellis of linear and nonlinear optical MSK signals is shown in Figure 3.31.

3.4.2 Single-Side Band Optical Modulators

An SSB modulator can be formed using a primary interferometer with two secondary MZM structures, the optical Ti-diffused waveguide paths that form a nested primary MZ, structure as shown in Figure 3.32. Each of the two primary arms contains an MZ structure. Two RF ports are for RF modulation and three DC ports are for biasing the two secondary MZMs and one primary MZM. The modulator consists of X-cut Y-propagation LiNbO$_3$ crystal, where you can produce an SSB modulation just by driving each MZ. DC voltage is supplied to produce the π phase shift between upper and lower arms. DC bias voltages are also supplied from DC$_B$ to produce the phase shift between 3rd and 4th arms. A DC bias voltage is supplied from DC$_C$ to achieve a $\pi/2$ phase shift between MZIM$_A$ and MZIM$_B$. The RF voltage

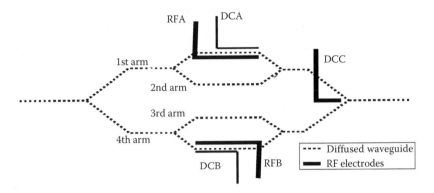

FIGURE 3.32
Schematic diagram (not to scale) of a single sideband (SSB) FSK optical modulator formed by nested MZ modulators.

applied $\Phi_1(t) = \Phi \cos \omega_m t$ and $\Phi_2(t) = \Phi \sin \omega_m t$ are inserted from RF_A and RF_B respectively by using a wideband $\pi/2$ phase shifter. Φ is the modulation level, while ω_m is the RF angular frequency.

3.4.3 Optical RZ-MSK

The RZ format of the optical MSK modulation scheme can also be generated. A bit is used to generate the ASK-like feature of the bit. The structure of such transmitter is shown in Figure 3.24. The encoder, as shown in the far left of the model, provides two outputs, one for MSK signal generation and the other for amplitude modulation for generation of the RZ or NRZ format. The phase of the RZ MSK must be nonzero so as to satisfy the continuity of the phase between the states.

3.4.4 Multi-Carrier Multiplexing Optical Modulators

Another modulation format that can offer much higher single-channel capacity and flexibility in dispersion and nonlinear impairment mitigation is the employment of multi-carrier multiplexing (MCM). When these sub-carrier channels are orthogonal, the term orthogonal frequency division multiplexing (OFDM) is used.

Our motivation in the introduction of OFDM is due to its potential as an ultra-high capacity channel for the next-generation Ethernet, the optical internet. The network interface cards for 1- and 10-Gb/s Ethernet are readily commercial available. Traditionally, the Ethernet data rates have grown in 10-Gb/s increments, so the data rate of 100 Gb/s can be expected as the speed of the next generation of Ethernet. The 100-Gb/s all-electrically time-division-multiplexed (ETDM) [17] transponders are becoming increasingly important because they are viewed as a promising technology that may be able to meet speed requirements of the new generation of Ethernet. Despite the recent progress in high-speed electronics, ETDM [17] modulators and photodetectors are still not widely available, so that alternative approaches to achieving a 100-Gb/s transmission using commercially available components and QPSK is very attractive.

OFDM is a combination of multiplexing and modulation. The data signal is first split into independent sub-sets and then modulated with independent sub-carriers. These sub-channels are then multiplexed for OFDM signals. OFDM is thus a special case of FDM, but instead like one stream, it is a combination of several small streams into one bundle.

A schematic signal flow diagram of an MCM is shown in Figure 3.33. The basic OFDM transmitter and receiver configurations are given in Figures 3.34a and 3.34b, respectively. Data streams (e.g., 1 Gb/s) are mapped into a two-dimensional signal point from a point signal constellation such as QAM. The complex-valued signal points from all sub-channels are considered the values of the discrete Fourier transform (DFT) of a multi-carrier OFDM signal. By selecting the number of sub-channels, sufficiently large, the OFDM symbol interval can be made much larger than the dispersed pulse width in a single-carrier system, resulting in an arbitrary small intersymbol interference (ISI). The OFDM symbol, shown in Figure 3.34, is generated under software processing as follows: input QAM symbols are zero-padded to obtain input samples for inverse fast Fourier transform (IFFT). The samples are inserted to create the guard band, and the OFDM symbol is multiplied by the window function (raised cosine function can be used). The purpose of cyclic extension is to preserve the orthogonality among sub-carriers even when the neighboring OFDM symbols partially overlap due to dispersion.

A system arrangement of the OFDM for optical transmission in laboratory demonstration is shown in Ref. [32]. Each individual channel at the input would carry the same data rate sequence. These sequences can be generated

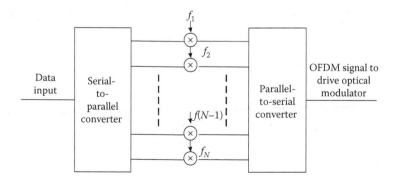

FIGURE 3.33

Multi-carrier FDM signal arrangement. The middle section is the RF converter as shown in Figure 3.26.

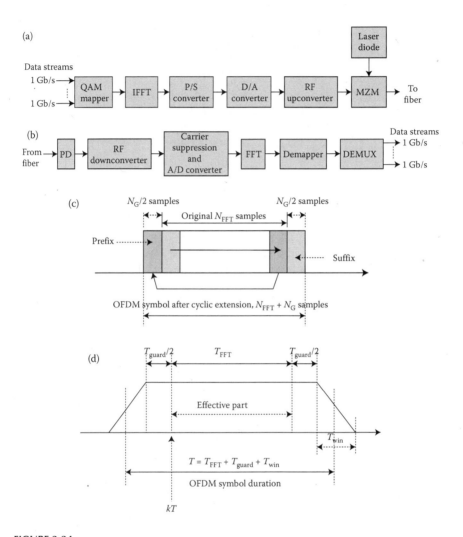

FIGURE 3.34
Schematic diagram of the principles of generation and recovery of OFDM signals. (a) Electronic processing and optical transmitter; (b) opto-electronic and receiver configurations; (c) OFDM symbol cyclic extension; (d) OFDM symbol after windowing.

from an arbitrary waveform generator. The multiplexed channels are then combined and converted to time domain using the IFFT module and then converted to the analog version via the two digital-to-analog converters (DACs). These orthogonal data sequences are then used to modulate I- and Q-optical waveguide sections of the electro-optica modulator to generate the orthogonal channels in the optical domain. Similar decoding of I and Q channels are performed in the electronic domain after the optical transmission and optical–electronic conversion via the photodetector and electronic amplifier.

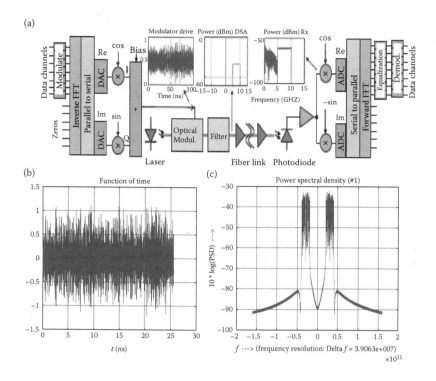

FIGURE 3.35

An optical FFT/IFFT based (a) OFDM system including representative waveforms and spectra (Extracted from M. Chacínski et al., *IEEE J. Sel. Top. Quant. Elect.*, 16(5), Sept/Oct. 2010, 1321–1327.); (b) typical time domain OFDM signals, (c) power spectral density of OFDM signal with 512 sub-carriers—a shift of 30 GHz for the line rate of 40 Gb/s and QPSK modulation.

In OFDM, the serial data sequence, with a symbol period of Ts and a symbol rate of 1/Ts, is split up into N-parallel sub-streams (sub-channels) (Figures 3.35 and 3.36).

3.4.5 Spectra of Modulation Formats

Utilizing this double-phase modulation configuration, different types of linear and nonlinear CPM phase shaping signals, including MSK, weakly nonlinear MSK, and linear-sinusoidal MSK can be generated. The third scheme was introduced by Amoroso [33] and its side lobes decay with a factor of 8 compared to 4 of MSK. The simulated optical spectra of DBPSK and MSK schemes at 40 Gb/s are contrasted in Figure 3.37. Table 3.4 outlines the characteristics and spectra of all different modulation schemes.

Figure 3.39 shows the power spectra of the DPSK modulated optical signals with various pulse shapes including NRZ, RZ33, and CSRZ types. For the convenience of the comparison, the optical power spectra of the RZ OOK counterparts are also shown in Figure 3.38.

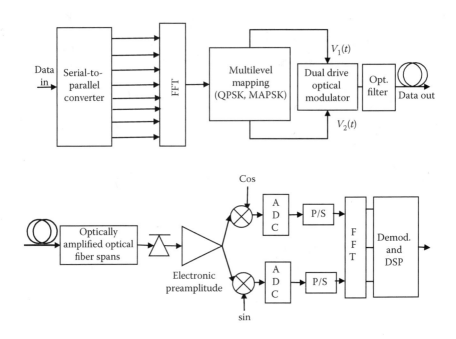

FIGURE 3.36
Schematic diagram of an optical FFT/IFFT-based OFDM system. S/P and P/S ~ serial to parallel conversion and vice versa.

Several key notes observed from and Figure 3.39 are outlined as follows: (i) The optical power spectrum of the OOK format has high power spikes at the carrier frequency or at signal modulation frequencies, which contribute significantly to the severe penalties caused by the nonlinear effects. Meanwhile, the DPSK optical power spectra do not contain these high-power

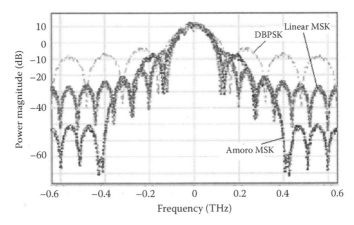

FIGURE 3.37
Spectra of 40 Gbps DBPSK, and linear and nonlinear MSK.

TABLE 3.4

Typical Parameters of Optical Intensity Modulators for Generation of Modulation Formats

Modulation Techniques	Spectra	Formats	Definition/Comments
Amplitude modulation—ASK-NRZ	DSB + carrier	ASK-NRZ	Biased at quadrature point or offset for pre-chirp
AM—ASK-RZ	DSB + carrier	ASK-RZ	Two MZIM s required—one for RZ pulse sequence and the other for data switching
ASK–RZ-Carrier suppressed	DSB – CSRZ	ASK-RZCS	Carrier suppressed, biased at π phase difference for the two electrodes
Single sideband	SSB + carrier	SSB NRZ	Signals applied to MZIM are in phase quadrature to suppress one sideband. Alternatively, an optical filter can be used
CSRZ DSB	DSB – carrier	CSRZ-ASK	RZ pulse carver is biased such that a π phase shift between the two arms of the MZM to suppress the carrier and then switched on and off or phase modulation via a data modulator
DPSK-NRZ DPSK-RZ, CSRZ-DPSK		Differential BPSK RZ or NRZ/RZ-carrier suppressed	
DQPSK		DQPSK-RZ or NRZ	Two bits per symbol
MSK	SSB equivalent	Continuous phase modulation with orthogonality	Two bits per symbol and efficient bandwidth with high side-lobe suppression
Offset-DQPSK		Oriented $\pi/4$ as compared to DQPSK constellation	Two bits/symbol
MCM (multi-carrier multiplexing- e.g., OFDM)	Multiplexed bandwidth— base rate per sub-carrier		
Duo-binary	Effective SSB		Electrical low-pass filer required at the driving signal to the MZM
FSK	Two distinct frequency peaks		
Continuous phase FSK	Two distinct frequency peaks with phase continuity at bit transition	Minimum shift keying	When the frequency difference equals a quarter of the bit rate, the signals for "1" and "0" are orthogonal and thus called MSK
Phase modulation (PM)	Chirped carrier phase		Chirpless MZM should be used to avoid inherent crystal effects, hence carrier chirp

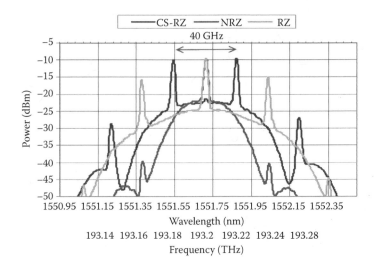

FIGURE 3.38
Spectra of CSRZ/RZ/NRZ–OOK modulated optical signals.

frequency components. (ii) RZ pulses are more sensitive to the fiber dispersion due to their broader spectra. In particular, RZ33 pulse type has the
broadest spectrum at the point of –20 dB down from the peak. This property
of the RZ pulses thus leads to faster spreading of the pulse when propagating
along the fiber. Thus, the peak values of the optical power of these CSRZ or
RZ33 pulses decrease much more quickly than the NRZ counterparts. As the

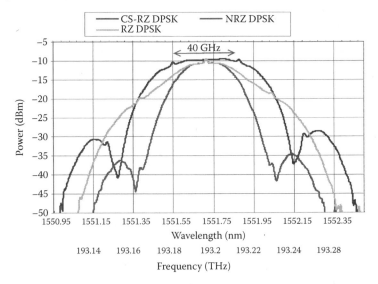

FIGURE 3.39
Spectra of CSRZ/RZ33/NRZ–DPSK modulated optical signals.

result, the peak optical power quickly turns to be lower than the nonlinear threshold of the fiber, which means that the effects of fiber nonlinearity are significantly reduced. (iii) However, NRZ optical pulses have the narrowest spectrum. Thus, they are expected to be most robust to the fiber dispersion. As a result, there is a trade-off between RZ and NRZ pulse types. RZ pulses are much more robust to nonlinearity but less tolerant to the fiber dispersion. The RZ33/CSRZ–DPSK optical pulses are proven to be more robust against impairments, especially self-phase modulation and PMD compared to the NRZ-DPSK and the CSRZ/RZ33-OOK counterparts.

Optical power spectra of three I–Q optical MSK modulation formats, which are linear, weakly nonlinear, and strongly nonlinear types are shown in Figure 3.40. It can be observed that there are no significant distinctions of the spectral characteristics between these three schemes. However, the strongly nonlinear optical MSK format does not highly suppress the side lobe as compared to the linear MSK type. All three formats offer better spectral efficiency compared to the DPSK counterpart, as shown in Figure 3.41. This figure compares the power spectra of three modulation formats: 80 Gb/s dual-level MSK, 40 Gb/s MSK, and NRZ-DPSK optical signals. The normalized amplitude levels into the two optical MSK transmitters comply with the ratio of 0.75/0.25.

Several key notes can be observed from Figures 3.40 and 3.41 as follows: (i) The power spectrum of the optical dual-level MSK format has identical characteristics to that of the MSK format. The spectral width of the main lobe is narrower than that of the DPSK. The base width takes a value of approximately ±32 GHz on either side compared to ±40 GHz in the case of the DPSK

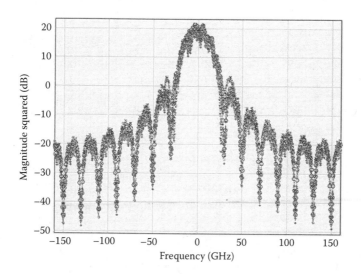

FIGURE 3.40
Optical power spectra of three types of I-Q optical MSK formats: linear (*), weakly nonlinear (o), and strongly nonlinear (x).

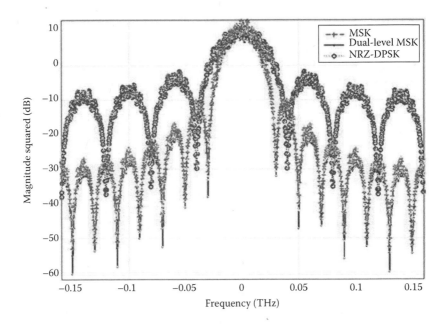

FIGURE 3.41
Spectral properties of three modulation formats: MSK (dash), dual-level MSK (solid), and DPSK (dot).

format. Hence, the tolerance to the fiber dispersion effects is improved. (ii) High suppression of the side lobes with a value of approximately greater than 20 dB in the case of 80 Gb/s dual-level MSK and 40 Gb/s optical MSK power spectra; thus, more robustness to inter-channel crosstalk between DWDM channels; and (iii) the confinement of signal energy in the main lobe of the spectrum leads to a better signal to noise ratio. Thus, the sensitivity to optical filtering can be significantly reduced. A summary of the spectra of different modulation formats is given in Table 3.4.

3.5 I–Q Integrated Modulators

3.5.1 Inphase and Quadrature Phase Optical Modulators

We have described in the above sections the modulation of QPSK and M-ary-QAM schemes using single-drive or dual-drive MZIM devices. However, we have also witnessed another alternative technique to generate the states of constellation of QAM, using I–Q modulators. I–Q modulators are devices in which the amplitude of the inphase and the quadrature components are modulated in synchronization, as illustrated in Figure 3.42b. These components are $\pi/2$ out of phase, thus we can achieve the inphase and quadrature of QAM. In optics, this phase quadrature at optical frequency can be implemented by a low-frequency electrode with an appropriate applied voltage as observed by the $\pi/2$ block in the lower optical path of the structure, given in Figure 3.42a. This type of modulation can offer significant advantages when multi-level QAM schemes are employed; for example, 16-QAM (see Figure 3.42c) or 64-QAM. Integrated optical modulators have been developed in recent years, especially in electro-optic structures such as LiNbO$_3$ for coherent QPSK and even with polarization division multiplexed optical channels. In summary, the I–Q modulator consists of two MZIMs performing ASK modulation and incorporating a quadrature phase shift.

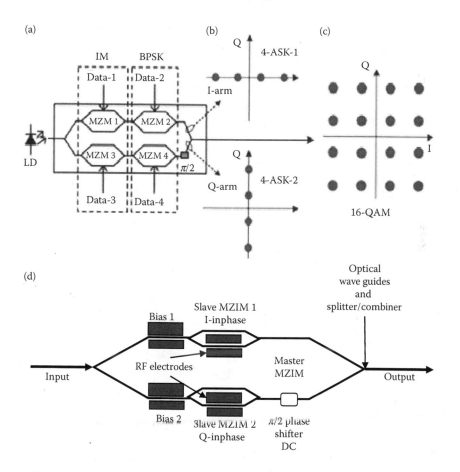

FIGURE 3.42
(a) Schematic of integrated IQ optical modulator and (b) amplitude modulation of a lightwave path in the inphase and quadrature components. (c) Constellation of a 16 QAM modulation scheme and (d) alternate structure of an I-Q modulator using two slave MZIM and one master MZIM.

Multi-level or multi-carrier modulation format such as QAM and orthogonal frequency division multiplexing (OFDM) have been demonstrated as the promising technology to support high capacity and high spectral efficiency in ultra-high-speed optical transmission systems. Several QAM transmitter schemes have been experimentally demonstrated using commercial modulators [18–20] and integrated optical modules [21,22] with binary or multi-level driving electronics. The integration techniques could offer a stable performance and effectively reduce the complexity of the electrics in a QAM transmitter with binary driving electronics. The integration schemes based on parallel structures usually require hybrid integration between LiNbO$_3$ modulators and silica-based planar lightwave circuits (PLCs). Except the DC electrodes for the bias control of each sub-MZM (child MZM), several additional electrodes are required to adjust the relative phase offsets among embedded sub-MZMs. Shown in Figure 3.42a is a 16-QAM transmitter using a monolithically integrated quad Mach–Zehnder in-phase/quadrature (QMZ-IQ) modulator. As distinguishable from the parallel integration, four sub-MZMs are integrated and arranged in a single IQ superstructure, where two of them are cascaded in each of the arms (I and Q arms). These two pairs of child MZMs are combined to form a master or parent MZ interferometer with a $\pi/2$ phase shift incorporated in one arm to generate the quadrature phase shift between them. Thus we have the inphase arm and the quadrature optical components. In principle, only one electrode is required to obtain orthogonal phase offset in these IQ superstructures, which makes the bias-control much easier to handle, and thus results in stable performance. A 16-QAM signal can be generated using the monolithically integrated Quadrature Mach–Zehnder Inphase and Quadraturee (QMZ-IQ) modulator with binary driving electronics, shown in Figure 3.42c, by modulating the amplitude of the lightwaves guided through both I (inphase) and Q (quadrature) paths as indicated in Figure 3.42b. Hence, we can see that the QAM modulation can be implemented by modulating the amplitude of these two orthogonal I–Q-components so that any constellation of Mary-QAM. For example, 16 QAM or 256 QAM, can be generated. Alternatively, the structure in Figure 3.42d gives an arrangement of two slave MZIMs and one master MZIM for the I–Q modulator, which is commonly used in Fujitsu type.

In addition, a number of electrodes would be incorporated so that biases can be applied to ensure that the amplitude modulation operating at the right point of the transfer curve. This can be commonly observed and simplified as in the IQ modulator manufactured by Fujitsu [23] shown in Figure 3.43 (top view only).

The arrangement of the electrodes of a high-speed IQ Mach–Zehnder modulator using Ti-diffused lithium niobate (LiNbO$_3$) is shown in Figure 3.43 in which the DC bias electrodes are separate from the traveling wave one. These electrodes allow adjustment of the bias independent with respect to the applied signals. This type of modulator can be employed for various modulation formats such as NRZ, DPSK, Optical Duo-Binary, DQPSK, DP-BPSK, DP-QPSK, and M-ary QAM, among others. It includes

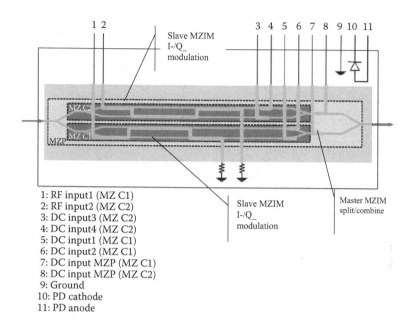

1: RF input1 (MZ C1)
2: RF input2 (MZ C2)
3: DC input3 (MZ C2)
4: DC input4 (MZ C2)
5: DC input1 (MZ C1)
6: DC input2 (MZ C1)
7: DC input MZP (MZ C1)
8: DC input MZP (MZ C2)
9: Ground
10: PD cathode
11: PD anode

FIGURE 3.43
Schematic diagram of Fujitsu PDM-IQ modulator with assigned electrodes.

built-in PD monitor and coupler function for auto bias control. Optical transmission equipment of 11 Gb/s (NRZ, DPSK, Optical Duo-Binary, DQPSK, DP-BPSK, PDM-QPSK) can be generated by this IQ modulator in which 4 wavelength carriers are used with 28–32 Gb/s per channel to form 100 Gb/s, including extra error coding bits and a payload of 25 Gb/s for each channel.

3.5.2 IQ Modulator and Electronic Digital Multiplexing for Ultra-High Bit Rates

Recently we [24–28] have seen reports on the development of electrical multiplexers with speeds reaching 165 Gb/s. These multiplexers will allow the interleaving of high-speed sequence so that we can generate a higher bit rate with the symbol rate, as shown in Figure 3.44. This type of multiplexing in the electrical domain has been employed for generation of superchannels in Ref. [29]. Thus, we can see that the data sequence can be generated from the DACs and then analog signals at the output of DACs can be condition to the right digital level of the digital multiplexer with assistance of a clock generator when the multiplexing occurs. Note that the multiplexer operates in digital mode so any pre-distortion of the sequence for dispersion pre-compensation will not be possible unless some pre-distortion can be done at the output of the digital multiplexer.

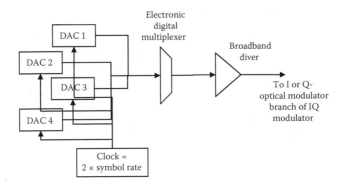

FIGURE 3.44
Time division multiplexing using high-speed sequences from DACs.

This time domain interleaving can be combined with the I–Q optical modulator to generate M-level QAM signal and thus further increase the bit rate of the channels. Several of these high bit-rate channels can be combined with pulse shaping; for example, the Nyquist shape to generate superchannels, which will be illustrated in Chapter 7. With a digital multiplexer operating higher than 165 Gb/s 128 Gb/s, bit rate can be generated with a 32 GSy/s data sequence. Thus, with polarization multiplexing and 16 QAM we can generate 8 × 128 Gb/s for one channel or 1.32 Tb/s per channel. If 8 of these 1.32 Gb/s form a superchannel, then the bit-rate capacity reaches higher than 10 Tb/s. At this symbol rate of 32 GBaud the required bandwidth for all electronic components as well as photonic devices of the transmitter and receiver can be satisfied with current high speed electronics and photonics.

3.6 DAC for DSP-Based Modulation and Transmitter

Recently we have witnessed the development of an ultra-high sampling rate DAC and ADC by Fujitsu and NTT of Japan. Figure 3.45a shows an IC layout of the InP-based DAC produced by NTT [30].

3.6.1 Fujitsu DAC

A differential input stage is structured with D-type flip flops and summing up circuits to produce analog signals as shown in Figure 3.45. These DACs allow the generation and programmable sampling and digitalized signals to form analog signals to modulate the I–Q modulator as described in Section 3.5. These DACs allow shaping of the pulse sequence for

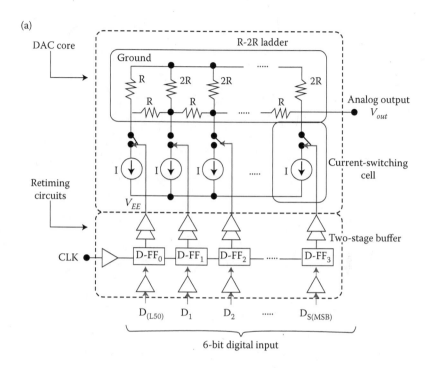

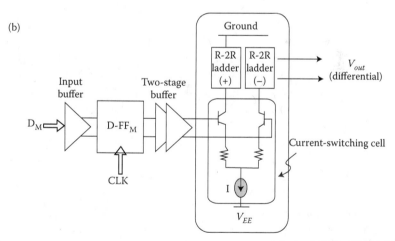

FIGURE 3.45
NTT InP-based DAC with 6-bit: (a) schematic; (b) differential input stage; and (c) IC layout. (Extracted from X. Liu et al., 1.5-Tb/s Guard-banded superchannel transmission over 56 GSymbols/s 16 QAM signals with 5.75-b/s/Hz net spectral efficiency, Paper Th.3.C.5.pdf ECOC Postdeadline Papers, ECOC 2012, Netherlands.)

(c) V_{out}
                   (differential)

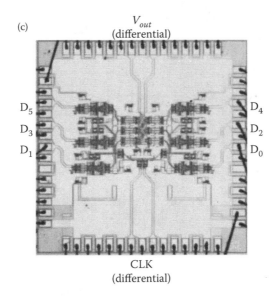

CLK
(differential)

FIGURE 3.45 (continued)
NTT InP-based DAC with 6-bit: (a) schematic; (b) differential input stage; and (c) IC layout.
(Extracted from X. Liu et al., 1.5-Tb/s Guard-banded superchannel transmission over 56
GSymbols/s 16 QAM signals with 5.75-b/s/Hz net spectral efficiency, Paper Th.3.C.5.pdf ECOC
Postdeadline Papers, ECOC 2012, Netherlands.)

pre-distortion to combat dispersion effects when necessary to add another
dimension of compensation in combination with that function implemented at the receiver. Test signals used are ramp waveform for assurance
of the linearity of the DAC at 27 GS/s as shown in Figure 3.46a and eye
diagrams of generated 16 QAM signals after modulation at 22 and 27 GSy/s
are shown in Figure 3.46b.

3.6.2 Structure

A DSP-based optical transmitter can incorporate a DAC for pulse shaping,
pre-equalization, and pattern generation as well as digitally multiplexing
for higher symbol rates. A schematic structure of the DAC and functional
blocks fabricated by Si–Ge technology is shown in Figures 3.47a and 3.47b,
respectively. An external sinusoidal signal is fed into the DAC so that multiple clock sources can be generated for sampling at 56–64 GSa/s. Thus the
noises and clock accuracy depends on the stability and noise of this synthe-
sizer/signal generator. Four DAC sub-modules are integrated in one IC with
4 pairs of 8 outputs of $(V_I^+, V_Q^+)(H_I^+, H_Q^+)\_and\_(V_I^-, V_Q^-)(H_I^-, H_Q^-)$.

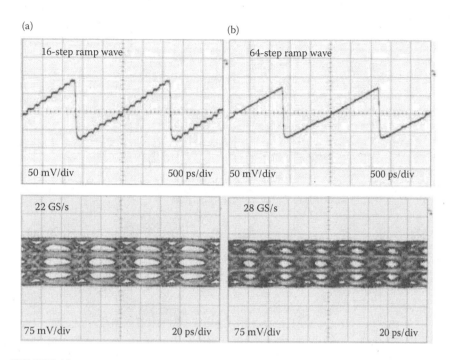

FIGURE 3.46
(a) Test signals and converted analog signals for linearity performance of the DAC; and
(b) 16 QAM 4-level generated signals at 22 and 27 GSy/s.

3.6.2.1 Generation of I and Q Components

The electrical outputs from the quad DACs are a mutual complementary pair of
positive and negative signs. Thus it would be able to form 2 sets of 4 output ports
from the DAC development board. Each output can be independently generated with offline uploading of pattern scripts. The arrangement of the DAC and
PDM-IQ optical modulator is depicted in Figure 3.48. Note that we require two
PDM IQ modulators for generation of odd and even optical channels.

As the Nyquist pulse-shaped sequences are required, a number of pressing steps can be implemented:

- Characterization of the DAC transfer functions
- Pre-equalization in the RF domain to achieve equalized spectrum in
  the optical domain, that is, at the output of the PDM IQ modulator

The characterization of the DAC is conducted by launching the DAC sinusoidal wave at different frequencies and measuring the waveforms at all
eight output ports. As observable in the insets of Figure 3.49, the electrical

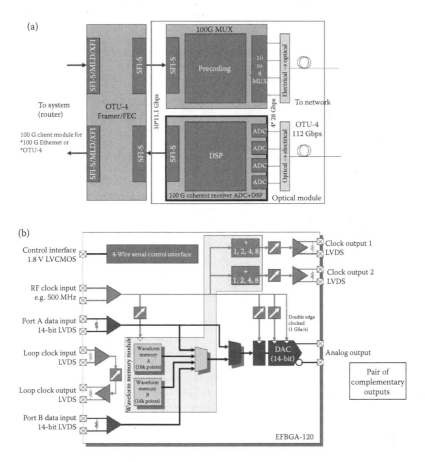

FIGURE 3.47
(a) Structure of the Fujitsu DAC; note 4 DACs are structured in one integrated chip. (b) Functional block diagram.

spectrum of the DAC is quite flat provided that pre-equalization is done in the digital domain launching to the DAC. The spectrum of the DAC output without equalization is shown in Figures 3.49a and 3.49b. The amplitude spectrum is not flat due to the transfer function of the DAC as given in Figure 3.49, which is obtained by driving the DAC with sinusoidal waves of different frequencies. This shows that the DAC acts as a low-pass filter with the amplitude of its passsband gradually decreasing when the frequency is increased. This effect can come from the number of samples being reduced when the frequency is increased, as the sampling rate can only be set in the range of 56–64 GSa/s. The equalized RF spectra are depicted in Figures 3.49c and 3.49d. The time domain waveforms corresponding to the RF spectra are shown in Figures 3.49e and 3.49f), and thence Figures 3.49g and 3.49h for the coherent detection after the conversion back to electrical domain from the

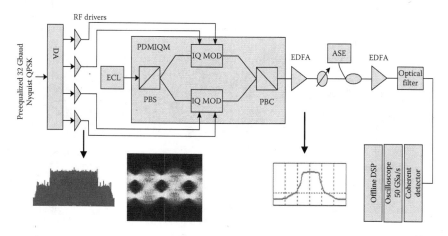

FIGURE 3.48

Experimental setup of the 128 Gb/s Nyquist PDM-QPSK transmitter and B2B performance evaluation.

optical modulator via the real-time sampling oscilloscope Tektronix DPO 73304A or DSA 720004B. Furthermore, the noise distribution of the DAC is shown in Figure 3.50b, indicating that the sideband spectra come from these noise sources (Figure 3.51).

It noted that the driving conditions for the DAC are very sensitive to the supply current and voltage levels, which are given in Appendix 2 with resolution of even down to 1 mV. With this sensitivity care must be taken when new patterns are fed to the DACs for driving the optical modulator. Optimal procedures must be conducted with the evaluation of the constellation diagram and BER derived from such constellation. However, we believe that the new version of the DAC supplied from Fujitsu Semiconductor Pty Ltd of England Europe has overcome somehow these sensitive problems. Still, we recommend that care should be taken and inspection of the constellation after the coherent receiver must be conducted to ensure the B2B connection is error free. Various time-domain signal patterns obtained in the electrical time domain generated by DAC at the output ports can be observed. Obviously the variations of the inphase and quadrature signals give rise to the noise, hence blurry constellations.

3.7 Remarks

Since the proposed dielectric waveguide and then the advent of an optical circular waveguide, the employment of modulation techniques has only been extensively exploited recently since the availability of optical amplifiers. The

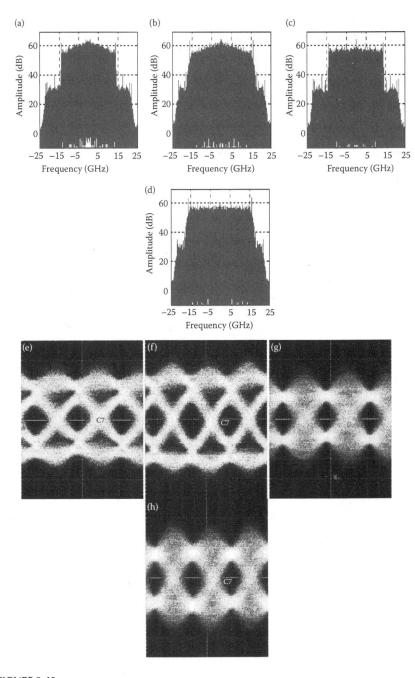

FIGURE 3.49
Spectrum (a) and eye diagram (e) of 28 Gbaud RF signals after DAC without pre-equalization,
(b) and (f) for 32 Gbaud; Spectrum (c) and eye diagram (g) of 28 Gbaud RF signals after DAC
with pre-equalization, (d) and (h) for 32 Gbaud.

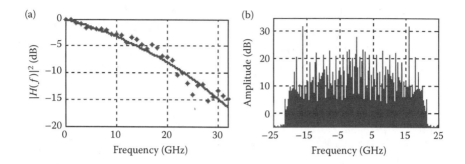

FIGURE 3.50

(a) Frequency transfer characteristics of the DAC. Note the near-linear variation of the magnitude as a function of the frequency. (b) Noise spectral characteristics of the DAC.

modulation formats allow the transmission efficiency, and hence the economy of ultra-high capacity information telecommunications. Optical communications have evolved significantly through several phases, from single-mode systems to coherent detection and modulation, which was developed with the main aim to improve on the optical power. The optical amplifiers defeated the main objective of modulation formats and allow the possibility of incoherent and all possible formats employing the modulation of the amplitude, phase, and frequency of the lightwave carrier.

Currently, photonic transmitters play a principal part in the extension of the modulation speed into several GHz range and make possible the modulation of the amplitude, phase, and frequency of the optical carriers and their multiplexing. Photonic transmitters using $LiNbO_3$ have been proven in laboratory and installed systems. The principal optical modulator is the MZIM, which can be a single or a combined set of these modulators to form binary or multilevel amplitude or phase modulation, and are even more effective for discrete or continuous PSK techniques. The effects of the modulation on transmission performance will be given in the coming chapters.

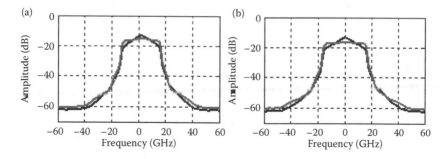

FIGURE 3.51

Optical spectrum after PDM IQM, black line for without pre-equalization, gray line for with pre-equalization: (a) 28 GSym/s, (b) 32 GSym/s.

Spectral properties of the optical 80 Gb/s dual-level MSK, 40 Gb/s MSK, and 40 Gb/s DPSK with various RZ pulse shapes are compared. The spectral properties of the first two formats are similar. Compared to the optical DPSK, the power spectra of optical MSK and dual-level MSK modulation formats have more attractive characteristics. These include the high spectral efficiency for transmission, higher energy concentration in the main spectral lobe, and more robustness to inter-channel crosstalk in DWDM due to greater suppression of the side lobes. In addition, the optical MSK offers the orthogonal property, which may offer a great potential in coherent detection, in which the phase information is reserved via I and Q components of the transmitted optical signals. In addition, the multi-level formats would permit the lowering of the bit rate and hence substantial reduction of the signal effective bandwidth and the possibility of reaching the highest speed limit of the electronic signal processing, the digital signal processing, for equalization and compensation of distortion effects. The demonstration of the ETDM receiver at 80G and higher speeds makes possible their applications in ultrahigh speed optical networks.

In recent years, research for new types of optical modulators using silicon waveguides has attracted several groups. In particular, graphene thin layer deposited [31] on silicon waveguides enables the improvement of the EA effects and enables the modulator structure to be incredibly compact and potentially perform at speeds up to ten times faster than current technology allows, reaching higher than 100 GHz and even to 500 GHz. This new technology will significantly enhance our capabilities in ultrafast optical communication and computing. This may be the world's smallest optical modulator, and the modulator in data communications is the heart of speed control. Furthermore, these grapheme Silicon modulators can be integrated with Si- or SiGe-based microelectronic circuits such as DAC, ADC, and DSP so that the operating speed of the electronic circuits can reach much higher than the 25–32 GGSy/s of today technology.

References

1. R.C. Alferness Optical guided-wave devices, *Science*, November 14, 1986: 234(4778), 825–829.
2. M. Rizzi and B. Castagnolo, Electro-optic intensity modulator for broadband optical communications, *Fiber Integr. Opt.*, 21:243–251, 2002.
3. H.Takara, High-speed optical time-division-multiplexed signal generation, *Opt. Quantum Electron.*, 33(7–10), 795–810, July 10, 2001.
4. E.L. Wooten, K.M. Kissa, A. Yi-Yan, E.J. Murphy, D.A. Lafaw, P.F. Hallemeier, D. Maack et al., A review of lithium niobate modulators for fiber-optic communications systems, *IEEE J. Sel. Topics Quant. Elect.*, 6(1), 69–80, Jan./Feb 2000.

5. K. Noguchi, O. Mitomi, H. Miyazawa, and S. Seki, A broadband Ti: LiNbO$_3$ optical modulator with a ridge structure, *J. Lightwave Technol.*, 13(6), June 1995.

6. J. Noda, Electro-optic modulation method and device using the low-energy oblique transition of a highly coupled super-grid, *IEEE/J. Lightwave Technol.*, LT-4, 1445–1453, 1986.

7. G.P. Agrawal, *Fiber-Optic Communication Systems*, 3rd ed., New York: Wiley, 2002.

8. A. Hirano, Y. Miyamoto, and S. Kuwahara, Performances of CSRZ-DPSK and RZ-DPSK in 43-Gbit/s/ch DWDM G.652 single-mode-fiber transmission, in *Proceedings of OFC'03*, 2, 454–456, 2003.

9. A.H. Gnauck, G. Raybon, P.G. Bernasconi, J. Leuthold, C.R. Doerr, and L.W. Stulz, 1-Tb/s (6/spl times/170.6 Gb/s) transmission over 2000-km NZDF using OTDM and RZ-DPSK format, *IEEE Photonics Technology Letters*, 15(11), 1618–1620, 2000.

10. M. Suzuki, Y. Noda, H. Tanaka, S. Akiba, Y. Kuahiro, and H. Isshiki, Monolithic integration of InGaAsP/InP distributed feedback laser and electroabsorption modulator by vapor phase epitaxy, *IEEE J. Lightwave Tech.*, LT-5(9), 127, Sept. 1987.

11. L.N. Binh, *Digital Optical Communications*, CRC Press, Taylor & Francis Group, Boca Raton, FL, 2008, Chapter 6.

12. L.N. Binh, Multi-amplitude minimum shift keying modulation formats for optical communications, *Optics Communications*, 281(17), 4245–4253, Sept. 2008.

13. T. Kawanishi, S. Shinada, T. Sakamoto, S. Oikawa, K. Yoshiara, and M. Izutsu, Reciprocating optical modulator with resonant modulating electrode, *Electronics Letters*, 41(5), 271–272, 2005.

14. R. Krahenbuhl, J.H. Cole, R.P. Moeller, and M.M. Howerton, High-speed optical modulator in LiNbO$_3$ with cascaded resonant-type electrodes, *J. Light. Technol.*, 24(5), 2184–2189, 2006.

15. O. Painter, P.C. Sercel, K.J. Vahala, D.W. Vernooy, and G.H. Hunziker, Resonant optical modulators US Patent No. WO/2002/050575, 27.06.2002.

16. F. Amoroso, Pulse and spectrum manipulation in the minimum frequency shift keying (MSK) format, *IEEE Trans. Commun.*, 24, 381–384, Mar. 1976.

17. M. Chacínski, W. Urban, S. Björn, D. Rachid, E.M. Robert, H. Volker, and G.S. Andreas, 100 Gb/s ETDM transmitter module, *IEEE J. Sel. Top. Quant. Elect.*, 16(5), Sept./Oct. 2010, 1321–1327.

18. P.J. Winzer, A.H. Gnauck, C.R. Doerr, M. Magarini, and L.L. Buhl, Spectrally efficient long-haul optical networking using 112-Gb/s polarization-multiplexed 16-QAM, *J. Lightwave Technol.*, 28, 547–556, 2010.

19. M. Nakazawa, M. Yoshida, K. Kasai, and J. Hongou, 20 Msymbol/s, 64 and 128 QAM coherent optical transmission over 525 km using heterodyne detection with frequency-stabilized laser, *Electron. Lett.*, 43, 710–712, 2006.

20. X. Zhou and J. Yu, 200-Gb/s PDM-16QAM generation using a new synthesizing method, ECOC 2009, paper 10.3.5, 2009.

21. T. Sakamoto, A. Chiba, and T. Kawanishi, 50-Gb/s 16-QAM by a quad-parallel Mach–Zehnder modulator, ECOC 2007, paper PDP2.8, 2007.

22. H. Yamazaki, T. Yamada, T. Goh, Y. Sakamaki, and A. Kaneko, 64QAM modulator with a hybrid configuration of silica PLCs and LiNbO3 phase modulators for 100-Gb/s applications, ECOC 2009, paper 2.2.1, 2009.

23. Fujitsu Optical Components Ltd., Japan, www.fujitsu.com

24. J. Hallin, T. Kjellberg, and T. Swahn, A 165-Gb/s 4:1 multiplexer in InP DHBT technology, *IEEE J. Solid-State Circ.*, 41(10), Oct. 2006, 2209–2214.

25. K. Murata, K. Sano, H. Kitabayashi, S. Sugitani, H. Sugahara, and T. Enoki, 100-Gb/s multiplexing and demultiplexing IC operations in InP HEMT technology, *IEEE J. Solid-State Circuits*, 39(1), 207–213, Jan. 2004.

26. M. Meghelli, 132-Gb/s 4:1 multiplexer in 0.13-μm SiGe-bipolar technology, *IEEE J. Solid-State Circuits*, 39(12), 2403–2407, Dec. 2004.

27. Y. Suzuki, Z. Yamazaki, Y. Amamiya, S. Wada, H. Uchida, C. Kurioka, S. Tanaka, and H. Hida, 120-Gb/s multiplexing and 110-Gb/s demultiplexing ICs, *IEEE J. Solid-State Circuits*, 39(12), 2397–2402, Dec. 2004.

28. T. Suzuki, Y. Nakasha, T. Takahashi, K. Makiyama, T. Hirose, and M. Takikawa, 144-Gbit/s selector and 100-Gbit/s 4:1 multiplexer using InP HEMTs, in *IEEE MTT-S Int. Microwave Symp. Dig.*, 117–120, Jun. 2004.

29. X. Liu, S. Chandrasekhar, P. J. Winzer, T. Lotz, J. Carlson, J.Yang, G. Cheren, and S. Zederbaum, 1.5-Tb/s Guard-banded superchannel transmission over 56 GSymbols/s 16QAM signals with 5.75-b/s/Hz net spectral efficiency, Paper Th.3.C.5.pdf ECOC Postdeadline Papers, ECOC 2012, Netherlands.

30. M. Nagatani and H. Nosaka, High-speed low-power digital-to-analog converter using InP heterojunction bipolar transistor technology for next-generation optical transmission systems, *NTT Techn. Rev.*, 9(4), 2011.

31. M. Liu, X. Yin, E. Ulin-Avila1, B. Geng, T. Zentgraf, L. Ju, F. Wang, and X. Zhang, A graphene-based broadband optical modulator, *Nat., Lett.*, | *Nature*, 474(2), 64, Jun. 2011.

32. W. Shieh and Djordjevic, *OFDM for Optical Communications*, Academic Press, Boston, USA, 2010.

33. Amoroso, F., Simplified MSK signaling technique, *IEEE Trans. Comm.*, 25(4), 433–441, 1977.

4

Optical Coherent Detection and Processing Systems

Detection of optical signals can be carried out at the optical receiver by direct conversion of optical signal power to electronic current in the photodiode (PD) and then electronic amplification. This chapter gives the fundamental understandings of coherent detection of optical signals that requires the mixing of the optical fields of the optical signals and that of the local oscillator (LO), a high-power laser, so that its beating product would result in the modulated signals preserving both its phase and amplitude characteristics in the electronic domain. Optical preamplification in coherent detection can also be integrated at the front end of the optical receiver.

4.1 Introduction

With the exponential increase in data traffic, especially due to the demand for ultrabroad bandwidth driven by multimedia applications, cost-effective ultra-high-speed optical networks have become highly desired. It is expected that Ethernet technology will not only dominate in access networks but also will become the key transport technology of next-generation metro/core networks. The next logical evolutionary step after 10 Gigabit Ethernet (10 GbE) is 100 Gigabit Ethernet (100 GbE). Based on the anticipated 100 GbE requirements, 100-Gbit/s data rate of serial data transmission per wavelength is required. To achieve this data rate while complying with current system design specifications such as channel spacing, chromatic dispersion, and polarization mode dispersion (PMD) tolerance, coherent optical communication systems with multilevel modulation formats will be desired, since it can provide high spectral efficiency, high receiver sensitivity, and potentially high tolerance to fiber dispersion effects [1–6]. Compared to conventional direct detection in intensity-modulation/direct-detection (IMDD) systems, which only detect the intensity of the light of the signal, coherent detection can retrieve the phase information of the light, and therefore, can tremendously improve the receiver sensitivity.

Coherent optical receivers are important components in long-haul optical fiber communication systems and networks to improve the receiver sensitivity and thus extra transmission distance. Coherent techniques were considered for optical transmission systems in the 1980s when the extension of repeater distance between spans is pushed to 60 km instead of 40 km for single-mode optical fiber at a bit rate of 140 Gb/s. However, in the late 1980s, the invention of optical fiber amplifiers has overcome this attempt. Recently, interests in coherent optical communications have attracted significant research activities for ultra-bit rate DWDM optical systems and networks. The motivation has been possible due to the following: (i) the uses of optical amplifiers in cascade fiber spans have added significant noises and thus limit the transmission distance; (ii) the advances of DSPs with sampling rates that can reach few tens of Giga-samples/s, allowing the processing of beating signals to recover the phase or phase estimation (PE); (iii) the availability of advanced signal processing algorithms such as Viterbi and Turbo algorithms; and (iv) the differential coding and modulation and detection of such signals may not require optical phase lock loop (OPLL), hence self-coherent and DSP to recover transmitted signals.

As is well known, typical arrangement of an optical receiver is that the optical signals are detected by a PD (a pin diode, avalanche photodiode [APD], or a photon counting device), electrons generated in the photodetector are then electronically amplified through a front-end electronic amplifier. The electronic signals are then decoded for recovery of original format. However, when the fields of incoming optical signals are mixed with those of a LO with a frequency that can be identical or different to that of the carrier, the phase and frequency property of the resultant signals reflect those of the original signals. Coherent optical communication systems have also been reviving dramatically due to electronic processing and availability of stable narrow linewidth lasers.

This chapter deals with the analysis and design of coherent receivers with OPLL and the mixing of optical signals and that of the LO in the optical domain thence detected by the optoelectronic receivers following this mixing. Thus, both the optical mixing and photodetection devices act as the fundamental elements of a coherent optical receiver. Depending on the frequency difference between the lightwave carrier of the optical signals and that of the LO, the coherent detection can be termed as heterodyne or homodyne detection. For heterodyne detection, there is a difference in the frequency and thus the beating signal falls in a passband region in the electronic domain. Thus, all the electronic processing at the front end must be in this passband region. While in homodyne detection there is no frequency difference and thus the detection is in the base band of the electronic signal, both cases would require a locking of the LO and carrier of the signals. An OPLL is thus treated in this chapter.

This chapter is organized as follows: Section 4.2 gives an account of the components of coherent receivers; Section 4.3 outlines the principles of optical coherent detection under heterodyne, homodyne, or intradyne techniques; and Section 4.4 gives details of OPLL, which are very important to the development of modern optical coherent detection.

4.2 Coherent Receiver Components

The design of an optical receiver depends on the modulation format of the signals and transmitted through the transmitter. The modulation of the optical carrier can be in the form of amplitude, phase, or frequency. Furthermore, phase shaping also plays a critical role in the detection and bit error rate (BER) of the receiver, and thence the transmission systems. In particular, it is dependent on the modulation in analog or digital, Gaussian or exponential pulse shape, on–off keying, multiple levels, and so on.

Figure 4.1 shows the schematic diagram of a digital coherent optical receiver, which is similar to the direct detection receiver but with an optical mixer at the front end. However, the phase of the signals at base or pass-band of the detected signals in the electrical domain would remain in the generated electronic current and voltages at the output of the electronic preamplifier. An optical front end is an optical mixer combining the fields of the optical waves of the local laser and the optical signals whereby several spectral components of the modulated optical signals can beat with the local oscillator frequency to give products with summation of the frequencies and with difference of the frequencies of the lightwaves. Only the lower frequency term, which falls within the absorption range of the photodetector, is converted into the electronic current, preserving both the phase and amplitude of the modulated signals.

Thence, an optical receiver front end, very much the same as that of the direct detection, is connected following the optical processing front end, consisting of a photodetector for converting lightwave energy into electronic currents; an electronic preamplifier for further amplification of the generated electronic current (see Figure 4.2), followed by an electronic equalizer for bandwidth extension, usually in voltage form, a main amplifier for further voltage amplification; a clock recovery circuitry for regenerating the timing sequence; and a voltage-level decision circuit for sampling

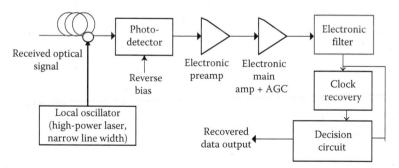

FIGURE 4.1
Schematic diagram of a digital optical coherent receiver with an additional LO mixing with the received optical signals before detection by an optical receiver.

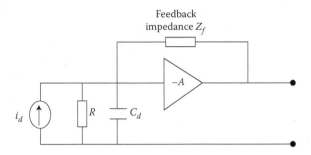

FIGURE 4.2
Schematic of the equivalent circuit of the electronic preamplifier of an optical transimpedance
electronic amplifier at the front end of an optical receiver. The current source represents the
electronic current generated in the photodetector due to the beating of the LO and the optical
signals. Cd, photodiode capacitance.

the waveform for the final recovery of the transmitted and received digital
sequence. Therefore, the optoelectronic preamplifier is followed by a main
amplifier with an automatic control to regulate the electronic signal voltage to be filtered and then sampled by a decision circuit synchronized by a
clock recovery circuitry.

An in-line fiber optical amplifier can be incorporated in front of the photodetector to form an optical receiver with an optical amplifier front end to
improve its receiving sensitivity. This optical amplification at the front end of
an optical receiver will be treated in the chapter dealing with optical amplification processes.

The structure of the receiver thus consists of four parts, the optical mixing front, front end section, the linear channel of the main amplifier and
automatic gain control (AGC) if necessary, and the data recovery section.
The optical mixing front end performs the summation of the optical fields
of the LO and that of the optical signals. Polarization orientation between
these lightwaves is very critical to maximize the beating of the additive
field in the PD. Depending on the frequency range and the magnitude of
the noises, the difference can be appreciable or not between these fields.
Hence, the resulting electronic signals derived from the beating in the
detector leads to noisy signals in the base band or in the passband depending on whether the detection technique is heterodyne or homodyne.

4.3 Coherent Detection

Optical coherent detection can be distinguished by the "demodulation"
scheme in communications techniques in association with the following
definitions: (i) Coherent detection is the mixing between two lightwaves or

optical carriers—one is information-bearing lightwaves and the other an LO with an average energy much larger than that of the signals. (ii) Demodulation refers to the recovery of baseband signals from the electrical signals.

A typical schematic diagram of a coherent optical communications employing guided wave medium and components is shown in Figure 4.1, in which a narrow-band laser incorporating an optical isolator cascaded with an external modulator is usually the optical transmitter. Information is fed via a microwave power amplifier to an integrated optic modulator; commonly used are LiNbO$_3$ or EA types. The coherent detection is a principal feature of coherent optical communications, which can be further distinguished with heterodyne and homodyne techniques depending on whether there is a difference or not between the frequencies of the LO and that of the signal carrier. An LO is a laser source with a frequency that can be tuned and approximately equivalent to a monochromatic source. A polarization controller would also be used to match its polarization with that of the information-bearing carrier. The LO and the transmitted signal are mixed via a polarization-maintaining coupler and then detected by coherent optical receiver. Most of previous coherent detection schemes are implemented in a mixture of photonic domain and electronic/microwave domain.

Coherent optical transmission returns into the focus of research. One significant advantage is the preservation of all the information of the optical field during detection, leading to enhanced possibilities for optical multilevel modulation. This section investigates the generation of optical multilevel modulation signals. Several possible structures of optical M-ary-PSK and M-ary-QAM transmitters are shown and theoretically analyzed. Differences in the optical transmitter configuration and the electrical driving lead to different properties of the optical multilevel modulation signals. This is shown by deriving general expressions applicable to every M-ary-PSK and M-ary-QAM modulation format and exemplarily clarified for square-16-QAM modulation.

Coherent receivers are distinguished between synchronous and asynchronous. Synchronous detection requires an OPLL, which recovers the phase and frequency of the received signals to lock the LO to that of the signal so as to measure the absolute phase and frequency of the signals relative to that of the LO. Thus, synchronous receivers allow direct mixing of the bandpass signals and the base band; thus, this technique is termed a homodyne reception. For asynchronous receivers, the frequency of the LO is approximately the same as that of the receiving signals and no OPLL is required. In general, the optical signals are first mixed with an intermediate frequency (IF) oscillator, whose magnitude is at least two to three times that of the signals in the 3 dB passband. The electronic signals can then be recovered using electrical phase lock loop (PLL) at a lower carrier frequency in the electrical domain. The mixing of the signals and an LO of an IF frequency is referred to as heterodyne detection.

If no LO is used for demodulating the digital optical signals, then differential or self-homodyne reception may be utilized, or classically termed as autocorrelation reception process or self-heterodyne detection.

Coherent communications have been an important technique in the 1980s and early 1990s, but then research came to interruption with the advent in the late 1990s of optical amplifiers that offer up to 20 dB gain without difficulty. Nowadays, however, coherent systems emerge again into the focus of interest, due to the availability of DSP and low-priced components, the partly relaxed receiver requirements at high data rates, and several advantages that coherent detection provides. The preservation of the temporal phase of the coherent detection enables new methods for adaptive electronic compensation of chromatic dispersion. When concerning WDM systems, coherent receivers offer tunability and allow channel separation via steep electrical filtering. Furthermore, only the use of coherent detection permits to converge to the ultimate limits of spectral efficiency. To reach higher spectral efficiencies, the use of multilevel modulation is required. Concerning this matter, coherent systems are also beneficial, because all the information of the optical field is available in the electrical domain. That way complex optical demodulation with interferometric detection—which has to be used in direct detection systems—can be avoided, and the complexity is transferred from the optical to the electrical domain. Several different modulation formats based on the modulation of all four quadratures of the optical field were proposed in the early 1990s, describing the possible transmitter and receiver structures and calculating the theoretical BER performance. However, a more detailed and practical investigation of multilevel modulation coherent optical systems for today's networks and data rates is missing thus far.

Currently, coherent reception has attracted significant interests due to the following reasons: (i) the received signals of the coherent optical receivers are in the electrical domain, which is proportional to that in the optical domain. This, in contrast to the direct detection receivers, allows exact electrical equalization or exact PE of the optical signals. (ii) Using heterodyne receivers, DWDM channels can be separated in the electrical domain by using electrical filters with sharp roll of the passband to the cutoff band. Presently, the availability of ultrahigh sampling rate DSP allows users to conduct filtering in the DSP in which the filtering can be changed with ease.

However, there are disadvantages that coherent receivers would suffer: (i) coherent receivers are polarization sensitive, which requires polarization tracking at the front end of the receiver; (ii) homodyne receivers require OPLL and electrical PLL for heterodyne that would need control and feedback circuitry, optical or electrical, which may be complicated; and (iii) for differential detection, the compensation may be complicated due to the differentiation receiving nature.

In a later chapter, when some advanced modulation formats are presented for optically amplified transmission systems, the use of photonic components are extensively exploited to take advantage of the advanced technology of integrated optics, planar lightwave circuits. Modulation formats of signals depend on whether the amplitude, the phase, or the frequency of the carrier is manipulated, as mentioned in Chapter 2. In this chapter, the detection is coherently converted to the IF range in the electrical domain and the signal envelope. The down-converted carrier signals are detected and then recovered. Both binary-level and multilevel modulation schemes employing amplitude, phase, and frequency shift keying modulation are described in this chapter.

Thus, coherent detection can be distinguished by the difference between the central frequency of the optical channel and that of the LO. Three types can be classified: (i) heterodyne, when the difference is higher than the 3 dB bandwidth of the base band signal; (ii) homodyne, when the difference is nil; and (iii) intradyne, where the frequency difference falls within the base band of the signal.

It is noted that to maximize the beating signals at the output of the photodetector, polarizations of the LO and the signals must be aligned. In practice, this can be implemented best by the polarization diversity technique.

4.3.1 Optical Heterodyne Detection

The basic configuration of optical heterodyne detection is shown in Figure 4.3. The LO, with a frequency that can be higher or lower than that of the carrier, is mixed with the information-bearing carrier, thus allowing down- or up-conversion of the information signals to the IF range. The down-converted electrical carrier and signal envelope is received by the photodetector. This combined lightwave is converted by the PD into electronic current signals, which are then filtered by an electrical bandpass filter (BPF) and then demodulated by a demodulator. A low-pass filter (LPF) is also used to

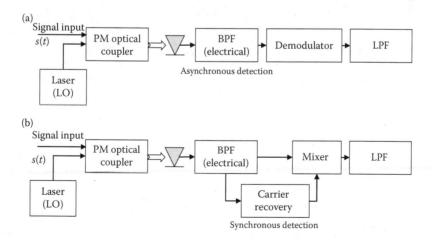

FIGURE 4.3
Schematic diagram of optical heterodyne detection. (a) Asynchronous and (b) synchronous receiver structures. LPF, low-pass filter; BPF, bandpass filter; PD, photodiode.

remove higher-order harmonics of the nonlinear detection photodetection process, the square-law detection. Under the use of an envelope detector, the process is asynchronous; hence the name asynchronous detection. If the down-converted carrier is recovered and then mixed with IF signals, then this is synchronous detection. It is noted that the detection is conducted at the IF range in the electrical domain; hence, the need for controlling the stability of the frequency spacing between the signal carrier and that of the LO. That means the mixing of these carriers would result into an IF carrier in the electrical domain prior to the mixing process or envelope detection to recover the signals.

The coherent detection thus relies on the electric field component of the signal and the LO. The polarization alignment of these fields is critical for the optimum detection. The electric field of the optical signals and the LO can be expressed as

$$E_s(t) = \sqrt{2P_s(t)} \cos\{\omega_s t + \phi_s + \varphi(t)\} \tag{4.1}$$

$$E_{LO} = \sqrt{2P_L} \cos\{\omega_{LO} t + \phi_{LO}\} \tag{4.2}$$

where $P_s(t)$ and P_{LO} are the instantaneous signal power and average power of the signals and LO, respectively; $\omega_s(t)$ and ω_{LO} are the signal and LO angular frequencies; ϕ_s and ϕ_{LO} are the phases, including any phase noise of the signal and the LO; and $\psi(t)$ is the modulation phase. The modulation can be amplitude with the switching on and off (amplitude shift keying [ASK]) of the optical power or phase or frequency with the discrete or continuous variation of the time-dependent phase term. For discrete phase, it can be PSK, differential PSK (DPSK), or differential quadrature PSK—DQPSK. When the variation of the phase is continuous, we have frequency shift keying if the rate of variation is different for the bit "1" and bit "0" as given in Chapter 4.

Under an ideal alignment of the two fields, the photodetection current can be expressed by

$$i(t) = \frac{\eta q}{h\upsilon}\left[P_S + P_{LO} + 2\sqrt{P_S P_{LO}} \cos\{(\omega_S - \omega_{LO})t + \phi_S - \phi_{LO} + \varphi(t)\}\right] \tag{4.3}$$

where the higher-frequency term (the sum) is eliminated by the photodetector frequency response, η is the quantum efficiency, q is the electronic charge, h is the Planck's constant, and v is the optical frequency.

Thus, the power of the LO dominates the shot noise process and at the same time boosts the signal level, hence enhancing the SNR. The oscillating term is the beating between the LO and the signal, that is proportional to the square root of the product of the power of the LO and the signal.

The electronic signal power S and shot noise N_s can be expressed as

$$S = 2\Re^2 P_s P_{LO}$$

$$N_s = 2q\Re(P_s + P_{LO})B$$

$$\Re = \frac{\eta q}{h\nu} = \text{responsivity}$$

(4.4)

with B is the 3 dB bandwidth of the electronic receiver. Thus, the optical signal-to-noise ratio (OSNR) can be written as

$$\text{OSNR} = \frac{2\Re^2 P_s P_{LO}}{2q\Re(P_s + P_{LO})B + N_{eq}}$$

(4.5)

where N_{eq} is the total electronic noise-equivalent power at the input to the electronic preamplifier of the receiver. From this equation, we can observe that if the power of the LO is significantly increased so that the shot noise dominates over the equivalent noise, at the same time increasing the SNR, the sensitivity of the coherent receiver can only be limited by the quantum noise inherent in the photodetection process. Under this quantum limit, the OSNR_{QL} is given by

$$\text{OSNR}_{QL} = \frac{\Re P_s}{qB}$$

(4.6)

4.3.1.1 ASK Coherent System

Under the ASK modulation scheme, the demodulator of Figure 4.3 is an envelope detector (in lieu of the demodulator) followed by a decision circuitry. That is, the eye diagram is obtained and a sampling instant is established with a clock recovery circuit. Under synchronous detection, it requires a locking between frequencies of the carrier and the LO. The LO is tuned to this frequency according to the tracking of the frequency of the central component of the signal band. The amplitude-demodulated envelope can be expressed as

$$r(t) = 2\Re\sqrt{P_s P_{LO}} \cos(\omega_{IF})t + n_x \cos(\omega_{IF})t + n_y \sin(\omega_{IF})t$$

$$\omega_{IF} = \omega_s - \omega_{LO}$$

(4.7)

The IF frequency ω_{IF} is the difference between those of the LO and the signal carrier, and n_x and n_y are the expected values of the orthogonal noise power components, which are random variables.

$$r(t) = \sqrt{[2\Re P_s P_{LO} + n_x]^2 + n_y^2} \cos(\omega_{IF}t + \Phi)t \quad \text{with } \Phi = \tan^{-1}\frac{n_y}{2\Re P_s P_{LO} + n_x}$$

(4.8)

4.3.1.1.1 Envelope detection

The noise power terms can be assumed to follow a Gaussian probability distribution and are independent of each other with a zero mean and a variance σ. The probability density function (PDF) can thus be given as

$$p(n_x, n_y) = \frac{1}{2\pi\sigma^2} e^{-(n_x^2 + n_y^2)/2\sigma^2} \tag{4.9}$$

With respect to the phase and amplitude, this equation can be written as [3]

$$p(\rho, \phi) = \frac{\rho}{2\pi\sigma^2} e^{-\rho^2 + A^2 - 2A\rho\cos\phi/2\sigma^2} \tag{4.10}$$

where

$$\rho = \sqrt{[2\Re\sqrt{P_s(t)P_{LO}} + n_s(t)]^2 + n_y^2(t)}$$

$$A = 2\Re\sqrt{P_s(t)P_{LO}} \tag{4.11}$$

The PDF of the amplitude can be obtained by integrating the phase amplitude PDF over the range of 0 to 2π and given as

$$p(\rho) = \frac{\rho}{\sigma^2} e^{-(\rho^2 + A^2)/2\sigma^2} I_0 \left\{ \frac{A\rho}{\sigma^2} \right\} \tag{4.12}$$

where I_0 is the modified Bessel's function. If a decision level is set to determine the "1" and "0" level, then the probability of error and the BER can be obtained, if assuming an equal probability of error between the "1s" and "0s", as

$$BER = \frac{1}{2}P_e^1 + \frac{1}{2}P_e^0 = \frac{1}{2}\left[1 - Q\left(\sqrt{2\delta}, d\right) + e^{-d^2/2}\right] \tag{4.13}$$

where Q is the magnum function given in Chapter 7 and δ is given by

$$\delta = \frac{A^2}{2\sigma^2} = \frac{2\Re^2 P_s P_{LO}}{2q\Re(P_s + P_{LO})B + i_{N_{eq}}^2} \tag{4.14}$$

When the power of the LO is much larger than that of the signal and the equivalent noise current power, then this SNR becomes

$$\delta = \frac{\Re P_s}{qB} \tag{4.15}$$

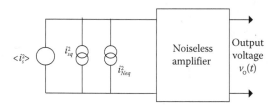

FIGURE 4.4
Equivalent current model at the input of the optical receiver, average signal current, and equivalent noise current of the electronic preamplifier as seen from its input port.

The physical representation of the detected current and the noises current due to the quantum shot noise and noise equivalent of the electronic preamplification are shown in Figure 4.4, in which the signal current can be derived from the output of a detection scheme of heterodyne or homodyne. This current can represent the electronic current which is generated from a single photodetector or from a back-to-back connected pair of photodetectors of a balanced receiver in the case of detecting the phase difference of DPSK, DQPSK, or CPFSK signals.

The BER is optimum when setting its differentiation with respect to the decision level δ, an approximate value of the decision level can be obtained as

$$d_{opt} \cong \sqrt{2 + \frac{\delta}{2}} \Rightarrow BER_{ASK-e} \cong \frac{1}{2}e^{-\delta/4} \tag{4.16}$$

4.3.1.1.2 Synchronous Detection

ASK can be detected using synchronous detection[*] and the BER is given by

$$BER_{ASK-S} \cong \frac{1}{2}erfc^{\sqrt{\delta}/2} \tag{4.17}$$

4.3.1.2 PSK Coherent System

Under the phase shift keying modulation format, the detection is similar to that of Figure 4.3 for heterodyne detection, but after the BPF, an electrical mixer is used to track the phase of the detected signal. The received signal is given by

$$r(t) = 2\Re\sqrt{P_s P_{LO}}\cos[(\omega_{IF})t + \varphi(t)] + n_x\cos(\omega_{IF})t + n_y\sin(\omega_{IF})t \tag{4.18}$$

---

[*] Synchronous detection is implemented by mixing the signals and a strong LO in association with the phase locking of the LO to that of the carrier.

The information is contained in the time-dependent phase term $\varphi(t)$.

When the phase and frequency of the voltage control oscillator (VCO) are matched with those of the signal carrier, then the received electrical signal can be simplified to

$$r(t) = 2\Re\sqrt{P_s P_{LO}}\,a_n(t) + n_x$$

$$a_n(t) = \pm 1$$

$$(4.19)$$

Under Gaussian statistical assumption, the probability of the received signal of a "1" is given by

$$p(r) = \frac{1}{\sqrt{2\pi\sigma^2}}e^{-(r-u)^2/2\sigma^2}$$

$$(4.20)$$

Furthermore, the probability of the "0" and "1" are assumed to be equal. We can obtain the BER as the total probability of the received "1" and "0" as

$$\mathrm{BER_{PSK}} = \frac{1}{2}erfc(\delta)$$

$$(4.21)$$

4.3.1.3 Differential Detection

As observed in the synchronous detection, there needs to be a carrier recovery circuitry, usually implemented using a PLL, which complicates the overall receiver structure. It is possible to detect the signal by a self-homodyne process by beating the carrier of one bit period to that of the next consecutive bit. This is called the differential detection. The detection process can be modified as shown in Figure 4.5, in which the phase of the IF carrier of one bit is compared with that of the next bit, and a difference is recovered to represent the bit "1" or "0." This requires a differential coding at the transmitter and an additional phase comparator for the recovery process. In a

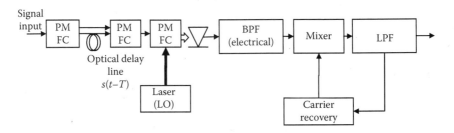

FIGURE 4.5
Schematic diagram of optical heterodyne and differential detection for PSK format.

later chapter on DPSK, the differential decoding is implemented in photonic domain via a photonic phase comparator in the form of an MZ delay interferometer (MZDI) with a thermal section for tuning the delay time of the optical delay line. The BER can be expressed as

$$\mathrm{BER}_{\mathrm{DPSK}-e} \cong \frac{1}{2}e^{-\delta} \tag{4.22}$$

$$r(t) = 2\Re\sqrt{P_s P_{LO}}\cos[\pi A_k s(t)] \tag{4.23}$$

where $s(t)$ is the modulating waveform and A_k represents the bit "1" or "0." This is equivalent to the baseband signal, and the ultimate limit is the BER of the baseband signal.

The noise is dominated by the quantum shot noise of the LO with its square noise current given by

$$i_{N-sh}^2 = 2q\Re(P_s + P_{LO})\int_0^\infty |H(j\omega)|^2\, d\omega \tag{4.24}$$

where $H(j\omega)$ is the transfer function of the receiver system, normally a transimpedance of the electronic preamp and that of a matched filter. As the power of the LO is much larger than the signal and integrating over the dB bandwidth of the transfer function, this current can be approximated by

$$i_{N-sh}^2 \simeq 2q\Re P_{LO}B \tag{4.25}$$

Hence, the SNR (power) is given by

$$\mathrm{SNR} \equiv \delta \simeq \frac{2\Re P_s}{qB} \tag{4.26}$$

The BER is the same as that of a synchronous detection and is given by

$$\mathrm{BER}_{\mathrm{homodyne}} \cong \frac{1}{2}erfc\sqrt{\delta} \tag{4.27}$$

The sensitivity of the homodyne process is at least 3 dB better than that of the heterodyne, and the bandwidth of the detection is half of its counterpart due to the double sideband nature of the heterodyne detection.

4.3.1.4 FSK Coherent System

The nature of FSK is based on the two frequency components that determine the bits "1" and "0." There are a number of formats related to FSK depending

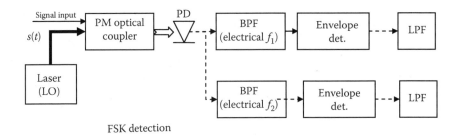

FSK detection

FIGURE 4.6
Schematic diagram of optical homodyne detection of FSK format.

on whether the change of the frequencies representing the bits is continuous or noncontinuous corresponding to either FSK or CPFSK modulation formats, respectively. For noncontinuous FSK, the detection is usually performed by a structure of dual-frequency discrimination as shown in Figure 4.6, in which two narrow band filters are used to extract the signals. For CPFSK, both the frequency discriminator and balanced receiver for PSK detection can be used. The frequency discrimination is indeed preferred as compared with the balanced receiving structures because it would eliminate the phase contribution by the LO or optical amplifiers, which may be used as an optical preamp.

When the frequency difference between the "1" and "0" equals a quarter of the bit rate, the FSK can be termed as the minimum shift keying (MSK) modulation scheme. At this frequency spacing, the phase is continuous between these states.

4.3.2 Optical Homodyne Detection

Optical homodyne detection matches the transmitted signal phases to that of the LO phase signal. A schematic of the optical receiver is shown in Figure 4.7. The field of the incoming optical signals is mixed with the LO, for which

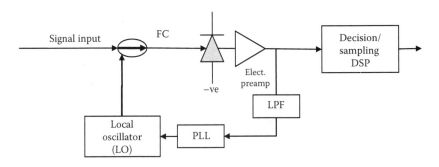

FIGURE 4.7
General structure of an optical homodyne detection system. FC, fiber coupler; LPF, low-pass filter; PLL, phase lock loop.

frequency and phase are locked with that of the signal carrier waves via a PLL. The resultant electrical signal is then filtered and thence a decision circuitry is formed.

4.3.2.1 Detection and OPLL

Optical homodyne detection requires the phase matching of the frequency of the signal carrier and that of the LO. In principles, this coherent detection would give a very high sensitivity that require only the energy of about nine photons per bit at the quantum limit. Implementation of such a system would normally require an OPLL, for which the structure of a recent development [4] is shown in Figure 4.8. The LO frequency is locked into the carrier frequency of the signals by shifting it to the modulated sideband component via the use of the optical modulator. A single sideband optical modulator is preferred. However, a double sideband may also be used. This modulator is excited by the output signal of a voltage-controlled oscillator for which frequency is determined by the voltage level of the output of an electronic BPF condition to meet the required voltage level for driving the electrode of the modulator. The frequency of the LO is normally tuned to the region such that the frequency difference with respect to the signal carrier falls within the passband of the electronic filter. When the frequency difference is zero, then there is no voltage level at the output of the filter, and thus the OPLL has reached the final stage of locking. The bandwidth of the optical modulator is important, so that it can extend the locking range between the two optical carriers.

Any difference in frequency between the LO and the carrier is detected and noise filtered by the LPF. This voltage level is then fed to a VCO to generate a sinusoidal wave, which is then used to modulate an intensity modulator modulating the lightwaves of the LO. The output spectrum of the modulator would exhibit two sidebands and the LO lightwave. One of these components

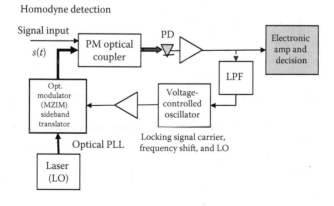

FIGURE 4.8
Schematic diagram of optical homodyne detection—electrical line (soft line) and optical line (heavy line) using an OPLL.

would then be locked to the carrier. A close loop would ensure a stable locking. If the intensity modulator is biased at the minimum transmission point and the voltage level at the output of the VCO is adjusted to $2V_\pi$ with driven signals of $\pi/2$ phase shift with each other, then we would have carrier suppression and sideband suppression. This eases the confusion of the closed-loop locking.

Under a perfect phase, matching the received signal is given by

$$i_s(t) = 2\Re\sqrt{p_s P_{LO}} \cos\{\pi a_k s(t)\} \tag{4.28}$$

where a_k takes the value ± 1 and $s(t)$ is the modulating waveform. This is a baseband signal; thus, the error rate is the same as that of the baseband system.

The shot-noise power induced by the LO and the signal power can be expressed as

$$i_{NS}^2 = 2q\Re(p_s + P_{LO})\int_0^\infty |H(j\omega)|d\omega \tag{4.29}$$

where $|H(j\omega)|$ is the transfer function of the receiver of which the expression, if under a matched filtering, can be

$$|H(j\omega)|^2 = \left[\frac{\sin(\omega T/2)}{\omega T/2}\right]^2 \tag{4.30}$$

where T is the bit period. Then, the noise power becomes

$$i_{NS}^2 = q\Re(p_s + P_{LO})\frac{1}{T} \simeq \frac{q\Re P_{LO}}{T} \quad \text{when} \quad p_s \ll P_{LO} \tag{4.31}$$

Thus, the SNR is

$$\text{SNR} = \frac{2\Re p_s P_{LO}}{q\Re P_{LO}/T} = \frac{2p_s T}{q} \tag{4.32}$$

thence, the BER is

$$P_E = \frac{1}{2}erfc(\sqrt{\text{SNR}}) \to \text{BER} = erfc(\sqrt{\text{SNR}}) \tag{4.33}$$

4.3.2.2 Quantum Limit Detection

For homodyne detection, a super quantum limit can be achieved. In this case, the LO is used in a very special way that matches the incoming signal field in polarization, amplitude, and frequency, and is assumed to be phase-locked to the signal. Assuming that the phase signal is perfectly modulated such that it acts in phase or in counter-phase with the LO, the homodyne detection would give a normalized signal current of

$$i_{sC} = \frac{1}{2T}\left[\mp\sqrt{2n_p} + \sqrt{2n_{LO}}\right]^2 \quad \text{for } 0 \le t \le T \qquad (4.34)$$

Assuming further that $n_p = n_{LO}$, the number of photons for the LO for the generation of detected signals, then the current can be replaced with $4n_p$ for the detection of a "1" and nothing for a "0" symbol.

4.3.2.3 Linewidth Influences

4.3.2.3.1 Heterodyne Phase Detection

When the linewidth of the light sources is significant, the IF deviates due to a phase fluctuation, and the PDF is related to the linewidth conditioned on the deviation $\delta\omega$ of the IF. For a signal power of p_s, we have the total probability of error as

$$P_E = \int_{-\infty}^{\infty} P_C(p_s, \partial\omega)p_{IF}(\partial\omega)\partial\omega \qquad (4.35)$$

The PDF of the IF under a frequency deviation can be written as [5]

$$p_{IF}(\partial\omega) = \frac{1}{\sqrt{\Delta \upsilon B T}}e^{-\partial\omega^2/4\pi\Delta\upsilon B} \qquad (4.36)$$

where $\Delta\upsilon$ is the full IF linewidth at FWHM of the power spectral density and T is the bit period.

4.3.2.3.2 Differential Phase Detection with Local Oscillator

4.3.2.3.2.1 *DPSK Systems* The DPSK detection requires an MZDI and a balanced receiver, either in the optical domain or in the electrical domain. If in the electrical domain, then the beating signals in the PD between the incoming signals and the LO would give the beating electronic current, which is then split and one branch is delayed by one bit period and then summed up. The heterodyne signal current can be expressed as [6]

$$i_s(t) = 2\Re\sqrt{P_{LO}p_s}\,\cos(\omega_{IF}t + \phi_s(t)) + n_x(t)\cos\omega_{IF}t - n_y(t)\sin\omega_{IF}t \qquad (4.37)$$

The phase $\phi_s(t)$ is expressed by

$$\phi_s(t) = \varphi_s(t) + \{\varphi_N(t) - \varphi_N(t+T)\} - \{\varphi_{pS}(t) - \varphi_{pS}(t+T)\} - \{\varphi_{pL}(t) - \varphi_{pL}(t+T)\}$$

(4.38)

The first term is the phase of the data and takes the value 0 or pi. The second term represents the phase noise due to shot noise of the generated current, and the third and fourth terms are the quantum shot noise due to the LO and the signals. The probability of error is given by

$$P_E = \int_{-\pi/2}^{\pi/2} \int_{-\infty}^{\infty} p_n(\phi_1 - \phi_2) p_q(\phi_1) \partial\phi_1 \partial\phi_1$$

(4.39)

where $p_n(.)$ is the PDF of the phase noise due to the shot noise, and $p_q(.)$ is for the quantum phase noise generated from the transmitter and the LO [7].

The probability or error can be written as

$$p_N(\phi_1 - \phi_2) = \frac{1}{2\pi} + \frac{\rho e^{-\rho}}{\pi} \sum_{m=1}^{\infty} a_m \cos\big(m(\phi_1 - \phi_2)\big)$$

$$a_m \sim \left\{ \frac{2^{m-1}\Gamma\!\left[\dfrac{m+1}{2}\right]\Gamma\!\left[\dfrac{m}{2}+1\right]}{\Gamma[m+1]} \left[I_{m-1/2}\frac{\rho}{2} + I_{(m+1)/2}\frac{\rho}{2} \right] \right\}^2$$

(4.40)

where $\Gamma(.)$ is the gamma function and is the modified Bessel function of the first kind. The PDF of the quantum phase noise can be given as [8]

$$p_q(\phi_1) = \frac{1}{\sqrt{2\pi D\tau}} e^{\frac{\phi_1^2}{2D\tau}}$$

(4.41)

With D as the phase diffusion constant, the standard deviation from the central frequency is given as

$$\Delta\upsilon = \Delta\upsilon_R + \Delta\upsilon_L = \frac{D}{2\pi}$$

(4.42)

The above is the sum of the transmitter and the LO FWHM linewidth. Substituting Equations 4.41 and 4.40 into Equation 4.39, we obtain

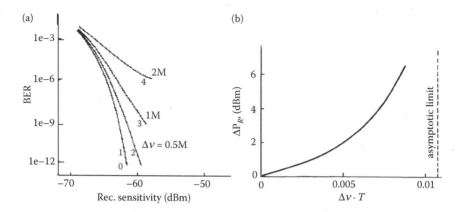

FIGURE 4.9
(a) Probability of error versus receiver sensitivity with linewidth as a parameter in megahertz.
(b) Degradation of optical receiver sensitivity at BER = 1e–9 for DPSK systems as a function of
the linewidth and bit period—bit rate = 140 Mb/s. (Extracted from G. Nicholson, *Electron. Lett.*,
20/24, 1005–1007, 1984.)

$$P_E = \frac{1}{2} + \frac{\rho e^{-\rho}}{2} \sum_{n=0}^{\infty} \frac{(-1)^n}{2n+1} e^{-(2n+1)^2 \pi \Delta \upsilon T} \left\{ I_{n-1/2} \frac{\rho}{2} + I_{(n+1)/2} \frac{\rho}{2} \right\}^2 \qquad (4.43)$$

This equation gives the probability of error as a function of the received
power. The probability of error is plotted against the receiver sensitivity and
the product of the linewidth, with the bit rate (or the relative bandwidth of
the laser line width and the bit rate) shown in Figure 4.9 for DPSK modulation format at 140 Mb/s bit rate and the variation of the laser linewidth from
0 to 2 MHz.

4.3.2.3.3 Differential Phase Detection under Self-Coherence

Recently, the laser linewidth requirement for DQPSK modulation and differential detection for DQPSK has also been studied, wherein no LO is used
means self-coherent detection. It has been shown that for the linewidth of up
to 3 MHz, the transmitter laser would not significantly influence the probability of error as shown in Figures 4.10 and 4.11 [8].

4.3.2.3.4 Differential Phase Coherent Detection of Continuous Phase FSK Modulation Format

The probability of error of CPFSK can be derived by taking into consideration the delay line of the differential detector, the frequency deviation, and
phase noise [10]. Similar to Figure 4.6, the differential detector configuration
is shown in Figure 4.12a and the conversion of frequency to voltage relationship in Figure 4.12b. If heterodyne detection is employed, then a BPF is used
to bring the signals back to the electrical domain.

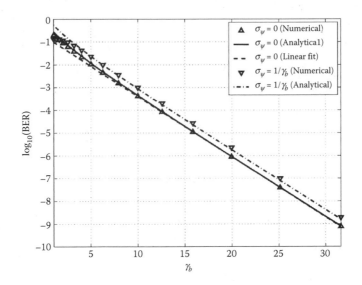

FIGURE 4.10
Analytical approximation (solid line) and numerical evaluation (triangles) of the BER for the cases of zero linewidth and that required to double the BER. The dashed line is the linear fit for zero linewidth. Bit rate 10 Gb/s per channel. (Extracted from S. Savory and T. Hadjifotiou, *IEEE Photonic Tech Lett.*, 16(3), 930–932, March 2004.)

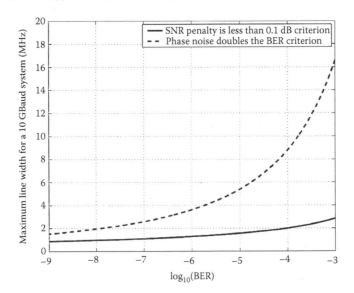

FIGURE 4.11
Criteria for neglecting linewidth in a 10-GSymbols/s system. The loose bound is to neglect linewidth if the impact is to double the BER, with the tighter bound being to neglect the linewidth if the impact is a 0.1-dB SNR penalty. Bit rate 10 GSymbols/s. (Extracted from Y. Yamamoto and T. Kimura, *IEEE J. Quantum Elect.*, QE-17, 919–934, 1981.)

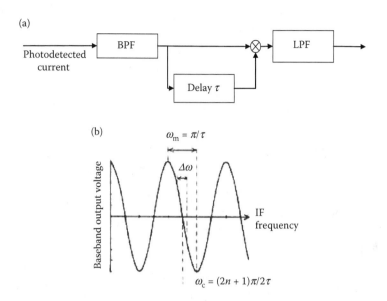

FIGURE 4.12
(a) Configuration of a CPFSK differential detection. (b) Frequency to voltage conversion relationship of FSK differential detection. (Extracted from K. Iwashita and T. Masumoto, LT-5/4, 452–462, 1987, Figure 1.)

The detected signal phase at the shot-noise limit at the output of the LPF can be expressed as

$$\phi(t) = \omega_c t + a_n \frac{\Delta\omega}{2}\tau + \varphi(t) + \varphi_n(t)$$

$$\text{with } \omega_c = 2\pi f_c = (2n+1)\frac{\pi}{2\tau}$$

(4.44)

where τ is the differential detection delay time, $\Delta\omega$ is the deviation of the angular frequency of the carrier for the "1" or "0" symbol, $\phi(t)$ is the phase noise due to the shot noise, and $n(t)$ is the phase noise due to the transmitter and the LO quantum shot noise, and takes the values of ±1, the binary data symbol.

Thus, by integrating the detected phase from $-(\Delta\omega/2)\tau \longrightarrow \pi - (\Delta\omega/2)\tau$, we obtain the probability of error as

$$P_E = \int_{-\frac{\Delta\omega}{2}\tau}^{\pi-\frac{\Delta\omega}{2}\tau/2} \int_{-\infty}^{\infty} p_n(\phi_1 - \phi_2)p_q(\phi_1)\partial\phi_1\partial\phi_1$$

(4.45)

Similar to the case of the DPSK system, substituting Equations 4.40 and 4.41 into Equation 4.45, we obtain

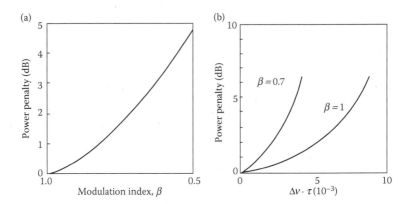

FIGURE 4.13
(a) Dependence of receiver power penalty at BER of 1e–9 on modulation index β (ratio between frequency deviation and maximum frequency spacing between f1 and f2). (b) Receiver power penalty at BER 1e–9 as a function of the product of beat bandwidth and bit delay time—effects excluding LD phase noise. (Extracted from Y. Yamamoto and T. Kimura, *IEEE J. Quantum Elect.*, QE-17, 919–934, 1981; K. Iwashita, and T. Masumoto, *IEEE J. Lightwave Tech.*, LT-5/4, 452–462, 1987, Figures 2 and 3.)

$$
P_E = \frac{1}{2}\frac{\rho e^{-\rho}}{2}\sum_{n=0}^{\infty}\frac{(-1)^n}{2n=1}e^{-(2n+1)^2\pi\Delta v\tau}\left\{I_{n-1/2}\frac{\rho}{2}+I_{(n+1)/2}\frac{\rho}{2}\right\}^2
$$

$$
\times e^{-(2n+1)^2\pi\Delta v\tau}\cos\left\{(2n+1)\alpha\right\} \tag{4.46}
$$

$$
\alpha = \frac{\pi(1-\beta)}{2} \quad \text{and} \quad \beta = \frac{\Delta\omega}{\omega_m} = 2\pi\tau/T_0
$$

where ω_m is the deviation of the angular frequency with *m* as the modulation index, and T_0 is the pulse period or bit period. The modulation index parameter b is defined as the ratio of the actual frequency deviation to the maximum frequency deviation. The variations of the power penalty at a specific BER with respect to the modulation index and the linewidth of the source are shown in Figure 4.13a and b, respectively.

4.3.3 Optical Intradyne Detection

Optical phase diversity receivers combine the advantages of the homodyne with minimum signal processing bandwidth and heterodyne reception with no optical phase locking required. The term "diversity" is well known in radio transmission links and describes the transmission over more than one path. In optical receivers, the optical path is considered to be a result of different polarization and phase paths. In intradyne detection, the frequency difference, the IF, or the LO frequency offset (LOFO), between the LO and the central carrier is

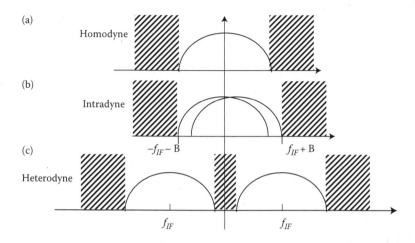

FIGURE 4.14
Spectrum of coherent detection: (a) homodyne, (b) intradyne, and (c) heterodyne.

nonzero, and lies within the signal bandwidth of the baseband signal as illustrated in Figure 4.14 [11]. Naturally, the control and locking of the carrier and the LO cannot be exact, sometimes due to jittering of the source and most of the time the laser frequency is locked stably by oscillating the reflection mirror; hence, the central frequency is varied by a few hundreds of kilohertz. Thus, the intradyne coherent detection method is more realistic. Furthermore, the DSP in the modern coherent reception system would be able to extract this difference without much difficulty in the digital domain [12]. Obviously, the heterodyne detection would require a large frequency range of operation of electronic devices while homodyne and intradyne receptions require simpler electronics. Either differential or nondifferential format can be used in DSP-based coherent reception. For differential-based reception, the differential decoding would gain advantage when there are slips in the cycles of bits due to walk-off of the pulse sequences over a very long transmission noncompensating fiber line.

The diversity in phase and polarization can be achieved by using a pi/2 hybrid coupler that splits the polarization of the LO and the received channels and mixed with pi/2 optical phase shift. Then the mixed signals are detected by balanced photodetectors. This diversity detection is described in the next few sections (see also Figure 4.19).

4.4 Self-Coherent Detection and Electronic DSP

The coherent techniques described above would offer significant improvement, but may face a setback due to the availability of a stable LO and an OPLL for locking the frequency of the LO and that of the signal carrier.

DSP have been widely used in wireless communications and play key roles in the implementation of DSP-based coherent optical communication systems. We can apply the DSP techniques in the receiver of coherent optical communication systems to overcome the difficulties of optical phase locking and also to improve the performance of the transmission systems in the presence of fiber degrading effects, including chromatic dispersion, PMD, and fiber nonlinearities.

Coherent optical receivers have the following advantages: (1) the shot-noise-limited receiver sensitivity can be achieved with a sufficient LO power; (2) closely spaced WDM channels can be separated with electrical filters, having sharp roll-off characteristics; and (3) the ability of phase detection can improve the receiver sensitivity compared with the IMDD system [13]. In addition, any kind of multilevel phase modulation formats can be introduced by using the coherent receiver. While the spectral efficiency of binary modulation formats is limited to 1 bit/s/Hz/polarization (which is called the Nyquist limit), multilevel modulation formats with N bits of information per symbol can achieve up to the spectral efficiency of N bit/s/Hz/polarization. Recent research has focused on M-ary phase-shift keying (M-ary PSK) and even quadrature amplitude modulation (QAM) with coherent detection, which can increase the spectral efficiency by a factor of $\log_2 M$ [14–16]. Moreover, for the same bit rate, since the symbol rate is reduced, the system can have higher tolerance to chromatic dispersion and PMD.

However, one of the major challenges in coherent detection is to overcome the carrier phase noise when using an LO to beat with the received signals to retrieve the modulated phase information. Phase noise can result from lasers, which will cause a power penalty to the receiver sensitivity. A self-coherent multisymbol detection of optical differential M-ary PSK is introduced to improve the system performance; however, higher analog-to-digital conversion resolution and more digital signal processing power are required as compared to a digital coherent receiver [17]. Further, differential encoding is also necessary in this scheme. As for the coherent receiver, initially, an OPLL is an option to track the carrier phase with respect to the LO carrier in homodyne detection. However, an OPLL operating at optical wavelengths in combination with distributed feedback (DFB) lasers may be quite difficult to implement because the product of laser linewidth and loop delay is too large [18]. Another option is to use electrical PLL to track the carrier phase after down converting the optical signal to an IF electrical signal in a heterodyne detection receiver as mentioned above. Compared to heterodyne detection, homodyne detection offers better sensitivity and requires a smaller receiver bandwidth [19]. On the other hand, coherent receivers employing high-speed ADCs and high-speed baseband DSP units are becoming increasingly attractive rather than using an OPLL for demodulation. A conventional block Mth power PE scheme is proposed in Refs. [13,18] to raise the received M-ary PSK signals to the Mth power to estimate the phase reference in conjunction with a coherent optical receiver. However, this scheme requires nonlinear operations, such as taking the Mth power and the $\tan^{-1}(\cdot)$, and resolving the $\pm 2\pi/M$

phase ambiguity, which results in a large latency to the system. Such non-linear operations would limit further potential for real-time processing of the scheme. In addition, nonlinear phase noises always exist in long-haul systems due to the Gordon–Mollenauer effect [20], which severely affects the performance of a phase-modulated optical system [21]. The results in Ref. [22] show that such Mth power PE techniques may not effectively deal with nonlinear phase noise.

The maximum likelihood (ML) carrier phase estimator derived in Ref. [23] can be used to approximate the ideal synchronous coherent detection in optical PSK systems. The ML phase estimator requires only linear computations; thus, it is more feasible for online processing for real systems. Intuitively, one can show that the ML estimation receiver outperforms the Mth power block phase estimator and conventional differential detection, especially when the nonlinear phase noise is dominant, thus significantly improving the receiver sensitivity and tolerance to the nonlinear phase noise. The algorithm of ML phase estimator is expected to improve the performance of coherent optical communication systems using different M-ary PSK and QAM formats. The improvement by DSP at the receiver end can be significant for the transmission systems in the presence of fiber-degrading effects, including chromatic dispersion, PMD, and nonlinearities for both single-channel and also DWDM systems.

4.5 Electronic Amplifiers: Responses and Noises

4.5.1 Introduction

The electronic amplifier as a preamplification stage of an optical receiver plays a major role in the detection of optical signals so that optimum SNR and thence the OSNR can be derived based on the photodetector responsivity. Under coherent detection, the amplifier noises must be much less than that of the quantum shot noises contributed by the high power level of the LO, which is normally about 10 dB above that of the signal average power.

This section provides an introduction to electronic amplifiers for wideband signals applicable to ultra-high-speed, high-gain, and low-noise transimpedance amplifiers (TIA). We concentrate on differential input TIA, but addressing the detailed design of a single input/single output with noise suppression technique in Section 4.7 with design strategy for achieving stability in the feedback amplifier as well as low noise and wide bandwidth. By electronic noise of the preamplifier stage, we define it as the total equivalent input noise spectral density, that is, all the noise sources (current and voltage sources) of all elements of the amplifier are referred to the input port of the amplifier and thence an equivalent current source is found and the current density is derived. Once this current density is found, the total equivalent noise current referred to the input can be found when the overall bandwidth of the receiver is known. When this current is known and with the average signal power we can obtain without difficulty the SNR at the input stage of the optical receiver, thence the OSNR. On the other hand, if the OSNR is required at the receiver, the signal average power or amplitude can be determined for any specific modulation format. Thus with the assumed optical power of the signal available at the front of the optical receiver and the responsivity of the photodetector, we can determine the maximum electronic noise spectral density allowable for the preamplification stage. Hence the design of the amplifier electronic circuit can be based on this requirement as a parameter.

The principal function of an optoelectronic receiver is to convert the received optical signals into electronic equivalent signals, then amplification, sampling, and processing to recover properties of the original shapes and sequence. So, at first, the optical domain signals must be converted to electronic current in the photodetection device, the photodetector of either p-i-n or APD, in which the optical power is absorbed in the active region and both electrons and holes generated are attracted to the positive and negative biased electrodes, respectively. Thus, the generated current is proportional to the power of the optical signals; hence, the name "square law" detection. The p-i-n detector is structured with a p+ and n+ :doped regions sandwiched within the intrinsic layer, where the absorption of optical signal occurs. A high electric field is established in this region by reverse biasing the diode, and then electrons and holes are attracted to either sides of the diode, thus generating current. Similarly, an APD works with the reverse biasing level close to the reverse breakdown level of the *pn* junction (no intrinsic layer) so that electronic carriers can be multiplied in the avalanche flow when the optical signals are absorbed.

This photogenerated current is then fed into an electronic amplifier in which transimpedance must be sufficiently high and generate low noise so that a sufficient voltage signal can be obtained and then further amplified by a main amplifier, a voltage gain type. For high-speed and wideband signals, transimpedance amplification type is preferred, as it offers wideband, much wider than high impedance type, though the noise level might be higher. With TIA, there are two types: the single input/single output port and differential inputs as well as differential outputs. The output ports can be differential with a complementary port. The differential input TIA offers much higher transimpenade gain (Z_T) and wider bandwidth as well. This is contributed to the use of a long-tail pair at the input and hence reasonable high input impedance that would ease the feedback stability [24–26].

In Section 4.8, a case study of a coherent optical receiver is described from the design to implementation, including the feedback control and noise reduction. Although the corner frequency is only a few hundreds of megahertz, with limited transition frequency of the transistors, this bandwidth is remarkable. The design is scalable to ultra-wideband reception subsystems.

4.5.2 Wideband TIAs

Two types of TIAs are described. They are distinguished by whether one single input or differential input, which are dependent on the use of a differential pair, long-tail pair, or a single transistor stage at the input of the TIA.

4.5.2.1 Single Input/Single Output

We prefer to present this section as a design example and experimental demonstration of a wideband and low-noise amplifier. This discussion appears in Section 4.7, located at the end of this chapter. So, in the next section, the differential input TIA is treated with large transimpedance and reasonably low noise.

4.5.2.2 Differential Inputs, Single/Differential Output

An example circuit of the differential input TIA is shown in Figure 4.15, in which a long-tail pair or differential pair is employed at the input stage. Two matched transistors are used to ensure the minimum common mode rejection and maximum differential mode operation. This pair has very high input impedance and thus the feedback from the output stage can be stable. Thus, the feedback resistance can be increased until the limit of the stability locus of the network pole. This offers the high transimpedance Z_T and wide bandwidth. Typical Z_T of 3000 to 6000 Ω can be achieved with 30 GHz 3 dB

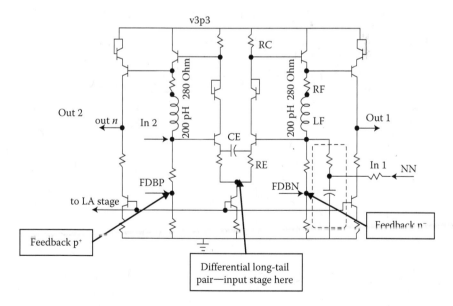

FIGURE 4.15
A typical structure of a differential TIA with differential feedback paths. (From H. Tran et al., *IEEE J. Solid State Circuits*, 39(10), 1680–1689, 2004.)

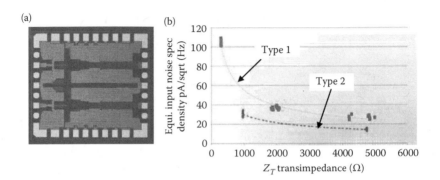

FIGURE 4.16
Differential amplifiers: (a) chip-level image; (b) referred input noise equivalent spectral noise density: Inphi *TIA* 3205 (type 1) and 2850 (type 2). (Courtesy of Inphi Inc., Technical information on 3205 and 2850 TIA device, 2012.)

bandwidth (see Figure 4.16), and can be obtained as shown in Figures 4.16 and 4.17. Also, the chip image of the TIA can be seen in Figure 4.16a. Such TIA can be implemented in either InP or SiGe material. The advantage of SiGe is that the circuit can be integrated with a high-speed Ge-APD detector and ADC and DSP. On the other hand, if implemented in InP, then high-speed p-i-n or APD can be integrated and RF interconnected with ADC and DSP. The differential group delay may be serious and must be compensated in the digital processing domain.

4.5.3 Amplifier Noise Referred to Input

There are several noise sources in any electronic systems, including thermal noises, shot noises, and quantum shot noises, especially in optoelectronic

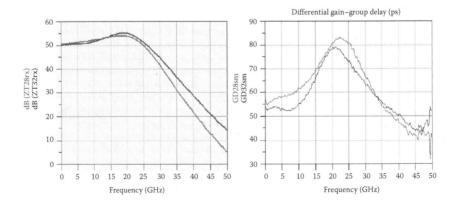

FIGURE 4.17
Differential amplifier: frequency response and differential group delay.

detection. Thermal noises result when the operating temperature is well above the absolute temperature, at which point no random movement of electrons and the resistance of electronic elements occur. This type of noise is dependent on temperate. Shot noises are due to the current flowing and random scattering of electrons; thus, this type of noise depends on the strength of the flowing currents such as biasing current in electronic devices. Quantum shot noises are generated due to the current emitted from opto-electronic detection processes, which are dependent on the strength of the intensity of the optical signals or sources imposed on the detectors. Thus, these types of noises are dependent on signals. In the case of coherent detection, the mixing of the LO laser and signals normally occur with the strength of the LO much larger than that of signal average power. Thus, the quantum shot noises are dominated by that from the LO.

In practice, an equivalent electronic noise source is the total noise as referred to in the input of electronic amplifiers, which can be determined by measuring the total spectral density of the noise distribution over the whole bandwidth of the amplification devices. Thence, the total noise spectral density can be evaluated and referred to the input port. For example, if the amplifier is a transimpedance type, then the transimpedance of the device is measured first, then the measure voltage spectral density at the output port can be referred to the input. In this case, it is the total equivalent noise spectral density. The common term employed and specified for transimpedence amplifiers is the total equivalent spectral noise density over the midband region of the amplifying device. The midband range of any amplifier is defined as the flat gain region from DC to the corner 3 dB point of the frequency response of the electronic device.

Figure 4.18 illustrates the meaning of the total equivalent noise sources as referred to the input port of a two-port electronic amplifying device. A noisy

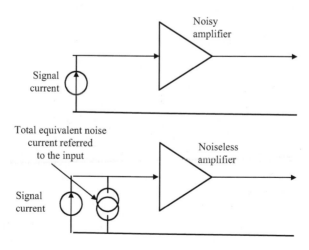

FIGURE 4.18
Equivalent noise spectral density current sources.

amplifier with an input excitation current source, typically a signal current generated from the PD after the optical to electrical conversion, can be represented with a noiseless amplifier and the current source in parallel with a noise source with a strength equal to the total equivalent noise current referred to the input. Thus, the total equivalent current can be found by taking the product of this total equivalent current noise spectral density and the 3 dB bandwidth of the amplifying device. One can find the SNR at the output of the electronic amplifier given by

$$\text{SNR} = \frac{\text{square of current generated}}{\text{square of current generated} + \text{total equivalent noise current power}} \tag{4.47}$$

From this SNR referred at the input of the electronic front end, one can estimate the eye opening of the voltage signals at the output of the amplifying stage, which is normally required by the ADC to sample and convert to digital signals for processing. One can then estimate the required OSNR at the input of the photodetector and hence the launch power required at the transmitter over several span links with certain attenuation factors.

Detailed analyses of amplifier noises and their equivalent noise sources as referred to input ports are given in the Appendix. It is noted that noises have no flowing direction, as they always add and do not subtract. Thus, the noises are measured as noise power and not as a current. Electrical spectrum analyzers are commonly used to measure the total noise spectral density, or the distribution of noise voltages over the spectral range under consideration, which is defined as the noise power spectral density distribution.

4.6 Digital Signal Processing Systems and Coherent Optical Reception

4.6.1 DSP-Assisted Coherent Detection

Over the years, since the introduction of optical coherent communications in the mid-1980s, the invention of optical amplifiers had left coherent reception behind until recently, when long-haul transmission suffered from nonlinearity of dispersion compensating fibers (DCF) and SSMF transmission line due to its small effective area. Furthermore, the advancement of DSP in wireless communication has also contributed to the application of DSP in modern coherent communication systems. Thus, the name "DSP-assisted coherent detection," that is, when a real-time DSP is incorporated after the optoelectronic conversion of the total field of the LO and that of the signals, the analog received signals are sampled by a high-speed ADC and then the digitalized signals are processed in a DSP. Currently, real-time DSP processors are intensively researched for practical implementation. The main difference between real-time and offline processing is that the real-time processing algorithm must be effective due to limited time available for processing.

When polarization division multiplexed (PDM) QAM channels are transmitted and received, polarization and phase diversity receivers are employed. The schematics of such receivers are shown in Figure 4.19a. Further, the structures of such reception systems incorporating DSP with the diversity hybrid coupler in the optical domain are shown in Figure 4.19b–d. The polarization diversity section with the polarization beam splitters at the signal and LO inputs facilitate the demultiplexing of polarized modes in the optical waveguides. The phase diversity using a 90° optical phase shifter allows the separation of the inphase (I) and quadrature (Q) phase components of QAM channels. Using a 2×2 coupler also enables the balanced reception using PDP connected back to back, and hence 3 dB gain in the sensitivity. Section 3.5 of Chapter 3 has described the modulation scheme QAM using I-Q modulators for single-polarization or dual-polarization multiplexed channels.

4.6.1.1 DSP-Based Reception Systems

The schematic of synchronous coherent receiver based on DSP is shown in Figure 4.20. Once the polarization and the I- and Q-optical components are separated by the hybrid coupler, the positive and negative parts of the I- and Q-optical components are coupled into a balanced optoelectronic receiver as shown in Figure 4.19b. Two PDs are connected back to back so that push–pull operation can be achieved, hence a 3 dB betterment as compared to a single PD detection. The current generated from the back-to-back connected PDs is fed into a TIA so that a voltage signal can be derived at the output. Further, a voltage-gain amplifier is used to boost these signals to the right level of the ADC so that sampling can be conducted and the analog signals converted to digital domain. These digitalized signals are then fetched into DSPs and processing in the "soft domain" can be conducted. Thus, a number of processing algorithms can be employed in this stage to compensate for linear and nonlinear distortion effects due to optical signal propagation through the optical guided medium, and to recover the carrier phase and clock rate for resampling of the data sequence. Chapter 6 will describe in detail the fundamental aspects of these processing algorithms. Figure 4.20 shows a schematic of possible processing phases in the DSP incorporated in a DSP-based coherent receiver. Besides the soft processing of the optical phase locking as described in Chapter 5, it is necessary to lock the frequencies of the LO and that of the signal carrier to a certain limit within which the algorithms for clock recovery can function, for example, within ±2 GHz.

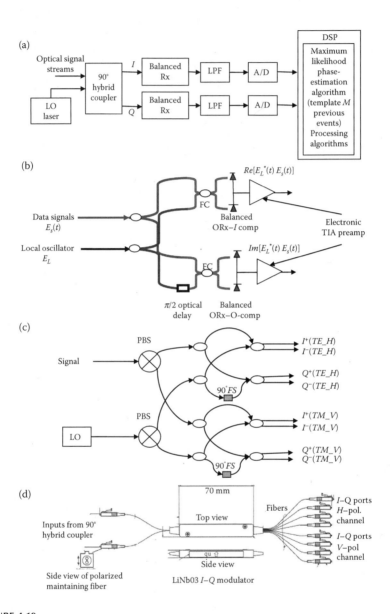

FIGURE 4.19
Scheme of a synchronous coherent receiver using DSP for PE of coherent optical communications. (a) Generic scheme; (b) detailed optical receiver using only one polarization phase
diversity coupler; (c) hybrid 90° coupler for polarization and phase diversity; (d) typical
view of a hybrid coupler with two input ports and eight output ports of structure in (c). FS,
phase shifter; PBS, polarization beam splitter; TM, transverse magnetic mode with polarization orthogonal to that of the TE mode; TE_V, TE_H, transverse electric mode with vertical
(V) or horizontal (H) polarized mode. (Adapted from (a) S. Zhang et al., *Photonics Global'08*,
Paper C3-4A-03. Singapore, December 2008; (b) S. Zhang, *OptoElectronics and Communications
Conference (OECC)'08*, Paper TuA-4, pp. 1–2, Sydney, Australia, July 2008.)

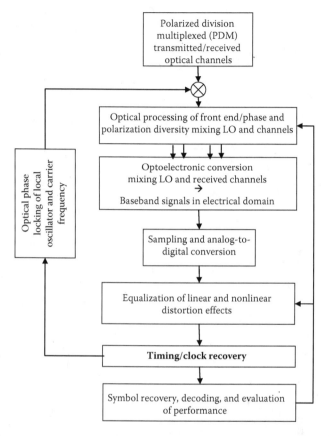

FIGURE 4.20
Flow of functionalities of DSP processing in a QAM-coherent optical receiver with possible feedback control.

4.6.2 Coherent Reception Analysis

4.6.2.1 Sensitivity

At an ultrahigh bit rate, the laser must be externally modulated; thus, the phase of the lightwave conserves along the fiber transmission line. The detection can be direct detection, self-coherent, or homodyne and heterodyne. The sensitivity of the coherent receiver is also important for the transmission system, especially the PSK scheme under both the homodyne and heterodyne transmission techniques. This section gives the analysis of the receiver for synchronous coherent optical fiber transmission systems. Consider that the optical fields of the signals and LO are coupled via a fiber coupler with two output ports 1 and 2 as shown in Figure 4.21. The output fields are then launched into two photodetectors connected back to back, and then the electronic current is amplified using a transimpedance type

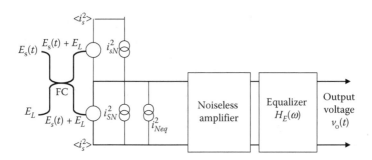

FIGURE 4.21
Equivalent current model at the input of the optical balanced receiver under coherent detection, average signal current, and equivalent noise current of the electronic preamplifier as seen from its input port and equalizer. FC, fiber coupler.

further equalized to extend the bandwidth of the receiver. Our objective is to obtain the receiver penalty and its degradation due to imperfect polarization mixing and unbalancing effects in the balanced receiver. A case study of the design, implementation, and measurements of an optical balanced receiver electronic circuit and noise suppression techniques is given in Section 4.7.

The following parameters are commonly used in analysis:

E_S	Amplitude of signal optical field at the receiver
E_L	Amplitude of LO optical field
P_s, P_L	Optical power of signal and LO at the input of the photodetector
$s(t)$	The modulated pulse
$\langle i^2_{NS}(t) \rangle$	Mean square noise current (power) produced by the total optical intensity on the photodetector
$\langle i^2_s(t) \rangle$	Mean square current produced by the photodetector by $s(t)$
$S_{NS}(t)$	Shot-noise spectral density of $\langle i^2_s(t) \rangle$ and LO power
$i^2_{Neq}(t)$	Equivalent noise current of the electronic preamplifier at its input
$Z_T(\omega)$	Transfer impedance of the electronic preamplifier
$H_E(\omega)$	Voltage transfer characteristic of the electronic equalizer followed by the electronic preamplifier

The combined field of the signal and LO via a directional coupler can be written to reflect their separate polarized field components as

$$E_{sX} = \sqrt{K_{sX}}\, E_S \cos(\omega_s t - \phi_{m(t)})$$
$$E_{sY} = \sqrt{K_{sY}}\, E_S \cos(\omega_s t - \phi_{m(t)} + \delta_s)$$
$$E_{LX} = \sqrt{K_{LX}}\, E_L \cos(\omega_L t) \tag{4.48}$$
$$E_{LY} = \sqrt{K_{LY}}\, E_L \cos(\omega_L t + \delta_L)$$
$$\phi_{m(t)} = \frac{\pi}{2} K_m s(t)$$

where $\phi_{m(t)}$ represents the phase modulation; K_m is the modulation depth; and K_{sX}, K_{sY}, K_{LX}, K_{LY} are the intensity fraction coefficients in the X- and Y-direction of the signal and LO fields, respectively.

Thus, the output fields at ports 1 and 2 of the FC in the X-plane can be obtained using the transfer matrix as

$$\begin{bmatrix} E_{R1X} \\ E_{R2X} \end{bmatrix} = \begin{bmatrix} \sqrt{K_{sX}(1-\alpha)}\cos(\omega_s t - \phi_{m(t)}) & \sqrt{K_{LX}\alpha}\sin(\omega_L t) \\ \sqrt{K_{sX}\alpha}\sin(\omega_s t - \phi_{m(t)}) & \sqrt{K_{LX}(1-\alpha)}\cos(\omega_L t) \end{bmatrix} \begin{bmatrix} E_s \\ E_L \end{bmatrix} \qquad (4.49)$$

$$\begin{bmatrix} E_{R1Y} \\ E_{R2Y} \end{bmatrix} = \begin{bmatrix} \sqrt{K_{sY}(1-\alpha)}\cos(\omega_s t - \phi_{m(t)}) & \sqrt{K_{LY}\alpha}\sin(\omega_L t + \delta_L) \\ \sqrt{K_{sY}\alpha}\sin(\omega_s t - \phi_{m(t)}) & \sqrt{K_{LY}(1-\alpha)}\cos(\omega_L t) + \delta_L \end{bmatrix} \begin{bmatrix} E_s \\ E_L \end{bmatrix} \qquad (4.50)$$

with α defined as the intensity coupling ratio of the coupler. Thus, the field components at ports 1 and 2 can be derived by combining the X- and Y-components from Equations 4.49 and 4.50; thence, the total power at ports 1 and 2 are given as

$$P_{R1} = P_s(1-\alpha) + P_L\alpha + 2\sqrt{P_sP_L\alpha(1-\alpha)K_p}\,\sin(\omega_{IF}t + \phi_{m(t)} + \phi_p - \phi_e)$$

$$P_{R2} = P_s\alpha + P_L(1-\alpha) + 2\sqrt{P_sP_L\alpha(1-\alpha)K_p}\,\sin(\omega_{IF}t + \phi_{m(t)} + \phi_p - \phi_e + \pi) \quad \text{with}$$

$$K_p = K_{sX}K_{LX} + K_{sY}K_{LY} + 2\sqrt{K_{sX}K_{LX}K_{sY}K_{LY}}\,\cos(\delta_L - \delta_s) \qquad (4.51)$$

$$\phi_p = \tan^{-1}\left[\frac{\sqrt{K_{sX}K_{LY}}\,\sin(\delta_L - \delta_s)}{\sqrt{K_{sX}K_{LX}} + \sqrt{K_{sY}K_{LY}}\,\cos(\delta_L - \delta_s)} \right]$$

where ω_{IF} is the intermediate angular frequency and equals to the difference between the frequencies of the LO and the carrier of the signals, ϕ_e is the phase offset, and $\phi_p - \phi_e$ is the demodulation reference phase error.

In Equation 4.51, the total field of the signal and the LO are added and then the product of the field vector and its conjugate is taken to obtain the power. Only the term with frequency falling within the range of the sensitivity of the photodetector would produce the electronic current. Thus, the term with the sum of the frequency of the wavelength of the signal and LO would not be detected and only the product of the two terms would be detected as given.

Now, assuming a binary PSK (BPSK) modulation scheme, the pulse has a square shape with amplitude +1 or −1, the PD is a p-i-n type, and the PD bandwidth is wider than the signal 3 dB-bandwidth, followed by an equalized electronic preamplifier. The signal at the output of the electronic equalizer or the input signal to the decision circuit is given by

$$\hat{v}_D(t) = 2K_H K_p \sqrt{P_s P_L \alpha(1-\alpha)K_p} \int\limits_{-\infty}^{\infty} H_E(f)df \int\limits_{-\infty}^{\infty} (t)dt \sin\left(\frac{\pi}{2}K_m\right)\cos(\phi_p - \phi_e)$$

$$\to \hat{v}_D(t) = 2K_H K_p \sqrt{P_s P_L \alpha(1-\alpha)K_p} \sin\left(\frac{\pi}{2}K_m\right)\cos(\phi_p - \phi_e)$$

(4.52)

$K_H = 1$ for homodyne; $K_H = 1/\sqrt{2}$ for heterodyne

For a perfectly balanced receiver $K_B = 2$ and $\alpha = 0.5$, otherwise $K_B = 1$. The integrals of the first line in Equation 4.52 are given by

$$\int\limits_{-\infty}^{\infty} H_E(f)df = \frac{1}{T_B} \quad \because H_E(f) = \text{sinc}(\pi T_B f)$$

(4.53)

$$\int\limits_{-\infty}^{\infty} s(t)dt = 2T_B$$

$V_D(f)$ is the transfer function of the matched filter for equalization, and T_B is the bit period. The total noise voltage as a sum of the quantum shot noise generated by the signal and the LO and the total equivalent noise of the electronic preamplifier at the input of the preamplifier at the output of the equalizer is given by

$$\langle v_N^2(t)\rangle = \frac{[K_B \alpha S'_{IS} + (2-K_B)S_{IX} + S_{IE}]\int\limits_{-\infty}^{\infty}|H_4(f)|^2 df}{K_{IS}^2}$$

(4.54)

or

$$\langle v_N^2(t)\rangle = \frac{[K_B \alpha S'_{IS} + (2-K_B)S'_{IX} + S_{IE}]}{K_{IS}^2 T_B}$$

For homodyne and heterodyne detection, we have

$$\langle v_N^2(t)\rangle = \Re q \frac{P_L}{\lambda T_B}[K_B \alpha S'_{IS} + (2-K_B)S_{IX} + S_{IE}]$$

(4.55)

where the spectral densities S'_{IX}, S'_{IE} are given by

$$S'_{IX} = \frac{S_{IX}}{S'_{IS}}$$

$$S'_{IE} = \frac{S_{IE}}{S'_{IS}}$$

(4.56)

Thus, the receiver sensitivity for binary PSK and equiprobable detection and Gaussian density distribution we have

$$P_e = \frac{1}{2} erfc\left(\frac{\delta}{\sqrt{2}}\right)$$

(4.57)

with δ given by

$$P_e = \frac{1}{2} erfc\left(\frac{\delta}{\sqrt{2}}\right) \quad \text{with } \delta = \frac{\hat{v}_D}{2\sqrt{\langle v_N^2 \rangle}}$$

(4.58)

Thus, using Equations 4.55, 4.58, and 4.52 we have the receiver sensitivity in a linear power scale as

$$P_s = \langle P_s(t) \rangle = \frac{\Re q \delta^2}{4\lambda T_B K_H^2} \frac{[K_B \alpha S'_{IS} + (2 - K_B)S_{IX} + S_{IE}]}{\eta K_p (1 - \alpha)\alpha K_B^2 \sin^2\left(\frac{\pi}{2} K_m\right)\cos^2(\phi_p - \phi_e)}$$

(4.59)

4.6.2.2 Shot-Noise-Limited Receiver Sensitivity

In case the power of the LO dominates the noise of the electronic preamplifier and the equalizer, then the receiver sensitivity (in linear scale) is given as

$$P_s = \langle P_{sL} \rangle = \frac{\Re q \delta^2}{4\lambda T_B K_H^2}$$

(4.60)

This shot-noise-limited receiver sensitivity can be plotted as shown in Figure 4.22.

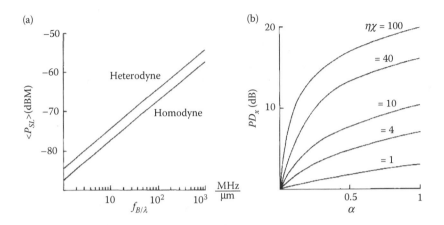

FIGURE 4.22
(a) Receiver sensitivity of coherent homodyne and heterodyne detection, signal power versus bandwidth over the wavelength. (b) Power penalty of the receiver sensitivity from the shot-noise-limited level as a function of the excess noise of the local oscillator. (Extracted from I Hodgkinson, *IEEE J. Lightw. Tech.*, 5(4), 573–587, 1987, Figures 1 and 2.)

4.6.2.3 Receiver Sensitivity under Nonideal Conditions

Under nonideal conditions, the receiver sensitivity departs from the shot-noise-limited sensitivity and is characterized by the receiver sensitivity penalty PD_T as

$$PD_T = 10 \log \frac{\langle P_s \rangle}{\langle P_{sL} \rangle} \, dB \tag{4.61}$$

$$PD_T = 10 \log_{10} \left[\frac{K_B \alpha S'_{IS} + (2 - K_B)S_{IX} + S_{IE}}{K_B \alpha} \right] - 10 \log_{10} \left[K_B(1 - \alpha) \right] \tag{4.62}$$

$$- 10 \log_{10} \left([\eta][K_p] \sin^2 \left(\frac{\pi}{2} K_m \right) \cos^2 (\phi_p - \phi_e) \right)$$

where η is the LO excess noise factor.

The receiver sensitivity is plotted against the ratio f_B/λ for the case of homodyne and heterodyne detection shown in Figure 4.22a, and the power penalty of the receiver sensitivity against the excess noise factor of the LO in Figure 4.22b. Receiver power penalty can be deduced as a function of the total electronic equivalent noise spectral density, and as a function of the rotation of the polarization of the LO can be found in Ref. [30]. Furthermore, in Ref. [31], the receiver power penalty and the normalized heterodyne center frequency can vary as a function of the modulation parameter and as a function of the optical power ratio.

4.6.3 Digital Processing Systems

A generic structure of the coherent reception and DSP system is shown in Figure 4.23, in which the DSP system is placed after the sampling and conversion from analog state to digital form. Obviously, the optical signal fields are beating with the LO laser, with which frequency would be approximately identical to the signal channel carrier. The beating occurs in the square law photodetectors, that is, the summation of the two fields are squared and the product term is decomposed into the difference and summation term; thus, only the difference term is fallen back into the baseband region and amplified by the electronic preamplifier, which is a balanced differential transimpedance type.

If the signals are complex, then these are the real and imaginary components that form a pair. Other pairs come from the other polarization mode channel. The digitized signals of both the real and imaginary parts are processed in real time or offline. The processors contain the algorithms to combat a number of transmission impairments such as the imbalance between the inphase and the quadrature components created at the transmitter, the recovery of the clock rate and timing for resampling, the carrier PE for estimation of the signal phases, adaptive equalization for compensation of propagation dispersion effects using MLSE, and so on. These algorithms are built into the hardware processors or memory and loaded to processing subsystems.

The sampling rate must normally be twice that of the signal bandwidth to ensure to satisfy the Nyquist criteria. Although this rate is very high for 25 G to 32 GSy/s optical channels, Fujitsu ADC has reached this requirement with a sampling rate of 56 G to 64 GSa/s as depicted in Figure 4.36.

The linewidth resolution of the processing for semiconductor device fabrication has progressed tremendously over the years in an exponential trend

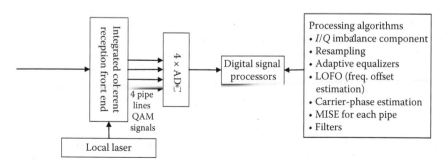

FIGURE 4.23
Coherent reception and the DSP system.

TABLE 4.1

Milestones of Progresses of Linewidth Resolution by Optical Lithography

Semiconductor Manufacturing Processes and Spatial Resolution (Gate Width)
10 µm—1971
3 µm—1975
1.5 µm—1982
1 µm—1985
800 nm (0.80 µm)—1989: UV lithography
600 nm (0.60 µm)—1994
350 nm (0.35 µm)—1995
250 nm (0.25 µm)—1998
180 nm (0.18 µm)—1999
130 nm (0.13 µm)—2000
90 nm—2002: electron lithography
65 nm—2006
45 nm—2008
32 nm—2010
22 nm—2012
14 nm—approx. 2014: x-ray lithography
10 nm—approx. 2016: x-ray lithography
7 nm—approx. 2018: x-ray lithography
5 nm—approx. 2020: x-ray lithography

as shown in Table 4.1. This progress was possible due to the successes in the lithographic techniques using optical sources at short wavelength such as the UV; electronic optical beam; and x-ray lithographic with appropriate photoresist, such as SU-80, which would allow the line resolution to reach 5 nm in 2020. So, if we plot the trend in a log-linear scale as shown in Figure 4.24, a linear line is obtained, meaning that the resolution is reduced exponentially. When the gate width is reduced, the electronic speed increases tremendously; at 5 mm, the speed of the electronic CMOS device in SiGe would reach several tens of gigahertz. Regarding the high-speed ADC and DAC, the clock speed is increased by parallel delay and summation of all the digitized lines to form a very high speed operation. For example, for Fujitsu 64 GSa/s DAC or ADC, the applied clock sinusoidal waveform is only 2 GHz. Figure 4.25 shows the progress in the speed development of Fujitsu ADC and DAC.

4.6.3.1 Effective Number of Bits

4.6.3.1.1 Definition

Effective number of bits (ENOB) is a measure of the quality of a digitized signal. The resolution of a DAC or an ADC is commonly specified by the number of bits used to represent the analog value, in principle, giving 2^N signal levels for an N-bit signal. However, all real signals contain a certain

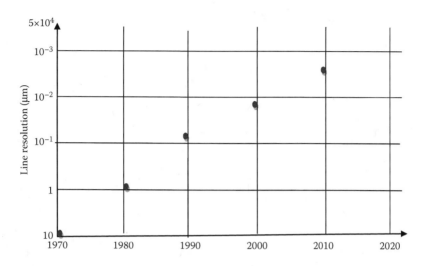

FIGURE 4.24
Semiconductor manufacturing with resolution of line resolution.

amount of noise. If the converter is able to represent signal levels below the system noise floor, the lower bits of the digitized signal only represent system noise and do not contain useful information. ENOB specifies the number of bits in the digitized signal above the noise floor. Often, ENOB is also used as a quality measure for other blocks, such as sample-and-hold amplifiers. This way, analog blocks can also be easily included in signal-chain calculations, as the total ENOB of a chain of blocks is usually below the ENOB of the worst block.

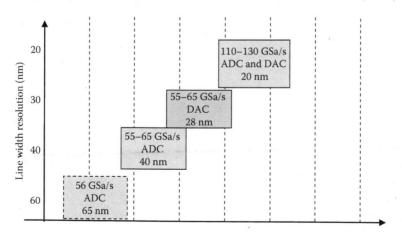

FIGURE 4.25
Evolution of ADC and DAC operating speed with corresponding linewidth resolution.

Thus, we can represent the ENOB of a digitalized system by writing

$$\text{ENOB} = \frac{\text{SINAD} - 1.76}{6.02} \qquad (4.63)$$

where all values are given in dB, and signal-to-noise and distortion ratio (SINAD) is the ratio of the total signal, including distortion and noise to the wanted signal; the 6.02 term in the divisor converts decibels (a $\log_{10}$ representation) to bits (a $\log_2$ representation), and the 1.76 term comes from the quantization error in an ideal ADC [32].

This definition compares the SINAD of an ideal ADC or DAC with a word length of ENOB bits with the SINAD of the ADC or DAC being tested.

Indeed, the SINAD is a measure of the quality of a signal from a communications device, often defined as

$$\text{SINAD} = \frac{P_{\text{sig}} + P_{\text{noise}} + P_{\text{distortion}}}{P_{\text{noise}} + P_{\text{distortion}}} \qquad (4.64)$$

where P is the average power of the signal, noise, and distortion components. SINAD is usually expressed in dB and is quoted alongside the receiver sensitivity, to give a quantitative evaluation of the receiver sensitivity. Note that with this definition, unlike SNR, a SINAD reading can never be less than 1 (i.e., it is always positive when quoted in dB).

When calculating the distortion, it is common to exclude the DC components. Owing to widespread use, SINAD has collected a few different definitions. SINAD is calculated as one of: (i) the ratio of (a) total received power, that is, the signal to (b) the noise-plus-distortion power. This is modeled by the above equation. (ii) The ratio of (a) the power of original modulating audio signal, that is, from a modulated radiofrequency carrier to (b) the residual audio power, that is, noise-plus-distortion powers remaining after the original modulating audio signal is removed. With this definition, it is now possible for SINAD to be less than 1. This definition is used when SINAD is used in the calculation of ENOB for an ADC.

Example: Consider the following measurements of a 3-bit unipolar D/A converter with reference voltage $V_{\text{ref}} = 8$ V:

Digital input	000	001	010	011	100	101	110	111
Analog output (V)	−0.01	1.03	2.02	2.96	3.95	5.02	6.00	7.08

The offset error in this case is −0.01 V or −0.01 LSB, as 1 V = 1 LSB in this example. The gain error is

$$\frac{7.08 + 0.01}{1} - \frac{7}{1} = 0.09 \, \text{LSB}$$

LSB stands for least significant bits. Correcting the offset and gain error, we obtain the following list of measurements: (0, 1.03, 2.00, 2.93, 3.91, 4.96, 5.93, 7) LSB. This allows the integral nonlinearity (INL) and differential nonlinearity (DNL) to be calculated: INL = (0, 0.03, 0, −0.07, −0.09, −0.04, −0.07, 0) LSB, and DNL = (0.03, −0.03, −0.07, −0.02, 0.05, −0.03, 0.07, 0) LSB.

Differential nonlinearity: For an ideal ADC, the output is divided into 2N uniform steps, each with a Δ width as shown in Figure 4.26. Any deviation from the ideal step width is called DNL and is measured in number of counts (LSBs). For an ideal ADC, the DNL is 0 LSB. In a practical ADC, DNL error comes from its architecture. For example, in an SAR ADC, DNL error may be caused near the midrange due to the mismatching of its DAC.

INL is a measure of how closely the ADC output matches its ideal response. INL can be defined as the deviation in LSB of the actual transfer function of the ADC from the ideal transfer curve. INL can be estimated using DNL at each step by calculating the cumulative sum of DNL errors up to that point. In reality, INL is measured by plotting the ADC transfer characteristics.

INL is popularly measured using either (i) best-fit (best straight line) method or (ii) end-point method.

4.6.3.1.1.1 Best-Fit INL The best-fit method of INL measurement considers offset and gain error. One can see in Figure 4.27 that the ideal transfer curve considered for calculating best-fit INL does not go through the origin. The

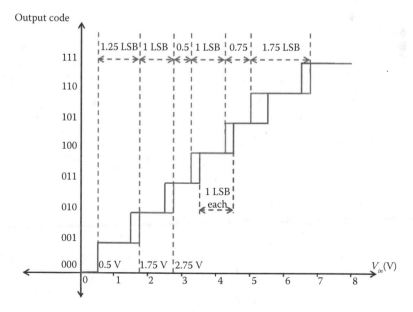

FIGURE 4.26
Representation of DNL in a transfer curve of an ADC.

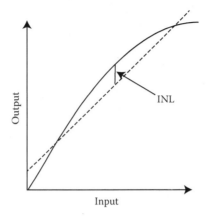

FIGURE 4.27
Best-fit INL.

ideal transfer curve here is drawn such that it depicts the nearest first-order approximation to the actual transfer curve of the ADC.

The intercept and slope of this ideal curve can lend us the values of the offset and gain error of the ADC. Quite intuitively, the best-fit method yields better results for INL. For this reason, many times, this is the number present on ADC datasheets.

The only real use of the best-fit INL number is to predict distortion in time-variant signal applications. This number would be equivalent to the maximum deviation for an AC application. However, it is always better to use the distortion numbers than INL numbers. To calculate the error budget, end-point INL numbers provide a better estimation. Also, this is the specification generally provided in datasheets. So, one has to use this versus end-point INL.

4.6.3.1.1.2 End-Point INL The end-point method provides the worst-case INL. This measurement passes the straight line through the origin and maximum output code of the ADC (refer to Figure 4.28). As this method provides

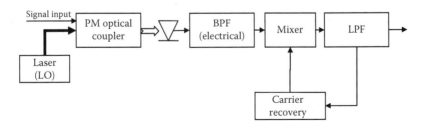

FIGURE 4.28
Schematic diagram of optical heterodyne detection for PSK format.

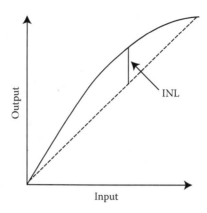

FIGURE 4.29
End-point INL.

the worst-case INL, it is more useful to use this number as compared to the one measured using best fit for DC applications. This INL number would typically be useful for error budget calculation. This parameter must be considered for applications involving precision measurements and control.

The absolute and relative accuracy can now be calculated. In this case, the ENOB absolute accuracy is calculated using the largest absolute deviation D, in this case, 0.08 V:

$$D = \frac{V_{ref}}{2^{ENOB}} \rightarrow ENOB = 6.64 \, bits \qquad (4.65)$$

The ENOB with relative accuracy is calculated using the largest relative (INL) deviation d, in this case, 0.09 V.

$$d = \frac{V_{ref}}{2^{ENOB}} \rightarrow ENOB_{rel} = 6.47 \, bits \qquad (4.66)$$

For this kind of ENOB calculation, note that the ENOB can be larger or smaller than the actual number of bits. When the ENOB is smaller than the ANOB, this means that some of the least significant bits of the result are inaccurate. However, one can also argue that the ENOB can never be larger than the ANOB, because you always have to add the quantization error of an ideal converter, which is ±0.5 LSB as shown in Figure 4.29. Note that different designers may use different definitions of ENOB!

4.6.3.1.2 High-Speed ADC and DAC Evaluation Incorporating Statistical Property

The ENOB of an ADC is the number of bits that an analog signal can convert to its digital equivalent multiplied by the number of levels represented by the module-2 levels, which are reduced due to noises contributed by electronic components in such convertor. Thus, only an effective number of equivalent bits can be accounted for. Hence, the term ENOB is proposed.

As shown in Figure 4.30b, a real ADC can be modeled as a cascade of two ideal ADCs and additive noise sources and an AGC amplifier [31]. The quantized levels are thus equivalent to a specific ENOB as far as the ADC is operating in the linear nonsaturated region. If the normalized signal amplitude/power surpasses unity, the saturated region, then the signals are clipped. The decision level of the quantization in an ADC normally varies following a normalized Gaussian PDF; thus, we can estimate the RMS noise introduced by the ADC as

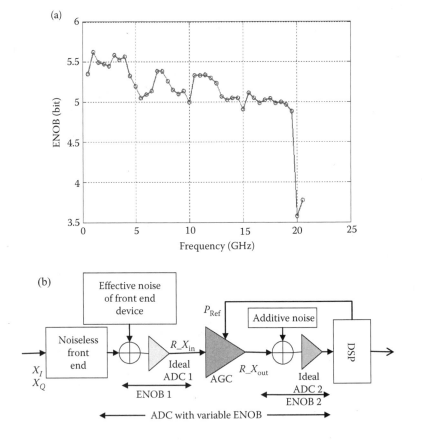

FIGURE 4.30
(a) Measured ENOB frequency response of a commercial real-time DSA of 20 GHz bandwidth and sampling rate of 50 GSa/s. (b) Deduced ADC model of variable ENOB based on experimental frequency response of (a) and spectrum of broadband signals.

$$\text{RMS\_noise} = \sqrt{\int_{\frac{\text{LSB}}{2}}^{\frac{\text{LSB}}{2}} \int_{-\infty}^{\infty} x^2 \frac{\frac{1}{\sqrt{2\pi\sigma^2}} \exp\left(-\frac{(x-y)^2}{2\sigma^2}\right)}{\text{LSB}} dxdy}$$

$$= \sqrt{\text{LSB}^2/12 + \sigma^2}$$

(4.67)

where σ is the variance, x is the variable related to the integration of decision voltage, and y that for the integration inside one LSB. Given the known quantity $\text{LSB}^2/12$, σ^2 can be determined via the Gaussian noise distribution. We can thus deduce the ENOB values corresponding to the levels of Gaussian noise as

$$\text{ENOB} = N - \log_2\left(\frac{\sqrt{\text{LSB}^2/12 + (A\sigma)^2}}{\text{LSB}/\sqrt{12}}\right)$$

(4.68)

where A is the RMS amplitude derived from the noise power. According to the ENOB model, the frequency response of ENOB of the DSA is shown in Figure 4.30a, with the excitation of the DSA by sinusoidal waves of different frequencies. As observed, the ENOB varies with respect to the excitation frequency, in the range from 5 to 5.5. Knowing the frequency response of the sampling device, what is the ENOB of the device when excited with broadband signals? This indicates the different resolution of the ADC of the transmission receiver operating under different noisy and dispersive conditions; thus an equivalent model of ENOB for performance evaluation is essential. We note that the amplitudes of the optical fields arriving at the receiver vary depending on the conditions of the optical transmission line. The AGC has a nonlinear gain characteristic in which the input-sampled signal power level is normalized with respect to the saturated (clipping) level. The gain is significantly higher in the linear region and saturated in the high level. The received signal R_X_{in} is scaled with a gain coefficient according to $R\_X_{out} = R\_X_{in}/\sqrt{P_{in\_av}/P_{Ref}}$, where the signal averaged power P_{in-av} is estimated and the gain is scaled relative to the reference power level P_{Ref} of the AGC; then a linear scaling factor is used to obtain the output-sampled value R_X_{out}. The gain of the AGC is also adjusted according to the signal energy, via the feedback control path from the DSP (see Figure 4.30b). Thus, new values of ENOB can be evaluated with noise distributed across the frequency spectrum of the signals, by an averaging process. This signal-dependent ENOB is now denoted as ENOB_s.

4.6.3.2 Impact of ENOB on Transmission Performance

Figure 4.31a shows the BER variation with respect to the OSNR under back-to-back (B2B) transmission using the simulated samples at the output of the 8-bit ADC with $ENOB_S$ and full ADC resolution as parameters. The difference is due to the noise distribution (Gaussian or uniform). Figure 4.31b depicts the variation of BER versus OSNR, with $ENOB_S$ as the variation parameter in case of offline data with ENOB of DSA shown in Figure 4.1a. Several more tests were conducted to ensure the effectiveness of our ENOB model. When the sampled signals presented to the ADC are of different amplitudes, controlled, and gain nonlinearly adjusted by the AGC, different degrees of clipping effect would be introduced. Thus, the clipping effect can be examined for the ADC of different quantization levels but with identical $ENOB_S$, as shown in Figure 4.32a for B2B experiment. Figure 4.32b and c, and d and e shows, with BER as a parameter, the contour plots of the variation of the adjusted reference power level of the AGC and $ENOB_S$ for the cases of 1500 km long-haul transmission of full CD compensation and non-CD compensation operating in the linear (0 dBm launch power in both links) and nonlinear regimes of the fibers, with the launch power of 4 and 5 dBm, respectively. When the link is fully CD compensated, the nonlinear effects further contribute to the ineffectiveness of the ADC resolution and hence moderate AGC freedom and the performance is achieved. In the case of non-CD compensation link (Figure 4.32d and e), the dispersive pulse sampled amplitudes are lower with less noise, allowing the resolution of the ADC to increase via the nonlinear gain of the AGC; thus, effective PE and equalization can be achieved. We note that

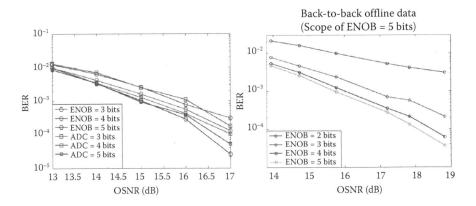

FIGURE 4.31

(a) B2B performance with different $ENOB_S$ values of the ADC model with simulated data (8-bit ADC). (b) OSNR versus BER under different simulated $ENOB_S$ of offline data obtained from an experimental digital coherent receiver. (From Mao B.N. et al., *Proc. ECOC 2011*, Geneva, 2011.)

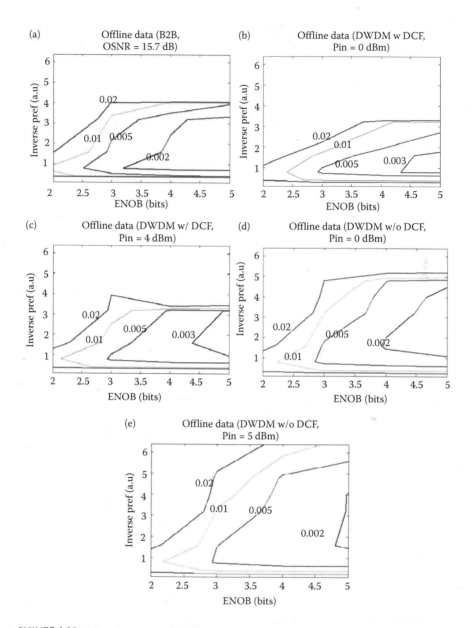

FIGURE 4.32
Comprehensive effects of AGC clipping (inverse of P_{Ref}) and $ENOB_S$ of a coherent receiver, experimental transmission under (a) B2B, DWDM linear operation with full CD compensation, and (b) linear region and (c) nonlinear region with 4 dBm launch power, and (d) non-CD compensation and linear region and (e) nonlinear region with launch power of 5 dBm.

the offline data sets, employed prior to processing with $ENOB_S$ to obtain the contours of Figure 4.32, produce the same BER contours of 2e–3 for all cases. Hence, a fair comparison can be made when the $ENOB_S$ model is used. The opening distance of the BER contours indicates the dynamic range of the $ENOB_S$ model, especially the AGC. It is obvious from Figure 4.32a–e that the dynamic range of the model is higher for noncompensating than for full CD-compensated transmission and even for the case of B2B. However, for nonlinear scenarios for both cases, the requirement for $ENOB_S$ is higher for the dispersive channel (Figure 4.32c and e). This may be due to the cross-phase modulation effects of adjacent channels; hence, more noise.

4.6.3.3 Digital Processors

The structures of the DAC and ADC are shown in Figures 4.33 and 4.34, respectively. Normally, there would be four DACs in an IC, in which each DAC section is clocked with a sequence derived from a lower-frequency sinusoidal wave injected externally into the DAC. Four units are required for the inphase and quadrature phase components of QAM-modulated polarized channels; thus, the notations of I_{DAC} and Q_{DAC} are shown in the diagram. Similarly, the optical received signals of PDM-QAM would be sampled by a four-phase sampler and then converted to digital form in four groups of I and Q lanes for processing in the DSP subsystem. Owing to the interleaving of the sampling clock waveform, the digitalized bits appear simultaneously at the end of a clock period that is sufficiently long, so that the sampling number is sufficiently large to achieve all samples. For example, in Figure 4.34, 1024 samples are achieved at a periodicity corresponding to 500 MHz cycle clock for an 8-bit ADC. Thus, the clock has been slowed down by a factor of 128, or alternatively, the sampling interval is $1/(128 \times 500 \text{ MHz}) = 1/64 \text{ GHz s}$. The sampling is implemented using a charged mode interleaved sampler (CHAIS).

Figure 4.35 shows a generic diagram of an optical DSP-based transceiver employing both DAC and ADC under QPSK modulated or QAM signals. The current maximum sampling rate of 64 GSa/s is available commercially. An IC image of the ADC chip is shown in Figure 4.36.

4.7 Concluding Remarks

This chapter has described the principles of coherent reception and associated techniques with noise considerations and main functions of the DSPs, for which algorithms will be described in separate chapters.

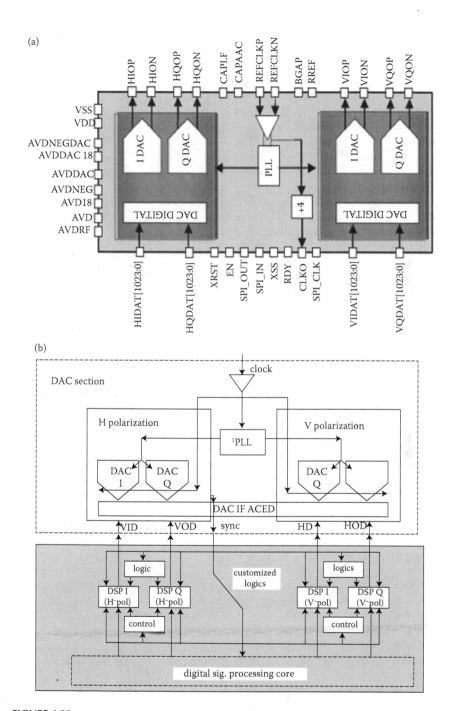

FIGURE 4.33
Fijitsu DAC structures for four-channel PDM_QPSK signals: (a) schematic diagram and (b) processing function.

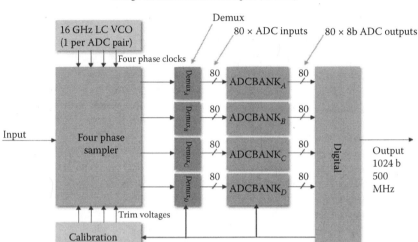

FIGURE 4.34
ADC principles of operations (CHAIS).

Furthermore, the matching of the LO laser and that of the carrier of the transmitted channel is very important for effective coherent detection, if not degradation of the sensitivity of results. The ITU standard requires that for a digital processing-based coherent receiver, the frequency offset between the LO and the carrier must be within the limit of ±2.5 GHz. Furthermore, in practice, it is expected that in network and system management, the tuning of the LO is to be done remotely and automatic locking of the LO with some prior knowledge of the frequency region to set the LO initial frequency. Thus, this action briefly describes the optical phase locking the LO source for an intradyne coherent reception subsystem.

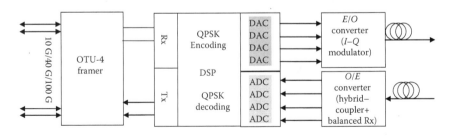

FIGURE 4.35
Schematic of typical structure of ADC and ADC transceiver subsystems for PDM-QPSK modulation channels.

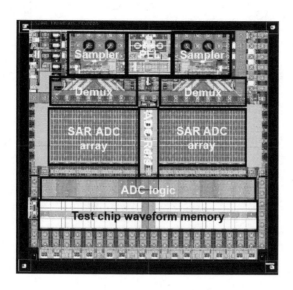

FIGURE 4.36
Fujitsu ADC subsystems with dual-converter structure.

4.8 Appendix: A Coherent Balanced Receiver and Method for Noise Suppression

It has been shown that a balanced optical receiver can suppress the excess noise intensity generated from the LO [34–36]. In a balanced receiver, an optical coupler is employed to mix a weak optical signal with an LO field, with average powers of E_S^2 and E_L^2, respectively. Figure 4.37 shows the generic

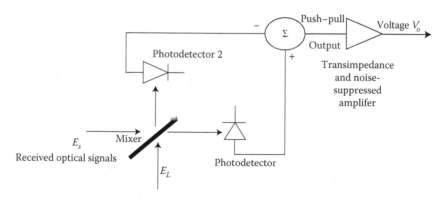

FIGURE 4.37
Generic block diagram of a coherent balanced detector optical receiver.

diagram of a balanced optical receiver under a coherent detection scheme, in which two photodetectors and amplifiers are operating in a push–pull mode.

Usually, the magnitude of the optical field of the LO E_L is much greater than E_S, the signal field. The fields at the output of a 2×2 coupler can thus be written as

$$\begin{bmatrix} \tilde{E_1} \\ \tilde{E_2} \end{bmatrix} = \begin{bmatrix} \sqrt{1-k} & \sqrt{k}e^{j\pi/2} \\ \sqrt{k}e^{j\mu/2} & \sqrt{1-k} \end{bmatrix} \begin{bmatrix} \tilde{E_S} \\ \tilde{E_L} \end{bmatrix} \qquad (4.69)$$

where E_S is the field of the received optical signal, E_L is the field of the LO (possibly a distributed feedback laser), and k is the intensity coupling coefficient of the coupler. For a 3 dB coupler, we have $k = 0.5$. E_S and E_L fields can then be written as

$$E_S = \sqrt{2}\,\tilde{E_S}\,e^{j(\omega_1 t + \phi_1)}$$

$$(4.70)$$

$$E_L = \sqrt{2}\,\tilde{E_L}\,e^{j(\omega_2 t + \phi_2)}$$

where $\tilde{E_S}, \tilde{E_L}$ are the magnitude of the optical fields of the signal and the LO laser. The electronic current generated from the PDs corresponding to the average optical power and the signals are given by

$$i_1(t) = \frac{\Re}{2}|E_1|^2 = \frac{\Re_1}{2}\left\{ \begin{array}{l} \frac{1}{2}\left(E_S^2 + E_L^2\right) + \\[2mm] E_S E_L \cos\left[(\omega_1 + \omega_2)t + \phi_1 + \phi_2 - \frac{\pi}{2} \right] \end{array} \right\} + N_1(t) \quad (4.71)$$

$$i_2(t) = \frac{\Re}{2}|E_2|^2 = \frac{\Re_2}{2}\left\{ \begin{array}{l} \frac{1}{2}\left(E_S^2 + E_L^2\right) \\[2mm] + E_S E_L \cos\left[\begin{array}{l} (\omega_1 + \omega_2)t \\ + \phi_1 + \phi_2 - \frac{\pi}{2} \end{array} \right] \end{array} \right\} + N_2(t) \qquad (4.72)$$

where ϕ_1, ϕ_2 and ω_1, ω_2 are the phase and frequency of the signal and LO, respectively. $N_1(t)$ and $N_2(t)$ are the noises resulting from the photodetection process, and $\Re_1, \Re_2$ are the responsivity of the PDs 1 and 2, respectively. It is assumed that the two PDs have the same quantum efficiency, thence, their responsivity. Thus, the total current of a back-to-back PD pair as referred to the input of the electronic amplifier is

$$i_{\text{Teq}}(t) = \begin{cases} I_{\text{dc1}} - I_{\text{dc2}} \\ \\ +(\Re_1 + \Re_2)E_S E_L \cos \begin{bmatrix} (\omega_1 - \omega_2)t \\ \\ +\phi_1 - \phi_2 - \dfrac{\pi}{2} \end{bmatrix} \end{cases} + N_1(t) + N_2(t) \qquad (4.73)$$

where the first two terms I_{dc1} and I_{dc2} are the detector currents generated by the detector pair due to the reception of the power of the LO, which is a continuous wave source; thus, these terms appear as the DC constant currents that are normally termed as the shot noises due to the optical power of the LO. The third term is the beating between the LO and the signal carrier, the signal envelope. The noise currents $N_{\text{1eq}}(t)$ and $N_{\text{2eq}}(t)$ are seen as the equivalent to the input of the electronic amplifier from the noise processes, which are generated by the PDs and the electronic amplifiers. Noise components are the shot noise due to bias currents, the quantum shot noise that is dependent on the strength of the signal and the LO, the thermal noise due to the input resistance of the amplifier, and the equivalent noise current referred to the input of the electronic amplifier contributed by all the noise sources at the output port of the electronic amplifier. We denote $N_{\text{1ep}}(t)$ as the quantum shot noise generated due to the average current produced by the PD pair in reception of the average optical power of the signal sequence.

The difference of the produced electronic currents can be derived using the following techniques: (i) the generated electronic currents of the PDs can be coupled through a 180° microwave coupler and then fed to an electronic amplifier; (ii) the currents are fed to a differential electronic amplifier; or (iii) a balanced PD pair connected back to back and then fed to a small-signal electronic amplifier [37]. The first two techniques require stringent components and are not normally preferred, as contrasted by the high performance of the balanced receiver structure of (iii).

Regarding the electronic pre-amplifier, a transimpedance configuration is selected due to its wide band and high dynamics. However, it suffers from high noises due to the equivalent input impedance of the shunt feedback impedance, normally around a few hundred ohms. In the next section, the theoretical analysis of noises of the optical balanced receiver is described.

4.8.1 Analytical Noise Expressions

In the electronic preamplifier, the selection of the transistor as the first-stage amplifying device is very critical. Either an FET or BJT could be used. However, the BJT is preferred for wideband application due to its sensitivity to noises. The disadvantages of using the BJT as compared with the FET are due to its small base resistance, which leads to high thermal noise. However, for the shunt feedback amplifier, the resistance of the Millers' equivalent resistance as referred to the input of the amplifier is much smaller and thus dominates over that of the base resistance. The advantage of the BJT over the FET is that its small-signal gain follows an exponential trend with respect to the small variation of the driving current derived from the photodetection as compared with parabolic for FET. The FET may also offer high input impedance between the gate and source of the input port, but may not offer much improvement in a feedback amplification configuration in term of noises. This section focuses on the use and design of BJT multistage shunt feedback electronic amplifiers.

Noises of electronic amplifying devices can be represented by superimposing all the noise current generators onto the small-signal-equivalent circuit. These noise generators represent the noises introduced into the circuit by physical sources/processes at different nodes. Each noise generator can be expressed by a noise spectral density current square or power as shown in Figure 4.38. This figure gives a general model of a small-signal-equivalent circuit, including noise current sources of any transistor, which can be represented by the transfer $[y]$ matrix parameters. Indeed, the contribution of the noise sources of the first stage of the electronic preamplifier are dominant, plus that of the shunt feedback resistance.

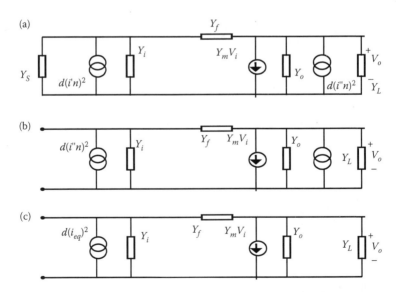

FIGURE 4.38
Small equivalent circuits, including noise sources: (a) Y-parameter model representing the ideal current model and all current noise sources at the input and output ports, (b) with noise sources at input and output ports, and (c) with a total equivalent noise source at the input.

The current generator $di_n'^2/df$ is the total equivalent noise generator contributed by all the noise sources to node 1, including the thermal noises of all resistors connected to the node. Similarly, for the noise source, $di_n''^2/df$ refers to the output node.

4.8.2 Noise Generators

Electrical shot noises are generated by the random generation of electron streams (current). In optical detection, shot noises are generated by (i) biasing currents in electronic devices and (ii) photo currents generated by the PD.

A biasing current I generates a shot noise power spectral density S_I given by

$$S_I = \frac{d\langle i_I^2 \rangle}{df} = 2qI \ \text{A}^2/\text{Hz} \qquad (4.74)$$

where q is the electronic charge. The quantum shot noise $\langle i_s^2 \rangle$ generated by the PD by an optical signal with an average optical power Pin is given by

$$S_Q = \frac{d\langle i_s^2 \rangle}{df} = 2q\langle i_s^2 \rangle \ \text{A}^2/\text{Hz} \qquad (4.75)$$

If the PD is an APD type, then the noise spectral density is given by

$$S_Q = \frac{d\langle i_s^2 \rangle}{df} = 2q\langle i_s^2 \rangle\langle G_n^2 \rangle \ \text{A}^2/\text{Hz} \qquad (4.76)$$

It is noted here again that the dark currents generated by the PD must be included in the total equivalent noise current at the input after it is evaluated. These currents are generated even in the absence of the optical signal, and can be eliminated by cooling the PD to at least below the temperature of liquid nitrogen (77 K).

At a certain temperature, the conductivity of a conductor varies randomly. The random movement of electrons generates a fluctuating current even in the absence of an applied voltage. The thermal noise spectral density of a resistor R is given by

$$S_R = \frac{d\langle i_R^2 \rangle}{df} = \frac{4k_B T}{R} \ \text{A}^2/\text{Hz} \qquad (4.77)$$

where k_B is the Boltzmann's constant, T is the absolute temperature (in K), R is the resistance in ohms, and i_R denotes the noise current due to resistor R.

4.8.3 Equivalent Input Noise Current

Our goal is to obtain an analytical expression of the noise spectral density equivalent to a source looking into the electronic amplifier, including the quantum shot noises of the PD. A general method for deriving the equivalent noise current at the input is given by representing the electronic device by a Y-equivalent linear network as shown in Figure 4.39. The two current noise sources $di_n'^2$ and $di_n''^2$ represent the summation of all noise currents at the input and output of the Y-network. This can be transformed into a Y-network circuit with the noise current referred to the input as follows:

The output voltages V_0 of Figure 4.38a can be written as

$$V_0 = \frac{i_N'(Y_f - Y_m) + i_N''(Y_i + Y_f)}{Y_f(Y_m + Y_i + Y_o + Y_L) + Y_i(Y_o + Y_L)} \tag{4.78}$$

and for Figure 4.38b and c

$$V_0 = \frac{(i_N')_{eq}(Y_f - Y_m)}{Y_f(Y_m + Y_i + Y_o + Y_L) + Y_i(Y_o + Y_L)} \tag{4.79}$$

Thus, comparing the above two equations, we can deduce the equivalent noise current at the input of the detector as

$$i_{Neq} = i_N' + i_N'' \frac{Y_i + Y_f}{Y_f - Y_m} \tag{4.80}$$

Then, reverting to mean square generators for a noise source, we have

$$d(i_{Neq})^2 = d(i_N')^2 + d(i_N'')^2 \left| \frac{Y_i + Y_f}{Y_f - Y_m} \right|^2 \tag{4.81}$$

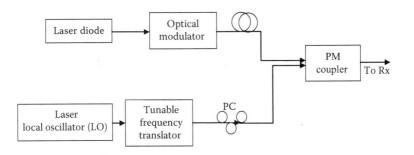

FIGURE 4.39

Typical arrangement of coherent optical communications systems. LD/LC is a very narrow linewidth laser diode as an LO without any phase locking to the signal optical carrier. PM coupler is a polarization-maintaining fiber-coupled device. PC, polarization controller. (Adapted from R.C. Alferness, *IEEE J. Quantum. Elect.*, QE-17, 946–959, 1981; W.A. Stallard, A.R. Beaumont, and R.C. Booth, *IEEE J. Lightwave Tech.*, LT-4(7), July 1986.)

It is therefore expected that if the Y-matrix of the front-end section of the amplifier is known, the equivalent noise at the input of the amplifier can be obtained by using Equation 4.81.

We propose a three-stage electronic preamplifier in AC configuration as shown in Figure 4.40. The details of the design of this amplifier are given in the next section. Small-signal and associated noise sources of this amplifier are given in Section 4.8.7. As it can be seen, this general configuration is a forward path of shut, series, and shut stages, which reduces the interaction between stages due to the impedance levels [6]. Shunt–shunt feedback is placed around the forward path; hence, stable transfer function is the transfer impedance, which is important for transferring the generated electronic photodetected current to the voltage output for further amplification and data recovery.

For a given source, the input noise current power of a BJT front end can be found by

$$i^2_{Neq} = \int_0^B d(i_{Neq})^2 = a + \frac{b}{r_E} + c r_E \tag{4.82}$$

where B is the bandwidth of the electronic preamplifier and r_E is the emitter resistance of the front-end transistor of the preamplifier. The parameters a, b, and c are dependent on the circuit elements and amplifier bandwidth. Hence, an optimum value of r_E can be found and an optimum biasing current can be set for the collector current of the BJT such that i^2_{Neq} is at its minimum, as

$$r_{Eopt} = \sqrt{\frac{b}{c}} \quad \text{hence} \rightarrow i^2_{Neq}\Big|_{r_E = r_{Eopt}} = a + 2\sqrt{bc} \tag{4.83}$$

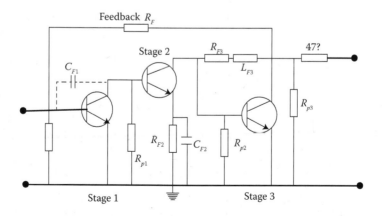

FIGURE 4.40
AC circuit model of a three-stage feedback electronic preamplifier.

If two types of BJT are considered, Phillips BFR90A and BFT24, then a good approximation of the equivalent noise power can be found as

$$a = \frac{8\pi B^3}{3}\left\{r_B C_s^2 + (C_s + C_f + C_{tE})\tau_T\right\}$$

$$b = \frac{B}{\beta_N} \qquad\qquad\qquad\qquad (4.84)$$

$$c = \frac{4\pi B^3}{3}(C_s + C_f + C_{tE})^2$$

The theoretical estimation of the parameters of the transistors can be derived from the measured scattering parameters as given by the manufacturer [38] as
For transistor type *BFR90A*

$$r_{Eopt} = 59\ \Omega \qquad\qquad \text{for } I_{Eopt} = 0.44\ \text{mA}$$

$$I_{eq}^2 = 7.3\times10^{-16}\,\text{A}^2 \quad \text{hence } \frac{I_{eq}^2}{B} = 4.9\times10^{-24}\ \text{A}^2/\text{Hz} \qquad (4.85)$$

$$I_{eq} = 27\ \text{nA} \qquad\qquad \text{thus } \frac{I_{eq}}{\sqrt{B}} = 2.21\ \text{pA}/\sqrt{\text{Hz}}$$

For transistor type *BFT24*

$$r_{Eopt} = 104\ \Omega \qquad\qquad \text{for } I_{Eopt} = 0.24\ \text{mA}$$

$$I_{eq}^2 = 79.2\times10^{-16}\,\text{A}^2 \quad \text{hence } \frac{I_{eq}^2}{B} = 6.1\times10^{-24}\,\text{A}^2/\text{Hz} \qquad (4.86)$$

$$I_{eq} = 30.2\ \text{nA} \qquad\qquad \text{thus } \frac{I_{eq}}{\sqrt{B}} = 2.47\ \text{pA}/\sqrt{\text{Hz}}$$

Note that the equivalent noise current depends largely on some poorly defined values, such as the capacitance, transit times, base spreading resistance, and the short-circuit current gain β_N. The term $I_{eq}/\sqrt{B}$ is usually specified as the noise spectral density equivalent referred to the input of the electronic preamplifier.

4.8.4 Pole-Zero Pattern and Dynamics

An AC model of a three-stage electronic preamplifier is shown in Figure 4.40, and the design circuit is shown in Figure 4.41. As briefly mentioned above, there are three stages with feedback impedance from the output to the input. The subscripts of the resistors, capacitors, and inductors indicate the order of

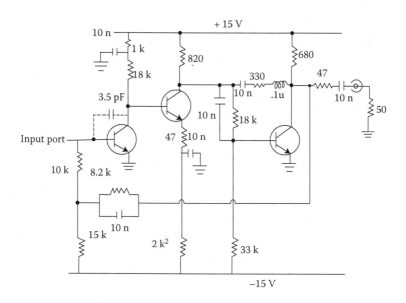

FIGURE 4.41
Design circuit of the electronic preamplifier for the balanced optical amplifier.

the stages. The first stage is a special structure of shunt feedback amplification in which the shunt resistance is increased to infinity. The shunt resistance is in order of hundreds of ohms for the required bandwidth, thus contributing to noises of the amplifier. This is not acceptable. The shunt resistance increases the pole of this stage and approaches the origin. The magnitude of this pole is reduced by the same amount of that of the forward path gain. Thus, the poles of the closed-loop amplifier remain virtually unchanged. As R_F is increased, the pole p_1 decreases, but G_1 is increased. Hence, $G_1 p_1 / s - p_1$ remains constant. Thus, the position of the root locus is almost unchanged.

A compensating technique for reducing the bandwidth of the amplifier is to add capacitance across the base-collector of the first stage. This may be necessary if oscillation occurs due to the phase shift becoming unacceptable at GH = 1, where G is the open loop gain and H is the feedback transfer function.

The second stage is a series feedback stage with feedback peaking. The capacitance C_{F2} is chosen such that at high frequencies it begins to bypass the feedback resistor. Thence, the feedback admittance partially compensates for the normal high-frequency drop in gain associated with the base and stray capacitances. Also, if the capacitance is chosen such that $1/R_F C_F = \tau_p$, the transfer admittance and input impedance become single-pole, which would be desirable [39]. The first and second stages are directly coupled.

The third stage uses inductive peaking technique. For a shunt stage with a resistive load, the forward path gain has only one pole and hence there is only one real pole in the closed-loop response. A complex pole pair can be

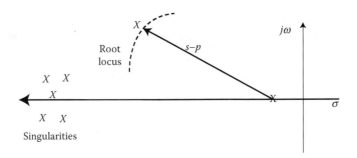

FIGURE 4.42
Effect of first-stage pole on root locus of a shunt feedback electronic amplifier. The root locus is given by $1 = (GH)_{\text{mid-band}} \Sigma(s - z_i/z_i)\Sigma(p_i/s - p_i)$.

obtained by placing a zero in Z_F2, and hence a pole in the feedback transfer function $H(\omega)$. Figure 4.43a–c shows the high-frequency singularity pattern of individual stages. Figure 4.43d shows the root locus diagram. It can be calculated that the poles take up the positions in Figure 4.43e when the lop gain GH is 220.

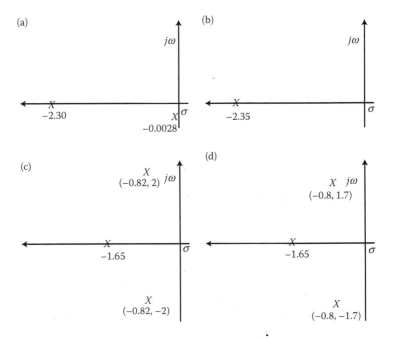

FIGURE 4.43
High-frequency root locus diagram. (a) Singularity pattern for first stage, (b) second stage, (c) third stage, and (d) root locus for close loop pattern for $GH = 220$.

In the low-frequency region, the loop gain has five poles and three zeroes on the negative axis, and two zeroes at the origin. The largest low-frequency pole is set near the input coupling capacitor, with the input resistance leading to a low-frequency cutoff of around tens of kilohertz.

The singularities of individual stages can be approximated by

Stage 1: The dominant pole at

$$p = -\frac{1}{\beta_N R_{L1} C_{F1}}$$ (4.87)

and the other pole located at

$$p = -\frac{C_{F1}}{C_{F1} + C_{L1}} \frac{1}{\tau_T + C_{F1}/g_m}$$

and the midband gain $= \beta_N R_{L1}$ with the load of stage 1 of R_{L1} (4.88)

Stage 2: The dominant pole at

$$p = -\frac{1}{r_B G_{T2} \tau_T} \text{ and the midband gain} = G_{T2} = 1/R_{F2}$$ (4.89)

Stage 3: The complex pole pair at

$$|p|^2 = -\frac{R_L}{\tau_T^2 R_{F3}}; \quad \sigma = -\frac{(R_L + R_{F3})}{2\tau_F R_{F3}}; \quad \tau_F = \frac{L_{F3}}{R_{F3}}; \quad \text{and a zero at} -\frac{1}{\tau_F};$$

with a gain $= -R_{F3}$

(4.90)

The feedback configuration of the circuit of the transimpedance electronic preamplifier has a 10 K or 15 K resistor, for which noises would contribute to the total noise current of the amplifier. The first and second stages are direct-coupled, hence eliminating the level shifter and a significant amount of stray capacitance. The peaking capacitor required for the second stage is in order of 0.5 pF; hence no discrete component can be used. This may not require any component, as the stay capacitance may suffice.

The equivalent noise current at the input of the TIA is approximately proportional to the square of the capacitance at the base of the transistor. Therefore, minimization of the stray capacitance at this point must be conducted by shortening the connection lead as much as possible. As a rough guide, critical points on the circuit where stray capacitance must be minimized are not at signal ground, and have impedance and small capacitance to ground. For instance, at the base of the output stage, the capacitance to the ground is more tolerable at this point, and therefore at the collector to the ground of the second transistor, than the collector of the last stage, which should ideally have no capacitance to ground.

An acceptable step response can be achieved by manipulating the values of $C_F 1, R_F 3, L_F 3$. Since $C_F 1$ contributes to the capacitance at the base of the first stage, $C_F 1$ is minimized. The final value of $C_F 1$ is 0.5 pF (about a half twist of two wires), $R_F 1$ is 330 ohms, and $L_F 1$ is about 0.1 μF (2 cm of wire). The amplifier is sensitive to parasitic capacitance between the feedback resistor and the first two stages. Thus, a grounded shield is placed between the 10 K resistor and the first and second stages.

The expressions for the singularities for the first two stages are similar, as described above. The base-collector capacitance of the third stage affects the position of the poles considerably. The poles of the singularities of this stage are as follows: the mid-band transresistance is $-R_{F3}$, with a zero@z $= 1/\tau_F$, and a complex pole pair given by

$$|p_p|^2 = \frac{1}{L_F C_F + R_F C_F \tau_T + \tau_T \tau_F \dfrac{R_F}{R_L}} ; \sigma = \frac{1}{2}|p_p|^2\left(R_F C_F + \tau_T\left(1 + \frac{R_F}{R_L}\right)\right) \tag{4.91}$$

In addition, a large pole given approximately at

$$p = -\frac{1}{L_F C_F \tau_T |p_p|^2} \tag{4.92}$$

Shown in Figure 4.44 are the open-loop singularities in (a), the root locus diagram in (b), and the closed-loop in (c). The two large poles can be ignored without any significant difference. Similarly, for the root locus diagram, the movement of the large poles is negligible, the pole and zero pair can be ignored, and the movement of the remaining three poles can be calculated by considering just these three poles as the open loop singularities.

4.8.5 Responses and Noise Measurements

4.8.5.1 Rise-Time and 3 dB Bandwidth

Based on the pole-zero patterns given in the previous section, the step responses can be estimated and contrasted with the measured curve as shown in Figure 4.45. The experimental set-up for the rise time measurement is shown in Figure 4.46, in the electrical domain without using photodetector. An artificial current source is implemented using a series of resistors with minimum stray capacitance. This testing in the electrical domain is

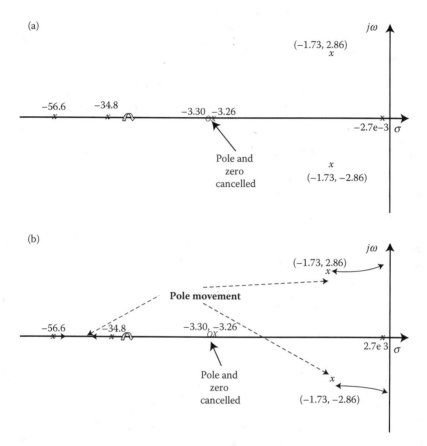

FIGURE 4.44
Pole-zero patterns in the s-plane and their dynamics of (a) open loop, (b) root locus, and (c) closed loop.

preferred over the optical technique, as the rise time of an electrical TDR is sufficiently short so that it does not influence the measurement of the rise time. If the optical method is used, the principal problem is that we must be able to modulate the intensity of the lightwave with a very sharp step function. This is not possible when the Mach–Zehnder intensity modulator (MZIM) is used, as its transfer characteristics follow a square of a cosine function. A very short pulse sequence laser such as mode-locked fiber laser can be used. However, this is not employed in this work, as the bandwidth of our amplifier is not very wide and the TDR measurement is sufficient to give us reasonable value of the rise time.

The passband of the TIA is also confirmed with the measurement of the frequency response by using a scattering parameter test set HP 85046A, the scattering parameter S_{21} frequency response is measured, and the 3 dB passband is confirmed at 190 MHz.

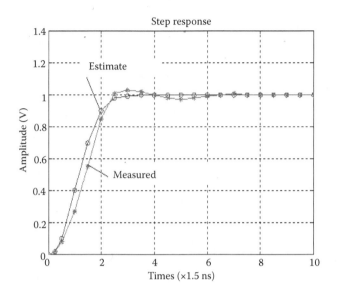

FIGURE 4.45
Step response of the amplifier: o, estimated; *, measured.

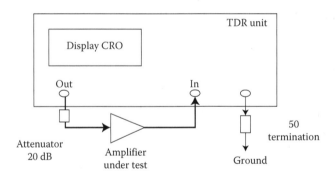

FIGURE 4.46
Experimental setup for measurement of the rise time in the electrical domain using a time domain reflectometer (TDR).

4.8.5.2 Noise Measurement and Suppression

Two pig-tailed PDs are used and mounted with back-to-back configuration at the input of the transistor of stage 1. A spectrum analyzer is used to measure the noise of the amplifier shown in Figure 4.47. The background noise of the analyzer is measured to be −88 dBm. The expected noise referred to the input is $9.87 \times 10^{-24} A^2/Hz$. When the amplifier is connected to the spectrum analyzer this noise level is increased to −85 dBm, which indicates the noise at the output of the amplifier is around −88 dBm.

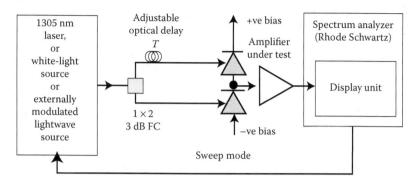

FIGURE 4.47
Experimental setup to measure excess noise impact and cancellation of a balanced receiver. FC, fiber coupler.

Since this power is measured as 50 Ω, using a spectral bandwidth of 300 kHz, the input current noise can be estimated as

$$\int di^2_{Neq} = \frac{v_N^2}{R_T^2} = \frac{10^{-11} \times 50}{(5k)^2} = 2 \times 10^{-17.5} A^2 \rightarrow \frac{di^2_{Neq}}{df} = \frac{2e^{-17.5} A^2}{3 \times 10^5}$$

$$= 1.06 \times 10^{-23} A^2/Hz \tag{4.93}$$

Thus, the measured noise is very close to the expected value.

4.8.5.3 Requirement for Quantum Limit

The noise required for near quantum limit operation can be estimated. The total shot noise referred to the input is

$$i^2_{T-shot} = (\Re_1 + \Re_2)|E_L|^2 \tag{4.94}$$

where $\Re_{1,2}$ is the responsivity of the photodetectors 1 and 2, respectively. The total excess noise referring to the input is

$$i^2_{NT-shot} = (\Re_1 + \Re_2)|E_L|^2 \; ; i^2_{N-excess} = 2q\gamma(\omega)(\Re_1 + \Re_2)^2|E_L|^4 \tag{4.95}$$

The excess noise from each detector is correlated so γ is a function of frequency, typically between 10^4 and 10^{10} A $^{-1}$. A receiver would operate within 1 dB of the quantum limit if the shot noise is about 6 dB above the excess noise and the amplifier noise. The amplifier noise power spectral density is 6.09×10^{-14} A^2/Hz. Assuming that a complete cancellation of the excess noise can be achieved, the required LO power is given as

$$i^2_{NT-shot} > 4i^2_{NTeq} \rightarrow \qquad |E_L|^2 > 0.24mW \approx -6.2dBm \tag{4.96}$$

Since there is bound to be excess noise due imperfection of the balancing of the two detectors, the power of the LO is expected to be slightly greater than this value.

The output voltage of the amplifier can be monitored so noises contributed by different processes can be measured. When the detectors are not illuminated we have

$$v_0^2 = Z_T^2 i_{Neq}^2 \tag{4.97}$$

where Z_T is the transimpedance and v_o is the output voltage of the amplifier. It is known that incoherent white light is free from excess intensity noise, so when the detectors are illuminated with white light sources we have

$$v_0^2 = Z_T^2 \left(i_{Neq}^2 + i_{N-shot}^2 \right) \tag{4.98}$$

When the LO is turned on, illuminating the detector pair, then

$$v_0^2 = Z_T^2 \left(i_{Neq}^2 + i_{N-shot}^2 + i_{N-excess}^2 \right) \tag{4.99}$$

4.8.5.4 Excess Noise Cancellation Technique

The presence of uncancelled excess noise can be shown by observing the output voltage of the electronic preamplifier using a spectrum analyzer. First, the detectors are illuminated with incoherent white light and measure the output noise. Then the detectors are illuminated with the laser LO at a power level equal to that of the white light source; thus the shot noise would be at the same magnitude. Thus, the increase in the output noise will be due to the uncancelled intensity noise. This, however, does not demonstrate the cancellation of the LO intensity noise.

A well-known method for this cancellation is to bias the LO just above its threshold [36]. This causes the relaxation resonance frequency to move over the passband. Thus, when the outputs of the single or balanced receivers are compared, the resonance peak is observed for the former case and not for the latter, in which the peak is suppressed. Since the current is only in the µA scale in each detector for an LO operating in the region near the threshold, the dominant noise is that of the electronic amplifier. This method is thus not suitable here. However, the relaxation resonance frequency is typically between 1 and 10 GHz. A better method can be developed by modulating the LO with a sinusoidal source and measuring the difference between single and dual photodetection receivers. The modulation can be implemented at either one of the frequencies to observe the cancellation of the noise peak power or by sweeping the LO over the bandwidth of the amplifier so that the cancellation of the amplifier can be measured. For fair comparison, the LO power would be twice that of the single detector case, so the same received power of both cases can be almost identical, and hence the SNR at the output of the amplifier can be derived and compared.

4.8.5.5 Excess Noise Measurement

An indication of excess noise and its cancellation within 0–200 MHz frequency range by the experimental set-up can be observed as shown in Figure 4.47. A laser of 1305 nm wavelength is used with an output power of −20 dBm, and is employed as the LO so that the dominant noise is that of the electronic transimpedance preamplifier as explained above. The LO is intensity-modulated by the sweep generator from the spectrum analyzer of 3 GHz frequency range (Rhode Schwartz model). The noise-suppression pattern of the same trend can be observed with suppression about every 28 MHz interval over 200 MHz, the entire range of the TIA.

Figure 4.48 shows the noise spectral density for the case of single and dual detectors. Note that the optical paths have different lengths. This path difference is used for cancellation of the excess noise as described in Refs. [40,41]. It can be shown that the optical delay line leads to a cancellation of noise following a relationship of $\{1 - \cos(2\pi fT)\}$, with T as the delay time of the optical path of one of the fibers from the fiber coupler to the detector. Hence, the maxima of the cancellation occur at $f = m/T$, where m is an integer. From Figure 4.47, we can deduce the optical path length as $f = 30$ MHz or $T = 31.7$ ns; thus $d = cT/n_{eff} = 6.8$ m with n_{eff} (~1.51 at 1550 nm) is the effective refractive index of the guided mode in the fiber. The total length of the fiber delay is equivalent to about 6.95 meters, in which the effective refractive index of the propagation mode is estimated to be 1.482 at the measured wavelength. Thus, there is a small discrepancy, which can account for the uncertainty of the exact value of the effective refractive index of the fiber.

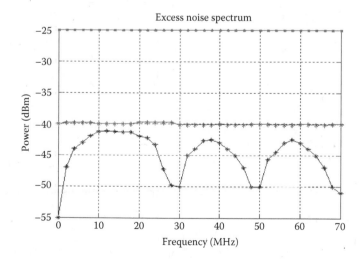

FIGURE 4.48
Excess noise for single and dual detectors with noise cancellation using delay and filter at the input port of the preamplifier.

Using this method, Abbas et al. [34] achieved a cancellation of 3.5 dB, while 8 dB is obtained in our set-up. This discrepancy can be accounted for due to the push pull operation of the photodetector pair which was tuned to be the same.

In our initial measurements we have observed some discrepancies due to (i) unmatched properties of the photodetection of the balanced detector pair, and (ii) the delay path has three extra optical connections and therefore more loss, so even when there is no delay, the excess noise would not be completely cancelled. Thus, at some stage the longer path was implemented with a tunable optical delay path so as to match the delay path to the null frequency of the RF wave to achieve cancellation of the excess noise. This noise cancellation can lead to an improvement of 16 dB in the OSNR due to the dual-detector configuration and thus an improvement of near 100 km length of standard single-mode fiber.

The noise cancellation is periodic following a relationship with respect to the delay time T imposed by the optical path of the interferometer T as $(1 - \cos 2\pi fT)$. The optical delay path difference is estimated by adjusting the optical delay line, which is implemented by a fiber path resultant of a fiber end to open air path and back to another fiber path via a pair of Selfoc lenses.

It is noted that the LO light source can be directly modulated (DD) with a sweep sinusoidal signal derived from the electrical spectrum analyzer (ESA). This light source has also been externally modulated using an MZIM bandwidth much larger than that of the electronic amplifier (about 25 GHz). The excess noise characteristics obtained using both types of sources are almost identical. Furthermore, with the excess noise reduction of 8 dB and the signal power for the balanced receiver operating in a push-pull mode, then the received electrical current would be double that of a single detector receiver. That means a gain of 3 dB in the SNR by this balanced receiver.

4.8.6 Remarks

In this Appendix, we consider the design and implementation of a balanced receiver using a dual-detector configuration and a BJT front end pre-amplifier. In light of recent growing interests in coherent optical transmission systems with electronic DSP [42,43], a noise-suppressed balanced receiver operating in the multi-GHz bandwidth range would be essential. Noise cancellation using optical delay line in one of the detection path leads to suppression of excess noise in the receiver.

In this section, we demonstrate, as an example, by design analysis and implementation, a discrete wideband optical amplifier with a bandwidth of around 190 MHz and a total input noise-equivalent spectral density of $10^{-23} A^2/Hz$. This agrees well with the predicted value using a noise model analysis. It is shown that if sensible construction of the discrete amplifier, minimizing the effects of stay capacitances and appropriate application of compensation techniques, a bandwidth of 190 MHz can be easily obtained for transistors with a few GHz transition frequencies. Further, by using an optimum emitter current for the first stage and minimizing the biasing and feedback resistance and capacitance at the input, a low-noise amplifier with an equivalent noise current of about 1.0 pA/$\sqrt{\text{Hz}}$ is achieved. Furthermore, the excess noise cancellation property of the receiver is found to give maximum SNR of 8 dB with a matching of the two photodetectors.

Although the bandwidth of the electronic amplifier reported here is only 190 MHz, which is about 1% of the 100 Gb/s target bit rate for modern optical Ethernet, the bandwidth achieved by our amplifier has reached close to the maximum region that can be achieved with discrete transistor stages and a transition frequency of only 5 GHz. The amplifier configuration reported here can be scaled up to a multi-GHz region for a 100 Gb/s receiver using integrated electronic amplification devices without much difficulty. We thus believe that the design procedures for determining the pole and zero patterns on the s-plane and their dynamics for stability consideration are essential so that any readers interested in their implementation using MMIC technology can use them. Should the amplifier be implemented in the multi-GHz range, then the microwave design technique employing noise Figure 4.2 method may be most useful when incorporating the design methodology reported in this section.

4.8.7 Noise Equations

Refer to the small-signal and noise model given in Figures 4.49 and 4.50. The spectral density of noises due to collector bias current I_C and small-signal transconductance g_m, and base conductance g_B generated in a BJT can be expressed as

$$di_1^2 = 4k_BTg_Bdf \, ; di_2^2 = 2qI_cdf + 2k_BTg_mdf$$

$$di_3^2 = 2qI_Bdf + 2k_BT(1-\alpha_N)g_mdf \, ; \text{and}$$

$$di_4^2 = \sum \text{shot noise of diodes and thermal noise of bias in } g_B \text{ resistors}$$

$$(4.100)$$

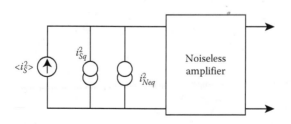

FIGURE 4.49

Equivalent noise current at the input and noiseless amplifier model. i_{Sq}^2 is the quantum shot noise, which is signal dependent; i_{Neq}^2 is the total equivalent noise current referred to the input of the electronic amplifier.

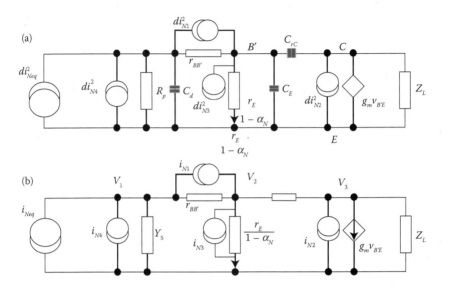

FIGURE 4.50
(a) Approximated noise-equivalent and small-signal model of a BJT front end. (b) Generalized
noise and small-signal model circuit. Note that $r_B = r_{sd} + r_{BB'}$ and $C_d = C_p + C_i$, where $r_{BB'}$ is the
base resistance, r_{sd} is the diode resistance, C_d is the PD capacitance, and C_i is the input capacitance.

From nodal analysis of the small-signal-equivalent circuit given in Figure
4.50 we can obtain the relationship

$$
\begin{bmatrix}
Y_s + g_B & -g_B & 0 \\
-g_B & g_B + y_1 + y_f & -y_f \\
0 & g_m - y_f & y_f + y_2
\end{bmatrix}
\begin{bmatrix}
V_1 \\ V_2 \\ V_3
\end{bmatrix}
=
\begin{bmatrix}
i_{eq} - i_{N1} \\
i_{N1} + i_{N3} \\
i_{N2}
\end{bmatrix}
\tag{4.101}
$$

Hence, V_3, V_2, V_1 can be found by using Euler's rule for the matrix relationship:

$$
V_3 = \frac{\Delta_{13}}{\Delta}(i_{eq} - i_{N1}) + \frac{\Delta_{23}}{\Delta}(i_{N1} + i_{N3}) + \frac{\Delta_{33}}{\Delta}(i_{N2})
\tag{4.102}
$$

and the noise currents as referred to the input are

$$
d(i_1'^2) = \left| \frac{Y_s}{g_B} \right|^2 dI_1^2 = \omega_2 C_s^2 r_B 4kTdf
\tag{4.103}
$$

$$
d(i_2'^2) = \left| (Y + y_1 + y_f) + \frac{Y_s}{g_B}(y_1 + y_f) \right|^2 \left| \frac{1}{y_f - g_m} \right|^2 = \left| \frac{1}{y_f - g_m} \right|^2 dI_2^2
\tag{4.104}
$$

$$
= \left(\frac{1}{\beta_N r_E^2} + \omega^2 \left(C_0^2 - \frac{2C_B r_B}{\beta_N r_E} \right) + \omega^4 2C_s r_B C_B^2 \right) 2kTr_E df
$$

and

$$d(i_3'^2) = \left| \frac{Y_s}{g_B} + 1 \right|^2 dI_3^2 = \left(\omega^2 C_s^2 r_B^2 + 1 \right) \frac{2kT}{\beta r_E} df \qquad (4.105)$$

in which we have assumed that

$$\omega C_f \ll g_m$$

$$C_0 = C_s + C_E + C_f + \frac{C_s r_s}{\beta r_E} \approx C_s + C_{tE} + C_f + \frac{\tau_t}{r_E} = C_x + \frac{\tau_t}{r_E} \qquad (4.106)$$

$$\text{with } C_x = C_E + C_f; \; C_E = C_{tE} + \frac{\tau_t}{r_E}$$

The total noise spectral density referred to the input of the amplifier as shown in Figures 4.49 and 4.50 are thus given as

$$\frac{di_{Neq}^2}{df^2} = \frac{di_{1eq}^2}{df^2} + \frac{di_{2eq}^2}{df^2} + \frac{di_{3eq}^2}{df^2}$$

$$= 2k_B T \left[\frac{1}{\beta r_E} \frac{\beta+1}{\beta} + \omega^2 \left(\frac{2}{r_B} + \frac{1}{\beta r_E} \right) + C_0 r_E - \frac{2C_s r_B C_x}{\beta} + \omega^2 C_s^2 r_B^2 C_x^2 r_E \right]$$

$$(4.107)$$

Thence, the total noise power referring to the input is given as

$$i_{Neq}^2 = \int_0^B di_{Neq}^2 = a + \frac{b}{r_E} + c r_E; \quad \text{with}$$

$$a \approx 2k_B T \left[\frac{8\pi^2 B^3}{3} (C_s r_B + C_x \tau_t + \frac{C_s^2 r_B^2}{\beta}) + \frac{(2\pi)^4}{5} B^5 C_s^2 r_B^2 2C_y \tau_t \right]$$

$$(4.108)$$

$$b \approx 2k_B T \left[\frac{B}{\beta} + \frac{4\pi^2 B^3}{3} (\frac{C_s r_B^2}{\beta} + \tau_t^2) + \frac{16\pi^4 B^5}{5} C_s^2 r_B^2 \tau_t^2 \right]$$

$$c \approx 2k_B T \left[\frac{4\pi^2 B^3 C_x^2}{3} + \frac{16\pi^4 B^5}{5} C_s^2 r_B^2 C_y^2 \right] \quad \text{with } C_y = C_{tE} + \frac{\tau_t}{r_E}$$

where the first term in each coefficient a, b, and c is dominant for a bandwidth of 190 MHz of the two transistor types, Phillip BFR90A and BFT24, under considerations.

If the dependence of C_E on I_E is ignored, then Equation 4.108 becomes

$$i_{Neq}^2 = 2k_BT\left[\frac{1}{r_E}\frac{B}{\beta} + r_E\left(\frac{4\pi^2B^3}{3}C_0^2\right) + \frac{16\pi^4B^5}{5}C_s^2r_B^2C_x^2 + \frac{8\pi^2B^3C_s^2r_B}{3}\right] \quad (4.109)$$

For the transistor BFR90A, the optimum emitter resistance is $r_{Eopt} = 49\ \Omega$ and the total equivalent noise power referred to the input is $4.3 \times 10^{-16}\ A^2$, which is moderately high.

Clearly, the noise power is a cubic dependence on the bandwidth of the amplifier. This is very critical for ultrawide band optical receiver. Thus, it is very important to suppress the excess noise of the optical receiver due to the quantum shot noise of the LO.

References

1. A.H. Gnauck and P.J. Winzer, Optical phase-shift-keyed transmission, *J. Lightwave Technol.* 23, 115–130, 2005.
2. R.C. Alferness, Guided wave devices for optical communication, *IEEE J. Quantum Electron.*, QE-17, 946–959, 1981.
3. W.A. Stallard, A.R. Beaumont, and R.C. Booth, Integrated optic devices for coherent transmission, *IEEE J. Lightwave Technol.*, LT-4(7), July 1986.
4. V. Ferrero and S. Camatel, Optical phase locking techniques: An overview and a novel method based on single sideband sub-carrier modulation, *Opt. Express*, 16(2), 818–828, January 21, 2008.
5. I. Garrett and G. Jacobsen, Theoretical analysis of heterodyne optical receivers for transmission systems using (semiconductor) lasers with nonnegligible line-width, *IEEE J. Lightwave Technol.*, LT-3/4, 323–334, 1986.
6. G. Nicholson, Probability of error for optical heterodyne DPSK systems with quantum phase noise, *Electron. Lett.*, 20/24, 1005–1007, 1984.
7. S. Shimada, *Coherent Lightwave Communications Technology*, Chapman and Hall, London, 1995, p. 27.
8. Y. Yamamoto and T. Kimura, Coherent optical fiber transmission system, *IEEE J. Quantum Electron.*, QE-17, 919–934, 1981.
9. S. Savory and T. Hadjifotiou, Laser linewidth requirements for optical DQPSK optical systems, *IEEE Photonic Tech. Lett.*, 16(3), 930–932, March 2004.
10. K. Iwashita and T. Masumoto, Modulation and detection characteristics of optical continuous phase FSK transmission system, *IEEE J. Lightwave Technol.*, LT-5/4, 452–462, 1987.
11. F. Derr, Coherent optical QPSK intradyne system: Concept and digital receiver realization, *IEEE J. Lightwave Technol.*, 10(9), 1290–1296, 1992.

12. G. Bosco, I.N. Cano, P. Poggiolini, L. Li, and M. Chen, MLSE-based DQPSK transmission in 43 Gb/s DWDM long-haul dispersion managed optical systems, *IEEE J. Lightwave Technol.*, 28(10), 1573–1581, 2010.

13. D-S. Ly-Gagnon, S. Tsukamoto, K. Katoh, and K. Kikuchi, Coherent detection of optical quadrature phase-shift keying signals with carrier phase estimation, *J. Lightwave Technol.*, 24, 12–21, 2006.

14. E. Ip and J.M. Kahn, Feedforward carrier recovery for coherent optical communications, *J. Lightwave Technol.*, 25, 2675–2692, 2007.

15. L.N. Binh, Dual-ring 16-star QAM direct and coherent detection in 100 Gb/s optically amplified fiber transmission: Simulation, *Opt. Quantum Electron.*, 40(10), 707–727, 2009.

16. L.N. Binh, Generation of multi-level amplitude-differential phase shift keying modulation formats using only one dual-drive Mach-Zehnder interferometric optical modulator, *Opt. Eng.*, 48(4), 045005-1–045005-10, 2008.

17. M. Nazarathy, X. Liu, L. Christen, Y.K. Lize, and A. Willner, Self-coherent multi-symbol detection of optical differential phase-shift-keying, *J. Lightwave Technol.*, 26, 1921–1934, 2008.

18. R. Noe, PLL-free synchronous QPSK polarization multiplex/diversity receiver concept with digital I&Q baseband processing, *IEEE Photon. Technol. Lett.*, 17, 887–889, 2005.

19. L.G. Kazovsky, G. Kalogerakis, and W.-T. Shaw, Homodyne phase-shift-keying systems: Past challenges and future opportunities, *J. Lightwave Technol.*, 24, 4876–4884, 2006.

20. J.P. Gordon and L.F. Mollenauer, Phase noise in photonic communications systems using linear amplifiers, *Opt. Lett.* 15, 1351–1353, 1990.

21. H. Kim and A.H. Gnauck, Experimental investigation of the performance limitation of DPSK systems due to nonlinear phase noise, *IEEE Photon. Technol. Lett.*, 15, 320–322, 2003.

22. S. Zhang, P.Y. Kam, J. Chen, and C. Yu, Receiver sensitivity improvement using decision-aided maximum likelihood phase estimation in coherent optical DQPSK system, *Conference on Lasers and Electro-Optics/Quantum Electronics and Laser Science and Photonic Applications Systems Technologies*, Technical Digest (CD), Optical Society of America, 2008, paper CThJJ2.

23. P.Y. Kam, Maximum-likelihood carrier phase recovery for linear suppressed-carrier digital data modulations, *IEEE Trans. Commun.*, COM-34, 522–527, June 1986.

24. E.M. Cherry and D.A. Hooper, *Amplifying Devices and Amplifiers*, J. Wiley, New York, 1965.

25. E. Cherry and D. Hooper, The design of wide-band transistor feedback amplifiers, *Proc. IEE*, 110(2), 375–389, 1963.

26. N.M.S. Costa and A.V.T. Cartaxo, Optical DQPSK system performance evaluation using equivalent differential phase in presence of receiver imperfections, *IEEE J. Lightwave Technol.*, 28(12), 1735–1744, June 2010.

27. H. Tran, F. Pera, D.S. McPherson, D.N. Viorel, and S.P. Voinigescu, 6-kΩ, 43-Gb/s differential transimpedance-limiting amplifier with auto-zero feedback and high dynamic range, *IEEE J Solid State Circuits*, 39(10), 1680–1689, 2004.

28. Inphi Inc., Technical information on 3205 and 2850 TIA device, 2012.

29. (a) S. Zhang, P.Y. Kam, J. Chen, and C. Yu, A comparison of phase estimation in coherent optical PSK system, *Photonics Global'08*, Paper C3-4A-03. Singapore,

December 2008; (b) S. Zhang, P.Y. Kam, J. Chen, and C. Yu, Adaptive decision-aided maximum likelihood phase estimation in coherent optical DQPSK system, *Optoelectronics and Communications Conference (OECC)'08*, Paper TuA-4, pp. 1–2, Sydney, Australia, July 2008.

30. I. Hodgkinson, Receiver analysis for optical fiber communications systems, *IEEE J. Lightwave Technol.*, 5(4), 573–587, 1987.
31. N. Stojanovic, An algorithm for AGC optimization in MLSE dispersion compensation optical receivers, *IEEE Trans. Circ. Syst. I*, 55, 2841–2847, 2008.
32. http://en.wikipedia.org/wiki/ENOB—cite_note-3, Access date: September 2011.
33. B.N. Mao et al., Investigation on the ENOB and clipping effect of real ADC and AGC in coherent QAM transmission system, *Proc. ECOC 2011*, Geneva, 2011.
34. G.L. Abbas, V. Chan, and T.K. Yee, A dual-detector optical heterodyne receiver for local oscillator noise suppression, *IEEE J. Lightwave Technol.*, LT_3(5), 1110–1122, October 1985.
35. B. Kasper, C. Burns, J. Talman, and K. Hall, Balanced dual detector receiver for optical heterodyne communication at Gb/s rate, *Electron. Lett.*, 22(8), 413–414, April 10, 1986.
36. S. Alexander, Design of wide-band optical heterodyne balanced mixer receivers, *IEEE J. Lightwave Technol.*, LT-5(4), 523–537, April 1987.
37. L.N. Binh, *Digital Optical Communications*, CRC Press, Boca Raton, FL, USA, 2008, Chapter 4.
38. Phillips Company, *Phillips Handbook of Semiconductors*, Vol. S10, Eindhoven, Netherland, 1987.
39. E.M. Cherry and D.E. Hooper, *Amplifying and Low Pass Amplifier Design*, John Wiley, New York, 1968, Chapters 4 and 8.
40. R. Sterlin, R. Battiig, P. Henchoz, and H. Weber, Excess noise suppression in a fiber optic balanced heterodyne detection system, *Opt. Quantum Electron.*, 18, 445–454, 1986.
41. J. Kahn and E. Ip, Principles of digital coherent receivers for optical communications, Paper OTuG5, *Proc. OFC 2009*, San Diego, USA, March 2009.
42. C. Zhang, Y. Mori, K Igarashi, K. Katoh, and K. Kikuchi, Demodulation of 1.28-Tbit/s polarization-multiplexed 16-QAM signals on a single carrier with digital coherent receiver, Paper OTuG3, *Proc. OFC 2009*, San Diego, USA, March 2009.
43. L.N. Binh, *Digital Optical Communications*, CRC Press, Taylors & Francis Group, Boca Raton, FL, USA, 2008, Chapter 11.

5

Optical Phase Locking

5.1 Overview of Optical Phase Lock Loop

Coherent receivers offer far more efficient detection and sensitivity of optical signals under various phase, amplitude, or frequency modulation scheme techniques as compared to simple direct detection. The critical component of a practical coherent receiver is the phase lock loop (PLL) in the optical domain, the optical PLL (OPLL). The PLL has found widespread applications in the electrical domain. However, the OPLL has yet to reach maturity due to its difficult setup and problems with stability. This chapter thus aims to improve the performance of the OPLL with the development of a novel FPGA-based loop filter, a digital phase locking state loop.

The OPLL and coherent optical reception are not recently demonstrated for DSP-based receivers, but significant development and research related to OPLL has been extensively conducted in the 1980s in the early days of single-mode fiber optical communications. Much of this research was performed aiming at the improvement of the receiver sensitivity, thereby increasing transmission distances. However, limitations in laser technology at that time limited the practical success of the OPLL. This can be coupled with the erbium-doped fiber amplifier (EDFA) in the late 1980s and in the first decade of this century that would provide simple and effective solutions to transmission distances.

An early OPLL pioneered by Kazovsky lists the following four important advantages of a coherent optical detection system [1]: (i) improved receiver sensitivity by 10–20 dB with respect to intensity modulation/direct detection; (ii) greatly enhanced frequency selectivity; (iii) conveniently tunable optical receivers; and (iv) the possibility of using alternative modulation formats such as FSK and PSK. Currently, when approaching the limit of what can be "wrung" out of traditional ASK-based systems, it is the fourth point of the list and the demand for installed systems and network management that has seen a resurgence in interest in the OPLL as an efficient method of decoding exotic modulation formats such as multilevel phase shift keying (M-ary PSK) and remote locking of the local oscillator (LO) laser to a reference carrier for maximizing the receiver sensitivity. In a subsequent paper, Kazovsky lists four improvements over a heterodyne-based receiver [2]: (i) 3 dB sensitivity improvement (signal power); (ii) higher bit rate with the same detector bandwidth; (iii) simpler postdetection processing since the IF part is eliminated; and (iv) a homodyne receiver that can be made from readily available direct detection receivers.

In their 2006 work, Ferrero and Camatel [3] gave an updated list of factors as to the attractiveness of the OPLL for modern coherent receivers: (i) Coherent detection would open the way for modulation formats currently successfully employed on other telecommunications fields such as M-PSK and M-QAM. These formats allow a much greater spectral efficiency over conventional intensity modulation by a factor $\log_2(M)$. For example, 8-PSK results in a bandwidth three times narrower than the intensity modulation. (ii) Coherent receivers allow phase information recovery at the receiver's end, opening new applications in receiver equalization for dispersion and PMD compensation. (iii) Coherent detection allows separating closely spaced DWDM channels without requiring narrow optical filters since channel selection is obtained directly at the base band by electrical filtering. In the early experiments performed by Kazovsky, the LO was tuned via the piezoelectric transducer (PZT) port of the tunable laser source. This provided a limited tuning sensitivity of 3 MHz/V [4]. In 2004, Ferrero and Camatel presented a novel subcarrier optical phase locked loop (SC-OPLL) [5], which presents a simpler and much more stable "optical VCO." Instead of tuning the laser directly, an electrical VCO was connected to an optical amplitude modulator modulated by the LO laser, which thus created two sidebands $f_{LO} \pm f_{VCO}$, one of which can be used for locking with the received signal (Rx). Because f_{VCO} is generated by a stable electrical VCO, the sideband used for locking the state is very stable and can have a tuning sensitivity of 100 MHz. The drawback of using this method is a 3 dB penalty, as one of the sidebands is not utilized. A blocking state diagram of their system is shown in Figure 5.1.

Note that the LPF in Figure 5.1 uses a first-order analog active filter. This is a typical design employed in PLL and forms a second-order closed-loop transfer function. This type of loop is unconditionally stable to any step change in phase and will settle to steady state for any step change in frequency. However, with this simple filter design, there exists a trade-off situation where the bandwidth of the filter is desired to be as small as possible to reduce noise filtering through to the output, but at the same time the bandwidth is wide enough allowing the locking onto the initial error signal and hence the real homodyne mixing in the Rx [6].

By digitizing the LPF and combining a control system to bring the error signal within the pull-in region of the suitably narrow digital LPF, the trade-off situation disappears. A further benefit of a digital implementation is the flexible and simple implementation of a monitoring system, in contrast to analog monitoring circuits, which are generally fairly complex. A particularly efficient example of implementing a digital LPF onboard an FPGA was demonstrated by Linn [7]. The natural frequency of PLL is generally

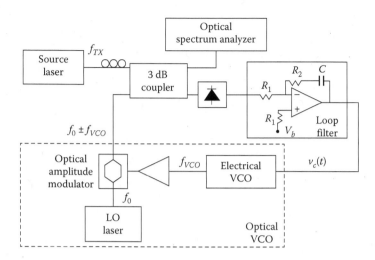

FIGURE 5.1
Subcarrier optical phase lock loop (SC-OPLL): analog feedback control technique.

relatively low. Indeed, the PLL itself can be thought of much like a low-pass filter; making this realization it follows that requirements on LPF locking state rate can also be reduced, and as such it is possible to implement multiplication with an iterative state machine, which sums and shifts partial products. This results in a more then 70% reduction in the gate count from a direct hardware implementation of the multiplication operation. The advantages of this architecture come at a cost of complexity; thus, at least our initial prototypes will simply utilize hardware multiplication blocking states.

This chapter thus focuses on the digital LPF, which acts to condition the signal as seen at the photodiode, such that the LO is driven to match or locking state in frequency and phase of the incoming optical signal. Experimental analog LPF of a simple first-order active filter loop, constructed using a single operational amplifier, is implemented to compare with the performance of the FPGA-digital-based phase locking. However, the LPF can be seen to be the "active" or controlling component of the loop and thus this project aims to "smarten" the LPF, thereby solving some of the observed problems of a traditional OPLL. The proposed LPF is implemented onboard an FPGA consisting of a first-order active digital filter, which also performs other functions such as conditioning the input signal, locking state monitoring, and sweeping for the incoming signal over the full range of the voltage-controlled oscillator (VCO). The locking range is much greater than the bandwidth of the internal filter, thereby simplifying the structure.

This chapter is arranged as follows: Section 5.1 reviews existing literature concerning OPLLs, along with a detailed section on standard phase lock loop theory. Various design considerations encountered in the implementation are discussed. Simulation and experimental results are presented, confirming the operation of the "Smart" Filter. Section 5.2 then gives an introduction of optical coherent detection and the role of optical phase locking state loop. Also in this section, we introduce the design of the digital phase locking state loop as part of the feedback loop with the analog-to-digital conversion, thence digital signal processing/filtering and then digital-to-analog conversion to generate a frequency shifting wave to drive the optical modulator to meet the locking of the modulated carrier and the frequency-shifted LO. Section 5.3 gives the performance of both analog and digital feedback loop techniques in the optical PLL. We then describe, in Section 5.4, a proposed technique for generating super-LO sources for mixing and betting with superchannels for Tbps coherent optical receivers. Finally, concluding remarks are stated.

5.2 Optical Coherent Detection and Optical PLL

The principle behind coherent detection is remarkably simple, as described in Chapter 4 and as summarized in Figure 5.2. A narrow linewidth laser is used as a LO and the frequency and phase are tuned to equal that of the Rx. Any phase information is then available by simply applying a low-pass filter.

Obviously the complexity in the coherent system is to tune or lock the LO to the incoming signal and moreover maintain that locking state. To solve this problem the phase locked loop (PLL) is essentially developed.

5.2.1 General PLL Theory

The PLL can be divided into three principal components arranged in a feedback loop: the phase detector (PD), the LPF, and the VCO as shown in Figure 5.3.

The basic principle of operation can be understood by moving around the loop. Starting at the PD, two input signals are compared and a signal is returned, which is proportional to the phase error. This signal is conditioned by the LPF before being applied to the VCO. The frequency (and thus phase)

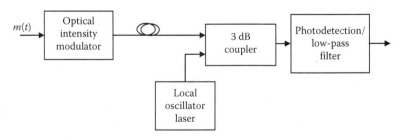

FIGURE 5.2
Typical arrangement of optical fiber coherent detection.

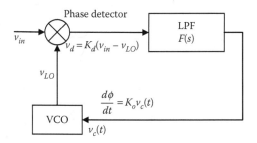

FIGURE 5.3
Generic PLL blocking state diagram based on analog technique.

output of VCO is proportional to the input voltage. In a well-designed loop, the input voltage applied to the VCO will reduce the phase error from the PD as we go around the loop again.

5.2.1.1 Phase Detector

The PD is generally a signal multiplier or coupler; however, other types exist, such as the sequential PD used in some digital PLLs. In this chapter we are concerned with a multiplier-type PD.

If we consider two signals $s_{in}(t)$ and $s_{LO}(t)$, assumed to be 90° apart, and multiply them together the output of the multiplier will be

$$s_3(t) = K_m A_{in} A_{LO} \sin[\omega_{in}t + \phi_{in}(t)]\cos[\omega_{LO}t + \phi_{LO}] \tag{5.1}$$

K_m is a gain factor associated with the multiplier, which is typically less than 1 for a passive coupler. Also, by applying a common trigonometric formula, this output signal can be written as

$$s_3(t) = \frac{K_m A_{in} A_{LO}}{2}\left[\sin(\omega_{in}t + \phi_{in}(t) + \omega_{LO}t + \phi_{LO}(t)) + \sin(\omega_{in}t + \phi_{in}(t)\right.$$

$$\left. - \omega_{LO}t - \phi_{in}(t))\right] \tag{5.2}$$

$$s_3(t) = \frac{K_m A_{in} A_{LO}}{2}\left[\sin((\omega_{in} + \omega_{LO})t + (\phi_{in} + \phi_{LO}(t))) + \sin((\omega_{in} - \omega_{LO})t\right.$$

$$\left. + (\phi_{in}(t) - \phi_{LO}(t)))\right] \tag{5.3}$$

If ω_{LO} is set such that it is equal (or even very close too) ω_{in} the above expression can be reduced to

$$s_3(t) = \frac{K_m A_{in} A_{LO}}{2}\left[\sin(2\omega_{in}t + (\phi_{in}(t) + \phi_{LO}(t))) + \sin(\phi_{in}(t) - \phi_{LO}(t))\right] \tag{5.4}$$

Notice that the second *sine* term is only a function of the phase error and the first term has a frequency of $2\omega_{in}$, which can be simply removed by low-pass filtering.

$$s_3(t) = \frac{K_m A_{in} A_{LO}}{2}\left[\sin\left(\phi_{in1}(t) - \phi_{LO}(t)\right)\right] \tag{5.5}$$

This is the phase error signal from the PD, which is seen by the LPF. When the phase error is zero, $\phi_1(t) - \phi_2$ is also zero. However, owing to the way that we have set up the above equations, we can recall that $\phi_1(t)$ is the phase of a sine function and ϕ_2 of a cosine; the actual phase difference between s_1 and s_2 is $\pi/2$. This was deliberately chosen to match the commonly observed behavior of the loop locking state in quadrature with the Rx.

5.2.1.2 Loop Filter

In its most basic form, the low pass filter (LPF) is passive, and effectively a low-pass filter, but commonly an active filter is used. This type of filter also incorporates a low-pass filtering action but also features an integrator. A common first-order LPF topology is shown in Figure 5.4 with an idealized magnitude response.

The transfer function of this filter for large open-loop gain can be written as

$$F(s) = -\frac{s\tau_2 + 1}{s\tau_1} \tag{5.6}$$

where s is the variable of the Laplace transform.

The magnitude response of this filter ($s = j\omega$) (Figure 5.4) can be intuitively determined by considering that if ω is small (low frequencies), the magnitude grows without bound according to $(1/s\tau_1)$. If ω is large (high frequencies), the magnitude response is constant according to $(s\tau_2/s\tau_1) = (\tau_2/\tau_1)$. The negative sign in front of the filter introduces a phase reversal of $\pi/2$.

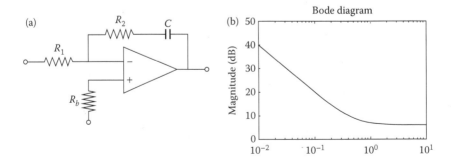

FIGURE 5.4
A simple LPF realization using analog circuit (a) and magnitude response (b).

5.2.1.3 Voltage-Controlled Oscillator

The VCO is the means by which the correction determined by the LPF is applied. The frequency generated at the output is described by

$$\omega_{ou} = \omega_c + K_o v_c(t) \tag{5.7}$$

where ω_c is the VCO central frequency, K_o is the VCO sensitivity in Hz/V, and v_c is the controlling voltage output from the LPF. Also note that phase can be expressed as the integral of frequency over time; thus, the output of VCO in terms of phase is

$$\phi_{LO}(t) = \int_{-\infty}^{\infty} \omega_c + K_o v_c(t)\, dt = \omega_c \int_{-\infty}^{\infty} dt + \int_{-\infty}^{\infty} K_o v_c(t)\, dt = \int_{-\infty}^{\infty} K_o v_c(t)\, dt \tag{5.8}$$

Note that we can discard the constant central frequency component ω_c due to the convention of phase wrapping around every 2π radians (or every complete cycle); thus it does not contribute to the output phase.

5.2.1.4 A Second-Order PLL

Now that all the basic components have been introduced, we consider the action of the loop as a whole. By rewriting the key loop equations, Equation 5.5 gives the output of the PD, which can be simplified by assuming the output is linear; this is true for small $\Delta\phi$, that is, when the loop is nearly locked.

$$s_3(t) = K_d\left[\phi_{in}(t) - \phi_{LO}(t)\right] \quad \text{with } K_d = \frac{K_m A_{in} A_{LO}}{2} \tag{5.9}$$

Taking the Laplace transform of Equation 5.9, we have

$$S_3(s) = K_d\left[\phi_{in}(s) - \phi_{LO}(s)\right] \tag{5.10}$$

Also, the output of the LPF is given by

$$V_c(s) = F(s)S_3(s) \tag{5.11}$$

where $F(s)$ is the LPF transfer function. We can also take the Laplace transform of Equation 5.8 to obtain

$$\phi_{LO}(s) = \frac{K_o V_c(s)}{s} \tag{5.12}$$

By substituting Equation 5.4a into Equation 5.5 and then substituting the result into Equation 5.10, we can write the closed-loop transfer function of a PLL as

$$\frac{\phi_{LO}(s)}{\phi_{in}(s)} = H(s) = \frac{K_o K_d F(s)}{s + K_o K_d F(s)} \tag{5.13}$$

If we substitute the first-order LPF developed previously into Equation 5.12, we can write

$$H(s) = \frac{\left(K_o K_d (s\tau_2 + 1)/\tau_1 \right)}{s^2 + s\left(K_o K_d \tau_2/\tau_1 \right) + K_o K_d/\tau_1} \tag{5.14}$$

or

$$H(s) = \frac{2\xi\omega_n s + \omega_n^2}{s^2 + 2\xi\omega_n s + \omega_n^2} \quad \text{where } \omega_n = \sqrt{\frac{K_o K_d}{\tau_1}} \quad \text{and } \xi = \frac{\tau_2 \omega_n}{2} \tag{5.15}$$

Also, we can write the phase error response as

$$\frac{\phi_e(s)}{\phi_{in}} = \frac{\phi_{LO} - \phi_{in}}{\phi_{in}} = 1 - H(s) = \frac{s^2}{s^2 + 2\xi\omega_n s + \omega_n^2} \tag{5.16}$$

Equation 5.14 is in the second-order canonical form and the parameters ω_n and ξ are known as natural frequency and damping factor, respectively, and are important parameters in characterizing the loop. Using the final value theorem to determine the steady-state phase error after a step change in the input $\Delta\phi_{in}$ is

$$\lim_{t \to \infty} \phi_e(t) = \lim_{s \to 0} \frac{s\Delta\phi_{in}}{s + K_o K_d F(s)} = 0 \tag{5.17}$$

That is, the second-order loop tracks out any step in phase applied to the input. This is an important result and demonstrates why the second-order loop is so popular.

For a ramp in phase (step change in input frequency $\Delta\omega_{in}$), the final value theorem results in a steady-state error of

$$\lim_{t \to \infty} \phi_e(t) = \lim_{s \to 0} \frac{\Delta\omega_{in}}{K_o K_d F(s)} = \frac{\Delta\omega_{in}}{K_o K_d F(0)} \tag{5.18}$$

This is also an important result, as it provides the constant DC voltage need to provide an offset from the VCO's central frequency. $F(0)$ is the LPF's DC gain.

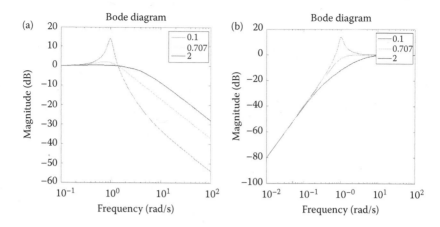

FIGURE 5.5
PLL response to (a) phase input and (b) phase error.

Plotting the magnitude responses given by Equations 5.15 and 5.16 over a normalized frequency range ω/ω_{in} furthers the understanding of the second-order loop. In Figure 5.5a, the PLL shows a low-pass filtering effect on phase inputs, while Figure 5.5b shows the high-pass filtering effect in phase error. In other words, the loop can track low-frequency errors but high-frequency errors pass through unaffected. This principle is used to allow data to be encoded in the incoming signal. The PLL will in effect "pass through" this high-frequency data while tracking any low-frequency drift in the LO.

5.2.2 PLL

The PLL is also designed to operate on the principles discussed above. Stability limitations in today's tunable lasers make directly tuning the LO laser impractical; however, a popular method of realizing an OPLL is through the use of a modulator. This type of loop is known as subcarrier modulated, whereby the loop actual locks onto a modulation sideband rather than the LO frequency.

It can be seen from Figure 5.6 that the OPLL shares many common features with the generic PLL presented in Figure 5.1. The 3 db coupler acts as the multiplier PD; however, we notice a significant difference in the addition of the amplitude modulator.

The optical amplitude/intensity modulator is biased at the minimum transmission point of the power-voltage transfer characteristics to suppress the carrier and forming two sidebands whose frequency is the output of the VCO. The electrical VCO has very good stability and tuning properties and thus the sidebands follow these properties providing a stable frequency that can be used to lock onto the Rx.

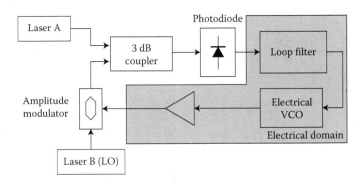

FIGURE 5.6
OPLL block diagram.

Hence, we see the frequency lock condition (see Figure 5.7) is

$$f_{LO} \pm f_{VCO} = f_{RX} \tag{5.19}$$

Other issues unique to the OPLL are linewidth requirements of the lasers. According to Kazovsky [8], for a typical second-order loop, at least 0.8 pW of signal power is need per kHz of laser linewidth. Furthermore, the loop bandwidth must be at least a few orders of magnitude larger than the line width of the Rx.

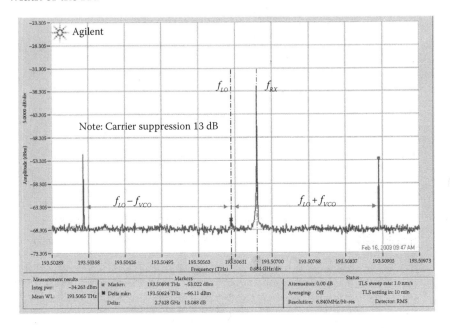

FIGURE 5.7
Input to the photodiode as see through a high-resolution spectrometer.

Other noise factors affecting OPLL performance such as laser phase noise and receiver shot noise are not considered in this report, but the reader is directed to the above-mentioned paper by Kazovsky [9] for detailed treatment.

5.2.3 OPLL

The OPLL follows much of the same principles and theory as the standard PLL discussed above, and thus the section above is justified. The principal difference in the OPLL is the realization of the VCO. Following the method developed by Ferrero and Camatel, this involves amplitude modulating the LO laser and creating two sideband frequencies at $f_{LO} \pm f_{VCO}$. The locking state is performed on one of the sidebands; however, the other sideband has the potential to interfere. Interference can be reduced by using a VCO with a high central frequency, ideally, much higher than the bandwidth of the optical receiver. Further noise reduction can be achieved by adding an electrical low-pass filter at the output of the receiver. By following these steps, the interference from the other sideband can be considered negligble [9] except for the division of signal power. However, it should be noted that in a WDM system, this spurious sideband will cause problems. One solution is to use a VCO with a very high central frequency (>40 GHz); however, this becomes expensive and difficult to work with. Ferrero presents another solution, which is single-side subcarrier modulation OPLL (SS-SC-OPLL) [10]. This method involves replacing the amplitude modulator with a QPSK modulator and applying a correct driving voltage such that the two arms of the modulator are $\pi/2$. This has the effect of producing a signal sideband that contains much of the power that was previously distributed between two sidebands. This implementation of the OVCO can be further investigated.

The FPGA-based PLL for OPLL requires the following:

5.2.3.1 Functional Requirements

The optical locking system requires that (i) the frequency of the LO be matched to the Rx within the largest laser linewidth. The precise linewidth of the lasers is estimated to be about 100 kHz. (ii) Once locking state has been acquired, the loop must track the slow frequency drift associated with the tuneable lasers. This should be over the entire VCO range. (iii) The digital LPF must be able to automatically acquire the signal if brought within the range of the VCO. (iv) The digital LPF can be determined if the locking state is acquired.

5.2.3.2 Nonfunctional Requirements

The nonfunctional requirements on the OPLL are that the LPF will be implemented onboard an FPGA. Initially, an Altera DE1* board for development

---

* See website http://www.altera.com/education/univ/materials/boards/de1/unv-de1-board.html.

with the aim of implementation onboard a smaller embeddable board. Thus, the entire filter must fit onto the FPGA chip.

5.2.4 Digital LPF Design

In this and subsequent sections, the term "LPF" is used to refer to the entire "smart" filter system, while the term "digital filter" refers to the digital filter that is incorporated into the LPF. Figure 5.8 provides an overview of the functional blocking states of the smart LPF as they will be implemented inside the FPGA. Note that this is an ideal theoretical view and does not exactly represent what was implemented, namely, the PID pathway was not implemented. In this respect, it actually suggests a system that achieves better performance and could perhaps be used as a basis for future work.

Referring to the blocking state diagram, three distinct data pathways can be seen; that is, the sweep unit, PID controller, and the digital filter. Only one path is chosen to go to the output via the 3:1 MUX. Each pathway will be described by referencing the startup procedure. Initially, the loop is not locked and the filter has no indication of where the receiver may be, hence, the sweep unit will be activated and will ramp up and down until a beat signal is detected. When the frequency counter recognizes a beat signal, the sweep unit will be disengaged and the PID controller will course tune the signal using the frequency counter output as an error signal. This will bring the signal into the locking state in range of the digital filter. Once this occurs, the output will be switched to the output of the digital filter. The frequency counter is continually monitoring the locking state and will output a control signal if action needs to be taken to correct the locking state. The control unit controls the MUX as well as provides diagnostic information to the outside world; its inputs are the frequency counter and external switches that can be used for manual control.

5.2.4.1 Fixed-Point Arithmetic

For efficient implementation of the digital filter as well as to keep hardware complexity down, it was quickly realized that a fixed point numbering system had to be used. The fixed point number system is often referred to as the $Q_{m.n}$ number format, where m represents the number of integer bits and n is the number of fractional bits. It is really a natural extension of the standard binary integer representation. The bits in the fractional section represent an unsigned integer (numerator) over an implicit denominator of 2^n. For example, in a $Q_{1.2}$ number format, the binary bit sequence 0.01 would represent an integer (decimal) value of 0 plus a fractional (decimal) value of 0.25, that is, $001_{Q1.2} \leftrightarrow 0.25_{10}$. The smallest (signed decimal) number a $Q_{m.n}$ format can represent is -2^m, while the largest is $2^m - 2^{-n}$. The resolution of such a system is 2^{-n}.

For the purposes of this filter, the only operations we are concerned with are multiplication and division. Signed fixed-point addition can actually

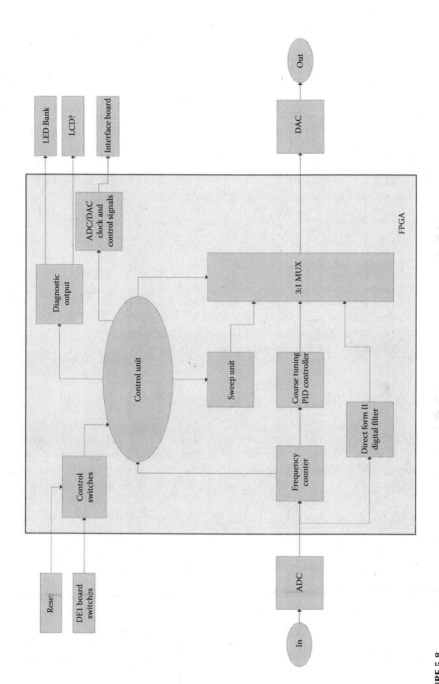

FIGURE 5.8
Module schematic of state diagram of the digital loop filter.

be handled transparently by an integer addition; no other operations are required. Fixed point multiplication is only slightly more involved in that the result of the standard integer multiplication must be shifted right by n bits (divide by 2^n) to correct for the implicit denominator. Note that the intermediary variable that holds the result of the integer multiplication must have at least $2 \times (m + n)$ to avoid overflow.

5.2.4.2 Digital Filter

In keeping with much of the existing literature, a second-order loop was the target for design. It results in a simple loop while also having the desired properties of zero steady-state error for a step change in phase and constant steady-state error for a ramp in phase. To achieve a second-order loop, the traditional LPF is commonly specified by the following transfer function:

$$-\frac{s\tau_2 + 1}{s\tau_1} \tag{5.20}$$

The single integrator in the denominator acts with the integrator inside the VCO to bring the total loop order to two. The analog filter described above can be converted to a digital filter using the bilinear transform with the equivalent digital transfer function shown below:

$$H(z) = \gamma \frac{1 - \beta z^{-1}}{1 - \alpha z^{-1}} \tag{5.21}$$

A plot of the two transfer functions (5.20) and (5.21) reveals very little difference. However, there are a number of sources of error when digitizing filters are used due to quantization and sampling. The number representation chosen for this filter is a $Q_{16.16}$ fixed point system. That is 16 bits for an integer section and 16 bits for a fractional section.

Using simulation as a starting point, the final coefficients were selected after significant experimentation and are presented in Table 5.1.

Note that for the coefficient α, exactly 1 was not used. This was to reduce the chance of overflow. However, the maximum fractional value approximates

TABLE 5.1

Quantization Errors for Filter Coefficients Using 16.16 Fixed Point

Coefficient	Decimal	Hexadecimal	Quantization Error (%)
$-\alpha$	1.0	0x0000FFFF	0.0015
β	−0.8	0xFFFF3333	−0.0003
γ	0.5	0x00008000	0

1 with a high level of accuracy. A sampling time of 100 ns (10 MHz) was chosen, thus with consideration of the Nyquist rate, the upper limit for the filter bandwidth is 5 MHz (equivalent to 200 ns rate). However, with practical considerations, this is more likely to be around 1 MHz. This upper limit is acceptable, as it leaves plenty of headroom while the filter tries to reduce the error down to less than 100 kHz. The 8 bit unsigned values (range [0,255]) returned from the ADC are denormalized from the bias voltage to create a signal centered at 0. Also the normalized values are shifted left by 4 to reduce the closed loop gain. Note that no precision has been lost at this stage due to the extra bits afforded by the 32 bit fixed point representation.

Values at the output of the digital filter are also 32 bit signed $Q_{16.16}$ format and are summed with the sweep signal, with the sum likewise being a signed 32 bit $Q_{16.16}$ representation. The integer section of a Q format number is represented in 2s compliment, and therefore the MSB is in effect a signed bit; as a negative VCO voltage is not applicable, a value of 0 was passed to the DAC if the MSB was 1 (i.e., the filter output is less than 0), else the next highest 16 bits. This assignment in effect introduces a shift left of 1 bit or a multiplication by 2, and must be kept in mind when setting the filter parameter γ. Note that by simply considering mapping of the ADC bits and the output mapping of the 16 DAC bits, we see the filter has a pass-through gain of

$$\frac{V_{out\_pk}}{V_{in\_pk}} \approx -29 \text{ dB} \tag{5.22}$$

Note that this does not take into account the large gain introduced by the integrator, and in fact, this seemingly large reduction was necessary in the simulation as well to dampen the action of the integrator. The next important consideration in regards to the digital filter is the topology shown in Figure 5.9. The critical consideration is propagation time, thus the two direct form filters are considered. It can be seen that the direct form type II filter uses only one register and thus is the most efficient to implement in hardware.

The drawback is that the single register is more prone to overflow as compared to the two registers required for the type I topology, especially

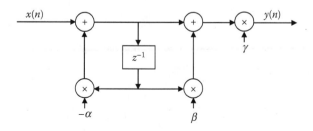

FIGURE 5.9
Direct form II digital filter topology.

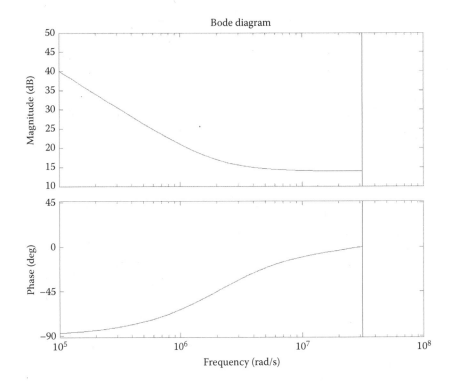

FIGURE 5.10
Bode plot of the digital filter described by Equation 5.18.

considering that the parameter $-\alpha$ is chosen very close to 1, thus approximating an ideal integrator. As such, the hardware adders were expanded to detect overflow and saturate rather then wrap around into the negative side of the signed integer. If an overflow is detected, the filter notifies the user by illuminating an LED and with the final filter configuration no overflows were observed. A Bode plot of the final digital filter is shown in Figure 5.10.

5.2.4.3 Interface Board

The purpose of the interface board is to allow a bidirectional flow of data between the loop and the FPGA. On the one side, the output of the optical receiver is sampled and digitized using an analog-to-digital converter (ADC) with the data input to the FPGA. The other side comprises a digital-to-analog converter (DAC), which takes the filtered output and converts it to an analog voltage to drive the electrical VCO. Other components are miscellaneous components required to drive the above-mentioned data converters.

The primary design consideration is the propagation delay through the LPF. This aspect governed many of the design choices involved in designing the interface board. The ADC (TLC5540) is implemented using a Flash architecture, which allows the output to follow the input with a much smaller delay as compared with a successive approximation device. The drawback with selecting this architecture is that they are only feasible up to 8 bits of resolution and thus will introduce some quantization error. However, as the signals that are expected to be the input to the ADC are sinusoidal and bounded in the range, the time delay was deemed a more significant factor. The selected sampling rate was the largest that was available, 40 Msps: this will allow digitization of signals of up to 20 MHz (theory). However, a more practical limit would be around 10 MSa/s. This is deemed adequate due to the sweeping action of the filter, that ensures that the signal passes through this window. A 5 MHz antialias low-pass filter ensures that there are no alias artifacts in the signal.

The input to the ADC is passed through a unity gain buffer with a very high slew rate of 3600 V/µs. This protects the ADC and provides a low output impedance to the ADC. Ideally, the buffer would incorporate bias circuitry. However, to increase flexibility and to meet manufacturing deadlines, this was left off the board. A separate bias board was constructed on a Vero board to bias the incoming signal at approximately 1 V while also providing 13 dB gain. This simple circuit should be incorporated into the interface board in future revisions. The lower ADC reference level was set at 0 V, while the upper was set at 2.2 V, using the ADC's internal voltage reference.

The prime consideration in the selection of the DAC (DAC8820) was accuracy, and hence a 16 bit DAC was chosen. The output reference set at 5 V translates to a quantization interval between each discrete voltage level of $5/2^{16} \approx 76$ µV, or with a VCO tuning sensitivity of approximately 100 MHz/V, a frequency resolution of 7.6 kHz.

Each of the data converter devices was chosen with a parallel bus to lower the time required to transfer the data to and from the FPGA. The PCB was laid out according to standard layout procedure. The top layer contains all the data signals. The bottom layer contains the power conversion components and significant distribution tracks. On both layers, a ground plane has been poured to minimize the introduction of external noise and reduce stray capacitance, and numerous links have been placed around the board to ensure a short path to ground from all corners of the board.

Another consideration was interfacing the data conversion components to the FPGA, which required LVTTL (3.3 V). The ADC provided output at 5 V TLL levels, which were converted down to 3.3 V using an 8 bit level translator (TXB0108). A second level translator was required to translate the control signals. The DAC accepted data input at 3.3 V and thus required no conversion: however, the control signals are converted to 5 V to fill the extra-level translator required for the ADC, but it may also help ensure a glitch-free operation.

A header is available for three external LEDs, two of which are connected to the FPGA. These LEDs are used to indicate that the board has power, when the FPAG has been correctly programmed and when signal locking state has been achieved. Currently, these have not been mounted, as the LEDs onboard the Altera DE1 board are used.

The entire interface board is powered by a single 12 V DC supply. This is regulated down to 10 V, after which a charge pump and regulator provide −10 and 5 V, respectively. When running without external indicator LEDs, the board draws 100 mA.

5.2.4.4 FPGA Implementation

The FPGA used is the EP2C20F484C7 from the Altera Cyclone II family. An Altera DE1 board was used for development, containing many useful peripherals used for debugging. However, the LPF was designed to be run on a smaller standalone microboard. The FPGA implementation was completed using the Verilog hardware description language and was constructed to be modular, allowing functions to be tested independently of the entire system. The entire LPF implementation uses only 18% of the total logic elements available on the selected FPGA, and thus may be used on smaller, cheaper FPGA chips. Figure 5.11 shows the FPGA development board employed for implementing the digital filter.

5.2.4.5 Indication of Locking State

The locking state indication is not merely useful for the purposes of monitoring. In fact, this signal controls the operation of the LPF, but it provides the trigger for the sweep unit to stop, and at the same time enables the digital filter. An initial frequency counter implementation was considered. However, owing to noise issues, it was not used in the final LPF. Instead, a simpler solution implemented involved comparing samples against a threshold. Once a certain number is reached, the system stops sweeping and enables the digital LPF, which will fine tune.

FIGURE 5.11
Illustration of the FPGA board implementing the digital PLL.

Once locking state is established, the system needs to determine if the locking state is lost and thus re-enable the sweep unit. The loop can be said to be unlocked if too many samples are above the threshold, that is, the beat frequency is close and within the range of the ADC, or if no samples above a threshold are detected. The latter case is the most likely indication of the system losing the locking state, as it can be presumed that the beat signal is outside the narrow field of view of the ADC. Short spurious beats that appear in the locking state case ensure the sweep unit is not enabled prematurely.

Another issue observed was if the gain of the digital filter was set too high when enabled, it would quickly force the signal beyond the range of the ADC. Thus what appears to be a "perfect" locking state, that is, one without spurious beats, is observed by the LPF. The assumption was made that such a perfect locking state was impossible in this system and thus the sweep unit will be enabled after a set time of no signal.

5.2.4.6 OPLL Hardware Details

The hardware setup used to create the OPLL consists of the following as shown in Table 5.2.

The optical power at the output of the coupler is approximately −5 dBm, or effectively is 3 dB less than the input power as shown in Table 5.3. Ideally, the electrical power available out of the O/E converter should be larger; this could be achieved by using a higher LO power or an O/E converter with a higher gain.

TABLE 5.2

Experimental OPLL Hardware Details

Component	Brand/Model	Comment
Input laser	Radians Innova Intun 1500	Tunable ECL
LO laser	Photonetics Tunics 1550	Tunable ECL
High-resolution spectrometer (HRS)	Agilent 83453 HRS	Optical HRS
Optical/electric converter (O/E)	Tektronix SA-64	Photodiode
O/E preamplifier	Mini-Circuits ZFL-500LN+	1.1–500 MHz min. gain 24 dB
O/E amplifier	Mini-Circuits ZX60-14012 L+	300 kHz to 14 GHz (wideband) Min. gain 9 dB
Low-pass filter	Mini-Circuits PLP-5+	5 MHz antialias filter
FPGA	Altera DE1 board	Can use smaller dedicated chip
VCO	Mini-Circuits ZX95-3080-S+	Range: 2920–3080 MHz (~200 MHz) Tuning sensitivity average 90 MHz/V
VCO amplifier	Mini-Circuits ZVA-183+	700 MHz to 18 GHz (wideband) Min. gain 24 dB
Amplitude modulator	Pirelli PIRMZM 153X	Bias at 3.1 V

TABLE 5.3

Optical Power Budget

Input laser	−1 dBm	At output
LO laser	5 dBm	At output
Amplitude modulator and polarization controller	−6.2 dBm	At output
Coupler	−4.9 dBm	At output

5.3 Performances: Simulation and Experiments

5.3.1 Simulation

A MATLAB Simulink model was constructed for the "smart" filter and was tested in a regular PLL loop without simulating the optical system. The PID pathway was not included in the final model. The model took into account quantization error and the limited bandwidth available to the interface board. However, several aspects such as fixed point arithmetic overflow and electrical noise sources are not considered in this work. Figures 5.12 through 5.14 presents the result of the simulation over the same time period (10 μs).

The simulation was performed with an initial frequency offset of 50 MHz and analog LPF parameters set as $\tau_1 = 10$ μs and $\tau_2 = 50$ μs, which describe the following digital filter. A sampling rate of 10 MHz was used.

$$H(z) = 5.5\frac{1 - 0.8182z^{-1}}{1 - z^{-1}} \tag{5.23}$$

In the first "half" of the plot, the "smart" filter is sweeping the LO searching for the Rx. Incrementally, it brings the beating "error" signal closer, which results in the input frequency being reduced, as can be seen in Figure 5.12. During this time, the frequency counter is constantly monitoring the input frequency. When this is below a set threshold (10 MHz in this case), the control unit enables the digital filter and halts the sweep unit. The action of the digital filter can be seen in the second "half" of the plots below. Figure 5.12a shows that the filter continues to reduce the frequency of the input signal until it becomes 0. At this point, the loop can be said to be in a locked state.

Figure 5.12b shows the voltage applied to the VCO. Initially, it consists of regular steps generated by the sweep unit. However, once the LPF is enabled, the signal quickly settles to the required voltage of 0.5 V that produces $\Delta f = 50$ MHz for a VCO sensitivity of 100 MHz/V. Figure 5.13 shows the input from the ADC and shows just how little information is needed to complete the locking state in the almost ideal conditions offered by the simulation environment. Figure 5.14 shows the output of the digital filter, particularly the integral action settling at a steady-state offset voltage.

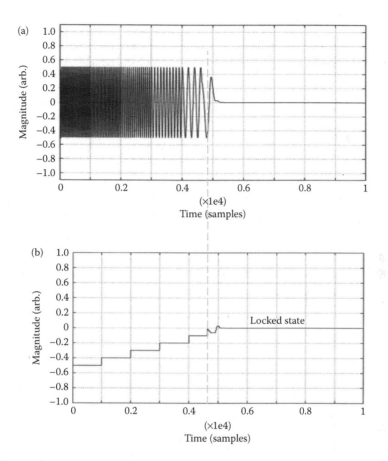

FIGURE 5.12

(a) LPF input (10 µs). (b) VCO drive voltage or LPF output (10 µs)—horizontal axis is number of samples and equivalent in time.

5.3.2 Experiment: Digital Feedback Control

The experimental setup can be seen in Figure 5.15a. The input signal is mixed with an externally modulated LO laser via a fiber coupler with a polarization controller. The beating signals in the photodetection diode are then amplified to the right level, digitally converted, and then low-pass filtered in an FPGA. The output level is then fed to a VCO whose output then acts as a sinusoidal wave which would be amplified and fed into the optical modulator to shift the optical carrier of the LO toward the frequency of the optical carrier of the optical signals. The FPGA allows the tuning and change of strategy for tuning, and locking the LO to the carrier. Figure 5.15b shows the prototype set-up, in the laboratory of the OPLL. Figure 5.16 shows the schematic of the setup, the evolution of the frequency of the LO, and a reference laser for testing the functionalities of the OPLL before feeding the real

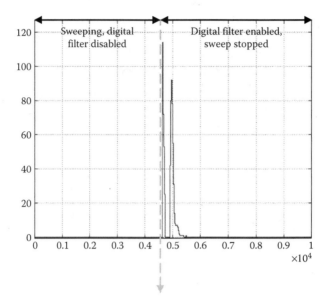

FIGURE 5.13
Output from ADC (10 µs full scale).

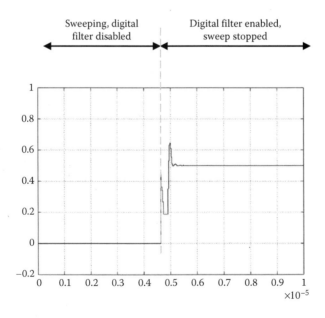

FIGURE 5.14
Output from the digital filter (10 µs full scale).

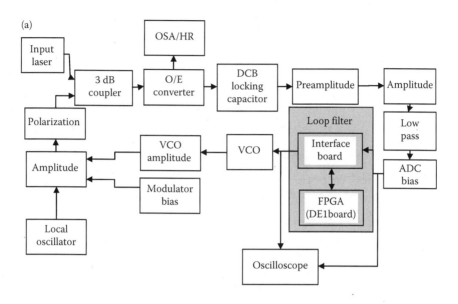

(a)

(b)

FIGURE 5.15

Laboratory implementation of the above blocking state diagram: (a) schematics and (b) experimental platform. ADC, analog to digital converter; HRS, high-resolution spectrum analyzer; OSA, optical spectrum analyzer.

channel. The coherent detection and frequency shifting with the digital feedback loop allow the locking and then digital control to generate the locked LO, which would be feed to the coherent receiver $\pi/2$ hybrid coupler or to a comb generator. The outputs of the comb generator would be mixed with the superchannel and then demuxed, to be detected coherently by a parallel balanced receiver and DSP interconnecting network for optical receiving structure. This subject is briefly described in the next section.

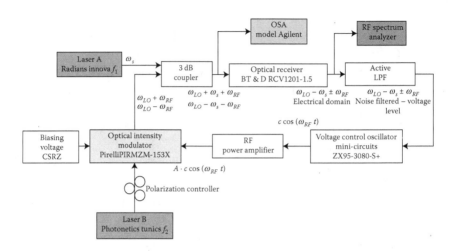

FIGURE 5.16
Experiment schematic and evolution of frequency components.

5.3.2.1 Noise Sources

Various noise sources can affect the performance of the PLL, the primary sources observed where electrical noise from the amplifiers after the O/E converter. Also, noises were observed to be superimposing to the interface board from the FPGA. When programmed and switching at 50 MHz; this could be solved by ensuring the interface board is electrically isolated from the FPGA using optical isolators.

Noise power is a major difficulty for any type of system, especially in a system incorporating a feedback loop as the loop gain may be continually amplifying the noise that would overflow the filter responses. The recording of the actual noise observed is shown in Figure 5.17, where the maximum and minimum deviations from the center are large. However, the standard deviation is relatively minute.

5.3.2.2 Quality of Locking State

Once the sweep unit was disabled and the digital filter enabled, the locking can be quickly obtained, often in less than half a millisecond. A perfect locking state as seen in simulation was not able to be achieved; however, a reasonable locking state quality was achieved and is shown in Figure 5.18a. Note the time between the spurious beats is in the order of tens of milliseconds and the beats themselves are in the order of a few kilohertz. This quality of locking state was sustained for at least 1 h and no doubt would have persisted longer had the experiment continued. During this time, the locking status was lost briefly a number of times; however, the intelligent sweep unit quickly recovered the locking state.

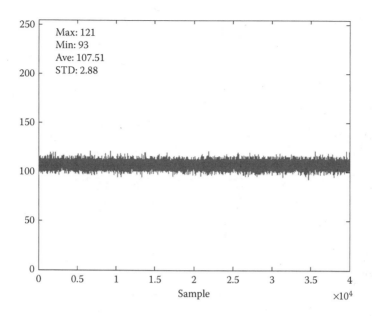

FIGURE 5.17
Noise observed after the ADC stage.

The fine-tuning action of the digital filter is shown in Figure 5.18b. Comparing this figure with Figure 5.14, a similar action can be observed. Once the sweep unit has been disabled, the filter quickly develops an offset voltage that holds this behavior as also seen in the simulation depicted in Figure 5.18a. An interesting feature of Figure 5.18 is the dip observed in the approximate center of the plot, where the filter acts to compensate for a sudden jump (beat) in one of

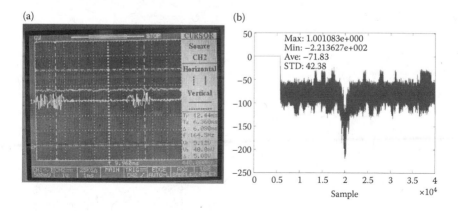

FIGURE 5.18
(a) O/E output when locking state (lower trace) and VCO voltage (upper trace). (b) Digital filter output under locking state.

the lasers. An offset average of approximately 70 indicates that a fine tuning of about 10.68 mV or 1.068 MHz is achieved. Also visible is the amplified noise as compared to Figure 5.17. This is an area for further work that would certainly have a positive influence on locking state quality.

5.3.2.3 Limitations

The greatest limitation observed with locking state quality maintained over time is the range of the VCO. The VCO employed was designed for an electrical PLL and had an approximate deviation from the central frequency of about 200 MHz, as shown in Figure 5.19a. It was noticed that over time, especially when the temperature is changing rapidly (dusk), the lasers became separated by a greater frequency deviation by the VCO and thus the locking state is irrecoverably lost, as shown in Figure 5.19b. Replacing the VCO with a travel of around 500 MHz to 1 GHz would ensure that the locking state is maintained over variations in temperature and over time.

It was also observed that occasional high-frequency jumps occurred, which the filter was not able to counter for due to its limited bandwidth. This is consistent with the operation of a PLL as discussed in Section 5.2.2. That is, high-frequency errors are passed through; however, since no data were being transmitted, these deviations are undesirable. A solution could be to upgrade the ADC sampling rate to allow a larger filter bandwidth; however, this increases the complexity of the interface board.

Finally, greater optical power would allow more aggressive filtering and require less amplification, thus reducing noise. This could be achieved with alternative lasers or perhaps through the use of an optical amplifier.

Figure 5.20 shows the spectrum of the LO and that of the channel at the initial setting stage. The LO is modulated using CSRZ modulation so that

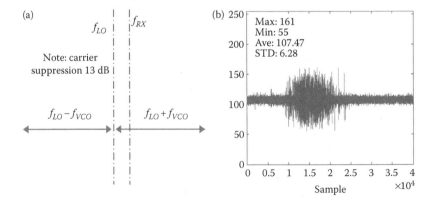

FIGURE 5.19
(a) Ranges of locking state of the LO laser and the optical signal carrier. (b) Manual sweeping for locking between the LO laser and the optical signal carrier.

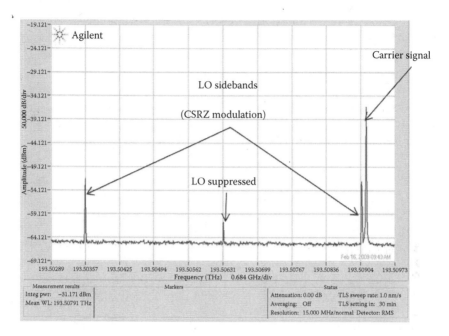

FIGURE 5.20
Spectrum of double sideband LO carrier and by CSRZ and channel carrier as observed by the high-resolution optical spectrum analyzer with vertical axis 5 dB/div and horizontal axis 684 MHz/div.

double sideband frequency doubling of the excitation sinusoidal signal is obtained. Figure 5.21 then shows near-locked and complete locking of the LO and the carrier.

The phase error of the OPLL and the variation of the ratio τ_2/τ_1 are also measured with respect to the variation of τ_2, as shown in Figure 5.22, in which a minimum phase error is observed at a time constant of about 0.425 μs.

5.3.3 Simulation and Experiment Test Bed: Analog Feedback Control

5.3.3.1 Simulation: Analog Feedback Control Loop

A MATLAB Simulink model was created to gain familiarity with the behavior of the loop. It allowed points of interest in the loop to be examined, thus allowing the investigation of loop parameters and their effect on loop performance. This Simulink model also allowed us to confirm some of the basic PLL theory discussed in the above section.

The scope of the simulation was limited with preference to building the loop in the lab and performing "real-life" experiments. As such, noise sources were assumed to be insignificant.

As can be seen from Figure 5.23, the result predicted in Section 5.2.1 is confirmed, that is, the loop will track out any step change in phase.

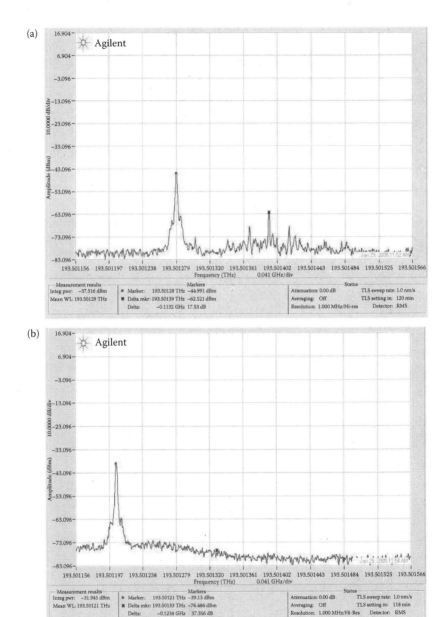

FIGURE 5.21
(a) The frequency of carrier signal and LO before locked. LOFO (LO frequency offset) is 193.501197 THz observed on a high-resolution optical spectrum analyzer (Agilent) at high-resolution mode. Vertical axis 10 dB/div and horizontal axis 41 MHz/div. (b) LOFO is zero—complete locking between LO and carrier.

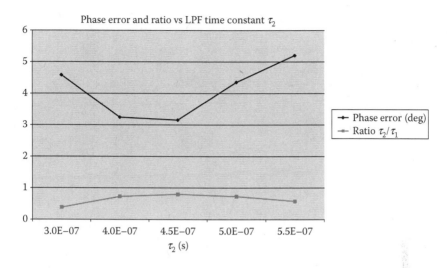

FIGURE 5.22
Phase error and ratio of time constants τ_1 and τ_2 of LPF.

The response to a frequency step predicted in Section 5.2.1 is confirmed, that is, a steady-state component will remain due to a frequency step (phase ramp). The locking error is also evaluated and monitored as shown in Figure 5.24.

As mentioned in the preceding section, the PLL tends to lock on in quadrature, that is, the Rx and LO signals are 90° out of phase. To demodulate phase-modulated data, the receiver and LO must have the same phase. This is achieved with a device known as a 90° optical hybrid. The properties of which are shown in Figure 5.25.

The results of transmitting a slow-changing chain of pulses can be seen as shown in Figure 5.26, as can be seen in the demodulated data, which closely follows the input data. However, some overshoot is clearly visible, and a noticeable rise time has been introduced. Figure 5.27 highlights these issues, where a small time delay is also visible.

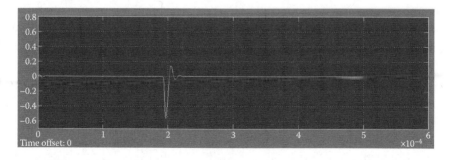

FIGURE 5.23
Error function due to a phase step of $\pi/2$ rad.

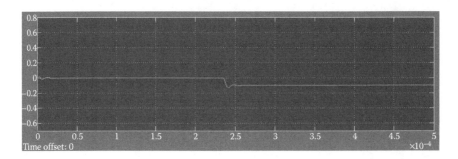

FIGURE 5.24
Error function due to a frequency step of $2\pi \times 10E6$ rad/s.

The design and modeling of the optical PLL have been described in previous sections. This section presents the experimental development of the OPLL. Figure 5.16 depicts the schematic of the test-bed of the OPLL. Two laser sources used in the experimental set shown in Figure 5.16 are Radians Innova [11] and Photonetics Tunics [12], the artificial carrier of received optical signals and the LO, respectively. The lasers are tunable external cavity-type with narrow line widths, approximately a few hundred megahertz or narrower. Although their actual line widths are not measured, they are expected at about 100 kHz line widths. Figure 5.28 illustrates the laser sources, the one on the left side is Photonetics Tunics, and the black box laser source on the right side is Radians Innova. Note that the Photonetics Tunics is selected due to its high power output of up to 5 dBm. The polarization of this laser can be rotated to match with that of the Ti:diffused $LiNbO_3$ channel waveguide of the optical modulator. The Photonetics Tunics laser source is used as the LO whose power suffers loss through propagation along the optical polarization controller and optical modulator (at least 6–10 dB). In the system, the insertion loss of the coupler, polarization controller, and the intensity modulator are measured at 3 dB, 1 dB, and 4 dB, respectively.

The spectrum of output lightwaves of the LO would exhibit two sidebands and the central carrier, normally about 10 dB above the optical sideband. Thus carrier suppression and single sideband (SSB) generation in the optical modulator may be necessary. In this experiment, the modulator is biased at

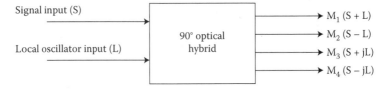

FIGURE 5.25
90° optical hybrid.

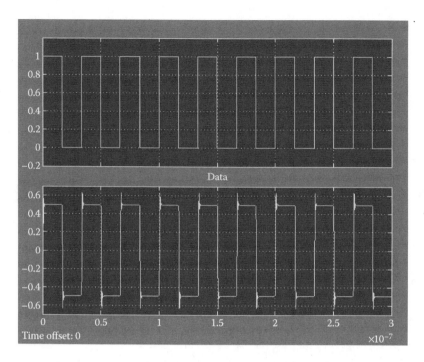

FIGURE 5.26
Transmitted data stream (top) and demodulated data (bottom). Modulation format is binary phase shift keying (BPSK).

the minimum transmission point and the voltage swing is moderate, thus carrier suppression double sideband (DSB) is used. The value V_π of the modulator is 3.3 V.

The lightwaves from the two lasers are mixed via the optical fiber coupler and then shined on a photodetector and optical receiver of sensitivity 0.6 A/W. The consequent electrical voltage output is then filtered by the active low pass filter, which has a cutoff frequency of 500 kHz and DC gain of 20 dB. After filtering, the electrical signal will drive the VCO. Signal output of the VCO, with a power of 3 dBm, is amplified by RF power amplifier. Then, the amplified signal will drive the optical intensity modulator. The optical PLL is a feedback loop so that any error in frequency and phase between laser A and laser B (refer to Figure 5.16) is responded to the loop and corrected by the VCO. The settings of the sub-v-systems of the OPLL are shown with corresponding output power, sensitivity, and gain/loss in Table 5.4. Figure 5.29 shows the overall bench-top set-up of the OPLL integrated in a coherent receiver.

The LO is modulated through the MZDI intensity modulator [13]. The frequency of the modulated sideband is controlled by the Mini-Circuits electrical VCO [14]. To correct the frequency and phase error between the LO and the Rx, they are mixed by 3 dB coupler. The product signal is then received

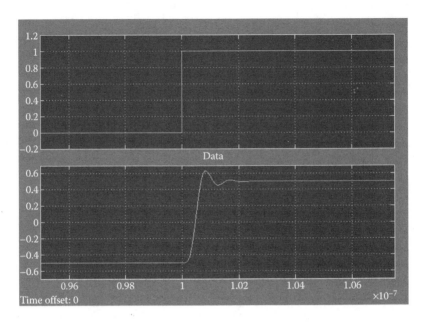

FIGURE 5.27
Close up of a rising edge from Figure 5.26.

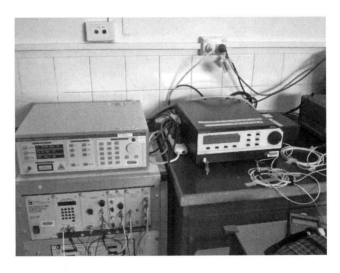

FIGURE 5.28
Laser sources. The white laser power box is used as an LO, and the black one is assumed as received optical signal.

TABLE 5.4

Configuration Settings of the Components

Component	Power Output	Gain/Loss	Sensitivity	Insertion Loss
Laser A	−1.5 dBm			
Laser B	5 dBm			
3 dB coupler (50:50)		$0.35 \cos(\omega_{LO} - \omega_s - \omega_{RF})t$		3 dB
Optical intensity modulator				6 dB
Polarization controller				0 dB
Power amplifier	21 dB			
Optical Rx			0.6 A/W—responsivity	
Active filter			Cut-off 3 dB point at 500 kHz gain 25 dB	
VCO			3 dBm at 0–5 V 100 MHz/V	

by the BT&D optical receiver [15]. The electrical output of the optical receiver is filtered by an appropriate loop filter. After filtering, the electrical signal will drive the electrical VCO.

Before closing the OPLL, we drive the VCO with a sinusoidal voltage from the output of a function generator so that the VCO output frequency can be tuned manually. The spectral property of the receiver output is also observed by an RF spectrum analyzer.

The following sections describe our procedures to achieve the locking of the optical phase by the OPLL. First, two laser sources are selected and

FIGURE 5.29
Experimental OPLL test-bed setup.

arranged for mix by a 3 dB fiber coupler. The summed beating signal in the photodetector/optical receiver is analyzed by an RF spectrum analyzer. An RF function generator is then used, manually controlled, to lock the frequency of the LO and the carrier of the Rx. Finally, the closed-loop optical PLL with the VCO and the active low-pass filter is implemented. The design of active low-pass filter is also described in this section.

5.3.3.2 Laser Beating Experiments

To observe the frequency components of mixed signal after 3 dB coupler, the Rx in electrical domain is analyzed by a 22 GHz RF spectrum analyzer.

Figure 5.30 shows the spectrum of beating signal, with a frequency of 700 MHz. First, the VCO is not implemented in the system; the driving frequency of the modulator is manually tuned by the function generator. Its bandwidth is up to 3 GHz. An oscilloscope is also used to observe the beating signal in the time domain. However, the bandwidth of the oscilloscope in the systems is limited at 150 MHz; it is thus essential to manually tune the function generator to reduce the beating frequency to the bandwidth of the oscilloscope. The result is observed in Figure 5.31.

By tuning the function generator carefully, the frequency of the LO is approximately locked into the carrier frequency of the Rx. Figure 5.32 shows the result observed in the RF spectrum analyzer. According to Figure 5.32, the Rx at 1.1 GHz is eliminated and shifted back to baseband. It means that the frequency of the LO is almost locked into the carrier frequency of the received optical signal.

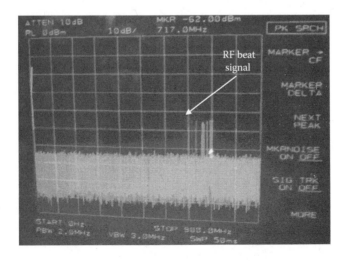

FIGURE 5.30
RF beat signal at 700 MHz, observed in RF spectrum analyzer (Model HP-8591E). Vertical axis: 10 dB/div, horizontal axis: center frequency 450 MHz span 900 MHz.

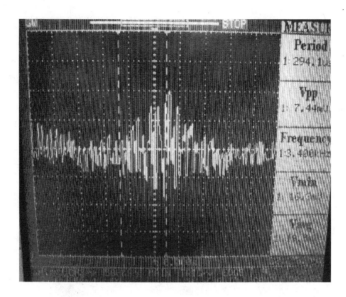

FIGURE 5.31
Screenshot of beating signal, observed in the time domain—oscilloscope model GDS-820C.

5.3.3.3 Loop Filter Design

The loop filter is built as described in previous sections. Figure 5.33 illustrates the loop filter we use for the optical PLL. The resistors are variable so that cutoff frequency and DC offset are able to tune. Because the circuit is built on prototype board, their bandwidth is limited at about 1 GHz. However, the operation amplifier limits itself at 10 MHz bandwidth. It is a bit different

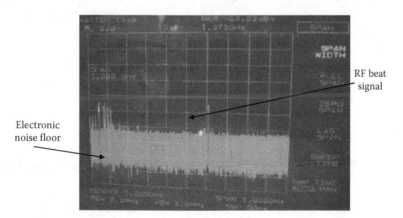

FIGURE 5.32
The frequency of the LO is almost locked into the carrier frequency of received optical signal by manually tuning the function generator as observed by HP spectrum analyzer (Model HP-8591E). Vertical axis: 10 dB/div, horizontal axis: 100 MHz/div, center frequency 1.0 GHz.

FIGURE 5.33
Active loop filter built on a typical prototype board to show low-frequency operation.

from theoretical design. In the optical PLL, the loop filter is meant to have a large amplification at frequencies close to DC. It is thus useful to put additional resistors in parallel with a positive feedback arm. Hence, according to Section 5.2.1, when the frequency is close to DC, the capacitance becomes open-circuit, so the gain factor is generally calculated as R_3/R_1.

The most important feature of an active loop filter compared to a passive filter is that the active loop filter is able to have significant DC offset. This DC voltage is essential to tuning the frequency of the VCO. General operation of the optical PLL is that the frequency of the LO is tuned such that it is locked to the carrier frequency of the Rx. Once this process finishes, the feedback loop will be able to recognize the phase error between the LO and the received optical signal. It is then corrected by the VCO.

5.3.3.4 *Closed-Loop Locking of LO and Signal Carrier: Closed-Loop OPLL*

In the previous section, the tuning of the shifting of the frequency/wavelength of the LO is manually controlled by tuning the RF signal generator. In this section, a closed-loop locking of the frequencies of the lasers is implemented. The RF signal generator is removed from driving the optical modulator. Instead, the output sinusoidal waves of the LPF and the VCO are now used to drive the modulator. Effectively, a closed-loop optical RF is formed as an adaptive OPLL.

The locking of the two lasers can be observed by two methods: (i) monitoring via an electrical spectrum analyzer to the beating of two optical signals (an indirect method) and (ii) monitoring the real optical spectra of two lasers at the output of the fiber coupler using a high-resolution spectrum analyzer (a direct method).

5.3.3.5 Monitoring of Beat Signals

An RF spectrum analyzer is connected to the optical receiver. When the frequency of one sideband of the laser B comes closer to the carrier frequency of laser A, on the RF spectrum analyzer, beat signal frequency will come closer to 0 Hz. Once the frequency is locked, beat signal frequency is eliminated, with the result observed in Figure 5.34.

The time domain oscilloscope is connect to the VCO to observe the VCO driving signal. As soon as the loop is closed, the frequency error is corrected by the VCO. After locking the frequency, the phase errors can be responded to by voltage shots. These shots are to change the frequency of the VCO so that phase error can be corrected. The result is observed in Figure 5.35.

The cutoff frequency of the active loop filter is 500 kHz in order to ease the electrical noise of the optical receiver. To observe the noise of the optical

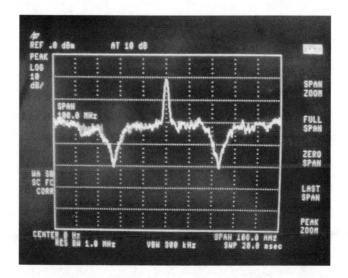

FIGURE 5.34
The frequency of laser B is clocked to the carrier frequency of laser A. RF spectrum analyzer (Model HP-8591E). Vertical axis: 10 dB/div; horizontal axis: center frequency 0 Hz, span 100 MHz.

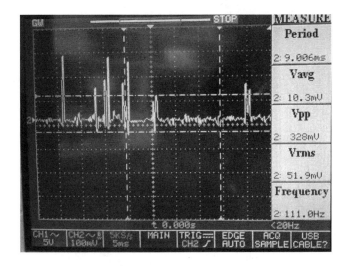

FIGURE 5.35
The voltage signal that drives the VCO, observed in time domain, oscilloscope (Model GDS-820C) channel 2, coupled with AC. Vertical axis: 100 mV/div; horizontal axis: 5 ms/div.

receiver, the oscilloscope is connected to the RF output port of the receiver and the optical input port is not connected. According to Figure 5.36, the noise voltage, which is about 550 mV, contributes to the driving voltage of the VCO. It is thus about 55 MHz error due to the noise of the receiver (sensitivity of the VCO is 100 MHz/V; see Table 5.4).

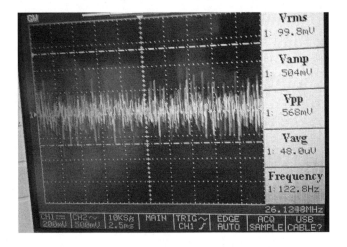

FIGURE 5.36
Electrical noise of the optical receiver, observed in oscilloscope. Channel 1, coupled with DC. Vertical axis set at 200 mV/div and horizontal axis at 2.5 ms/div.

5.3.3.6 High-Resolution Optical Spectrum Analysis

A high-resolution optical spectrum analyzer is connected to the fiber coupler. The carrier signal is tuned such that the frequency difference between the carrier signal and one sideband of the LO is less than 400 MHz. Figure 5.37 shows the spectrums of LO and carrier signal. The carrier of LO is suppressed to about 13 dB less than its sideband. Figure 5.38 shows the spectrums of the carrier signal and the LO under high-resolution operation. Their frequency difference is less than 120 MHz. Then, the OPLL is closed and the LO sideband is locked to the carrier signal (observed in Figure 5.39). Under lab conditions, the OPLL is able to maintain the lock for approximately 30 min.

5.3.3.7 Phase Error and LPF Time Constant

Referring to the time constants τ_1 and τ_2 of R_1C and R_2C, respectively according to Equation 5.3, changing the ratio of τ_1 and τ_2 will change the cutoff frequency of the active low-pass filter, so that the values of R_1 and R_2 are changed by tuning manually. The phase error is calculated as below [16]:

$$\sigma_{PE} = \frac{V_{RMS}}{K_d} \tag{5.24}$$

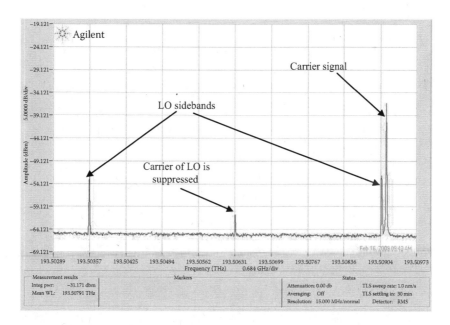

FIGURE 5.37
Spectrum components of LO and the carrier signal. Observed on a high-resolution optical spectrum analyzer (Agilent) wide mode. Vertical axis: 5 dB/div; horizontal axis: 684 MHz/div.

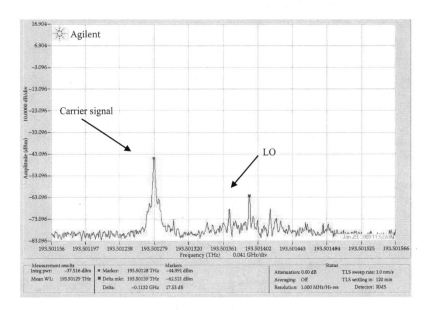

FIGURE 5.38
The frequency of carrier signal and LO before locked. Frequency error is 113.2 MHz. Observed on high-resolution optical spectrum analyzer (Agilent) high-resolution mode. Vertical axis: 10 dB/div; horizontal axis: 41 MHz/div.

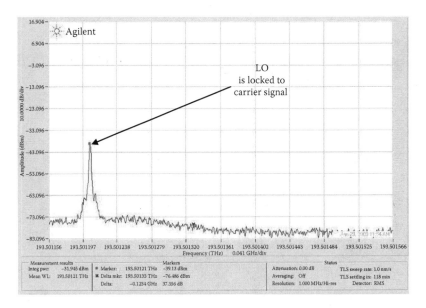

FIGURE 5.39
The frequency of the LO is locked to carrier frequency of the signal. Observed on a high-resolution optical spectrum analyzer (Agilent) high-resolution mode. Vertical axis: 10 dB/div; horizontal axis: 41 MHz/div.

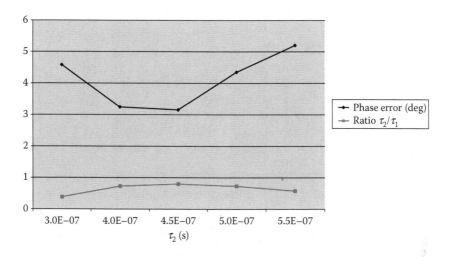

FIGURE 5.40
Phase error (deg) and ratio τ_2/τ_1 versus time constant τ_2.

where V_{RMS} is the RMS voltage measured at the electrical output port of the optical receiver. σ_{PE} is the standard deviation of the phase error. The variation of the phase error is plotted against the measured time constants of the LPF, plotted in Figure 5.40. According to Figure 5.40, as the ratio τ_2/τ_1 approaches 1, the phase error also decreases, agreeing with work reported in Ref. [16].

5.3.3.8 Remarks

Initial experimental results have demonstrated that they are in agreement with the modeling. Locking of the LO and the signal carrier can be tracked and converged by the LP-VCO and optical modulator loop. The time constant of the LPF is critical and further works are required in extending the locking range of the difference between the two laser sources.

The success of the locking loop for OPLL will influence our experimental demonstration of phase shift keying transmission in long-haul optically amplified fiber systems, which will be conducted in the near future.

The remaining issues/problems to be resolved are

- The rise time and fall time, as well as the roll off of the loop filter and effects on the phase locking
- The phase noises of the LO and its impact on the phase locking
- The quantum shot noises of the receiver due to high-level power of the LO; a noise analysis of the receiver should be done

5.4 OPLL for Superchannel Coherent Receiver

Superchannels are channels consisting of a number of subchannels that carry information data streams such that the total bit rate of all the subchannels would be equivalent to at least 1 Tbp. Naturally, the number of subchannels must be greater than 1, as at this rate the electronic and digital circuits would not be operating due to limited electronic speed. Thus, modulation and optical techniques must be employed as described in Chapter 3. The most common techniques are high-level QAM, such as QPSK, 16 QAM, OFDM, and so on, and for optical waves, the polarization and pulse shaping such as Nyquist filtering either in the electronic domain or optical domain by optical filtering.

At the superchannel receivers employing DSP-based coherent reception, there needs to be several LOs that can match with those of the subchannels so that they can be demultiplexed, and then mixed in the photodetectors to produce analog electric signals to be sampled by the ADC and then digitally processed in the DSP. Figure 5.41 shows the schematic of the optical reception subsystems employing an OPLL LO in association with a superchannel comb generator for producing several equally spaced LOs and mixing with the subchannels of the superchannels. Figure 5.42 gives details of the optical phase locked loop to lock the carrier of a subchannel of a superchannel and then lock it to the LO, thence using this locked source feeding into a comb generator to the number of sub-LO carriers, which are then fed into the phase and polarization diversity hybrid coupler. The outputs of the polarization and phase diversity coupler are then entered into a bank of photodetectors in which the currents produced are proportional to the inphase and quadrature components of the subchannels.

Figure 5.43 shows the parallel processing of the detected subchannels in the analog domain and then sampled by the bank of ADC to obtain a digitalized

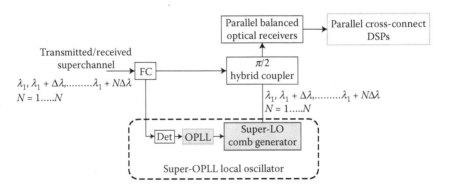

FIGURE 5.41
Schematic of generation of comb source for acting as LO of superchannel.

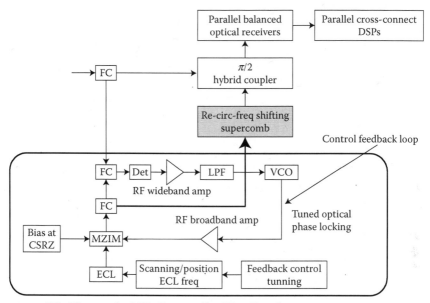

FC = Fiber coupler, LPF = Lowpass filter, PD = Photodetector,
VCO = Voltage-controlled oscillator, MZIM = Mach–Zehnder Int. Mod.
ECL = External cavity laser

FIGURE 5.42
Super LO using OPLL and comb generation.

sampled sequence of the subchannels. Interconnecting DSP processors then process the digital sampled sequences of subchannels.

Having described the optical phase locking techniques for generation of the LO for single channel, in the case of the superchannel, if a comb line LO can be generated and matched to the received subcarriers of the subchannels

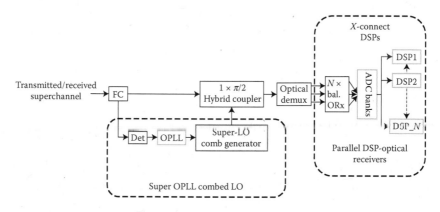

FIGURE 5.43
Super LO- and DSP-based superchannel coherent receiver system.

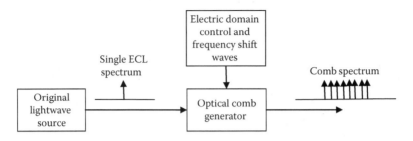

FIGURE 5.44
Schematic block diagram of a generic comb generator.

of the superchannel, then this would reduce significantly the complexity of the receiving system. Regarding the comb lightwave source for coherent detection of superchannels, combed optical source can be defined as a source in which the spectrum consists of equally spaced optical lines or subcarriers generated from an original narrow laser source. A typical schematic of a generic comb generator can be shown in Figure 5.44, in which there must be a lightwave source with a narrow-band single-frequency spectrum [17–27]. The comb generator receives the RF electric signals whose frequency determines the spacing of the comb-generated lines. The frequency shifter uses electro-optic or other trans-electric to optical conversion to translate the RF waves into the optical domain, and a frequency shift occurs. The frequency shifting can be a recirculating optical loop in which the lightwave is frequency shifted once every time they pass through the frequency shifter, hence multiple comb lines, or a nonlinear-conversion optical device through which several harmonics of the fundamental RF component can be generated. The frequency spacing in the comb lines can thus be adjusted to match with the spacing of the subchannels of the superchannel. The main matter is the matching of a reference frequency between the received channels and the first referenced comb line.

5.5 Concluding Remarks

Simulated results are presented supporting the operation of the "smart" digital LPF leading to the physical implementation of such a filter. An experimental OPLL is demonstrated for further exploration of the LPF and to confirm its "real-life" viability. Various design considerations pertaining to the hardware implementation of the LPF have been discussed, leading to the development of an interface board between the "smart" filter implemented onboard an FPGA and the OPLL. This is included in this chapter as a case study for generation of a LO for an intradyne reception system, especially the superchannel coherent receiver.

Experimental results have been presented displaying the correct operation of the LPF, and ultimately it can be concluded that a successful "proof of concept" was developed with the modified OPLL-maintained locking state for at least 1 h with limited spurious beats in the locking state. Further work remains to incorporate the OPLL into a coherent receiver and perform data transmission experiments.

Ultimately, an OPLL is of little use on its own, but only if it is incorporated into a coherent receiver to lock with the signal carrier and performing decoding experiments with phase encoded data. Any implementation using an analog filter is not very stable, and the LO frequency offset (LOFO) may wander over significant periods of time. Thus, a gap in the literature exists for data transmission employing a digital LPF.

References

1. L. Kazovsky, Balanced phase-locking stateed loops for optical homodyne receivers: Performance analysis, design considerations, and laser linewidth requirements, *IEEE J. Lightwave Technol.*, 4, 182–195, 1986.
2. L. G. Kazovsky, Decision-driven phase-locking stateed loop for optical homodyne receivers: Performance analysis and laser linewidth requirements, *IEEE Trans. Electron Dev.*, 32, 2630–2639, 1985.
3. S. Camatel and V. Ferrero, Homodyne coherent detection of ASK and PSK signals performed by a subcarrier optical phase- locking stateed loop, *IEEE Photonics Technol. Lett.*, 18, 142–144, 2006.
4. L. G. Kazovsky and D. A. Atlas, A 1320-nm experimental optical phase-locking stateed loop: Performance investigation and PSK homodyne experiments at 140 Mb/s and 2 Gb/s, *IEEE J. Lightwave Technol.*, 8, 1414–1425, 1990.
5. S. Camatel, V. Ferrero, R. Gaudino, and P. Poggiolini, Optical phase-locking stateed loop for coherent detection optical receiver, *Electron. Lett.*, 40, 384–385, 2004.
6. F. M. Gardner, *Phase Locking State Techniques*, 2nd ed., Wiley-Interscience, Boston, 1979.
7. Y. Linn, Efficient loop filter design in FPGAs for Phase locking state Loops in high-data rate wireless receivers—Theory and case study, in *Wireless Telecommunications Symposium, 2007. WTS 2007*, 2007, pp. 1–8.
8. L. Kazovsky, Decision-driven phase-locked loop for optical homodyne receivers: Performance analysis and laser linewidth requirements, *IEEE J. Lightwave Technol.*, 3(6), 1238–1247, December 1985.
9. S. Camatel and V. Ferrero, Design, analysis and experimental testing of BPSK homodyne receivers based on subcarrier optical phase-locking stateed Loop, *IEEE J. Lightwave Technol.*, 26, 552–559, 2008.
10. V. Ferrero and S. Camatel, Optical phase locking stateing techniques: An overview and a novel method based on single side sub-carrier modulation, *Opt. Express*, 16, 818–828, 2008.
11. Radians Innova Laser Diode, http://www.radians.se/

12. Photonetics Tunics Laser Diode, http://www.surpluseq.com/vdirs/info/Photonetics_Tunics-BT_info.pdf

13. Pirelli Inc., PIRMZM-153X, http://www.pirellilabs.com/web/default.page

14. Mini-Circuits Inc., ZX95–3080-S+, http://minicircuits.com/pdfs/ZX95–3080+.pdf

15. BT&D Technology Inc., RCV1201–1.5.

16. S. Camatel and V. Ferrero, Design, analysis and experimental testing of BPSK homodyne receivers based on subcarrier optical phase-locked loop, *IEEE J. Lightwave Technol.*, 26(5), 552–559, March 1, 2008.

17. B. Noll and F. Auracher, and A. Ebberg, Optical Comb Generator, US Patent No. 5,265,112, issued November 23, 1993.

18. M. Kourogi, W. Bambang, O. Nakamoto, S. Misawa, and Y. Nakayama, Optical frequency comb generator, US Patent No. US 7,239,442 B2, issued July 3, 2007.

19. T. Sakamoto, T. Kawanishi, M. Tsuchiya, and M. Izutsu, Super-stable optical comb and pulse generation using electro-optic modulation, in *IEEE Optical Fiber Conference*, 2007, paper MC3.1.

20. K.-P. Ho and J. M. Kahn, Optical frequency comb generator using phase modulation in amplified circulating loop, *IEEE Photonics Technol. Lett.*, 5(6), 712–725, June 1993.

21. M. Kourogi, K. Imai, and W. Bambang, Optical frequency comb generator and optical modulator, US Patent No. US 7,551,342 B2, issued June 23, 2009.

22. M. Fujiwara, K. Araya, M. Teshima, J. Kani, and K. Suzuki, Multiwavelength generating method and apparatus based on flattening of spectrum, US patent No. 7,068,412 B2, issued June 27, 2006.

23. T. Kawanashi et al. *IEICE Electron. Exp.*, 1(8), 217–231, 2004.

24. J. L. Hall, S. A. Diddams, L.-S. Ma, and J. Ye, Comb generating optical cavity that includes an optical amplifier and an optical modulator, US Patent No. 06201638B1, issued March 13, 2001.

25. I. Morohashi, T. Sakamoto, N. Yamamoto, H. Sotobayashi, T. Kawanishi, and I. Hosako, 1 Thz-bandwidth optical comb generation using Mach-Zehnder-modulator-based flat comb generator with optical feedback loop, in *Optical Fiber Conference 2011*, poster session (JTh4), California, USA, 2011.

26. N. Dupuis, C. R. Doerr, L. Zhang, L. Chen, N. J. Sauer, P. Dong, L. L. Buhl, and D. Ahn, InP-based comb generator for optical OFDM, *IEEE J. Lightwave Technol.*, 30(4), 466–472, 2012.

27. T. Healy, F. C. Garcia Gunning, A. D. Ellis, and J. D. Bull, Multi-wavelength source using low drive-voltage amplitude modulators for optical communications, *Opt. Express*, 15(6), 2981–2986, 2007.

6

Digital Signal Processing Algorithms and Systems Performance

6.1 Introduction

Digital signal processing (DSP) is the principal functionality of modern optical communications systems, employed at both the coherent optical receivers and the transmitter. At the transmitter, the DSP is employed to shape the pulse at the output of optical modulators, and compensate for the limited bandwidth of the digital-to-analog converter (DAC) as well as that of the transfer characteristics of the electro-optic external modulators, deskewing of the electrical path differences between the inphase and quadrature phase signals that occur often in ultra-high speed systems. It is noted that the mixing of the received signals and the local oscillator (LO) laser happens in the photo-detection stage.

Due to the square law process in the optoelectronic conversion, this mixing gives a number of terms; two square terms related to the intensity of the optical signals and the intensity of the LO laser, and two product terms that result from the beating between the optical signal beam and that of the LO. Only the differential product between these two optical frequencies, the signal frequency when the optical frequencies of the signals and the oscillator are the same, homodyne coherent detection. Thus, provided that perfect homodyne mixing is achieved in the photo-detection device, the detected signals, now in electrical domain, are in the base band superimposed by noises of the detection process. Thence after electronic preamplification and then sampled by an analog-to-digital converter (ADC), the signals are in the discrete domain and thus processed by digital signal processors, which can be considered to be similar to processing in wireless communications systems. The main difference is the physical processes of distortion, dispersion, nonlinear distortion, and clock recovery; broadband properties of optical modulated signals transmitted over long-haul optical fibers or short-reach scenarios.

On the other hand, at the receiver end, DSP follows the ADC with the digitized symbols and processes the noisy and distorted signals with algorithms to compensate for linear distortion impairments such as chromatic dispersion (CD), polarization mode dispersion (PMD), nonlinear self-phase modulation effects (SPM), clock recovery, and LO frequency offset (LOFO) with respect to the relieved channel carrier, among others. A possible flow of sequences in DSP can be illustrated in Figure 6.1 for modulated channel with the inphase and quadrature components and polarization multiplexing as recovered in the electronic domain after the optical processing front end as shown at the beginning,

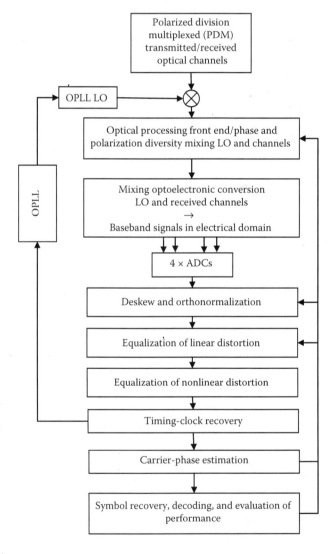

FIGURE 6.1
Flow of functionalities of DSP processing in a QAM-coherent optical receiver with possible feedback control.

the input boxes. Thus there are two pairs of electrical signals output from the two polarized channels and two for the inphase and quadrature components. Interprocessing of these pairs of signals can be implemented using 2×2 multiple input multiple output (MIMO) techniques, which are well known in the arena of digital processing of wireless communications signals [1].

After the opto-electronic detection of the mixed optical signals, the electronic currents of the PDM channels are then amplified with a linear transimpedance amplifier and a further amplification main stage if required to boost the electric signals to the level at which the ADC can digitalize the signals into their equivalent discrete states. Once the digitalized signals are obtained, the first task is to ensure that any delay difference between the channels and the inphase and quadrature components in the electronic and digitalization processes are eliminated by de-skewing. Thence all the equalization processing can be carried out to compensate for any distortion of the signals due to the propagation through the optical transmission line. Furthermore, nonlinear distortion effects on the signals can also be superimposed at this stage. Then the clock recovery and carrier phase detection can be implemented. Hence the processes of symbol recovery, decoding, and evaluation of performance can be implemented.

Regarding noises in the optical detection, coherent type then the noises mainly dominated by the shot noises generated by the LO power. However, if it is direct detection type then the quantum shot noises contributed by the signal power dominate. These noises will contribute to the convergence of algorithms.

This chapter is thus organized as follows. Section 6.2 gives a basic background of equalization using transversal filtering and zero-forcing with and without feedback; in the linear sense, the linear equalizer. When a decision is made in a nonlinear sense under some criteria to feedback to the input sequence, then the equalization process is nonlinear and the equalizer is classified as a nonlinear or decision-directed equalizer. Tolerance to noises is also given and simple quality factor of the system performance is deduced.

Clock recovery is important to determine the timing for sampling of the received sequence and the technique for obtaining this is also briefly given. Determining the reference phase for coherent reception is also critical so as to evaluate several quadrature-modulated transmitted signals. Thus carrier phase recovery is also described using the DSP technique, especially in the case of when there is significant offset between the LO and the lightwaves carrying the signals. This is described in Section 6.5.

When the maximum likelihood sequence estimation (MLSE) is used and the nonlinear equalizer is employed, the equalization can be considered optimal but consuming substantial memory. Section 6.3 is dedicated to this algorithm and illustrates an example of this MLSE. This section demonstrates the effectiveness of MLSE in MSK self-homodyne coherent reception transmission systems when under the influence of linear distortion effects with the minimum shift keying (MSK) modulation scheme. This is presented in Section 6.6.

TABLE 6.1

Functionalities of Subsystem Processing in a DSP-Based Coherent Receiver

De-skewing	Temporary alignment of inphase and quadrature components or polarized channel components or any different propagation time due to electrical connections
Orthogonalization	Ensuring independence or decorrelation between channels
Normalization	Standardized amplitude of components to the maximum value of unity
Equalization	Compensation of impairments due to physical effects or imperfections of sub-systems
Interpolation	Correction of timing error
Carrier phase estimation	Compensation of phase errors of the carrier
Frequency offset	Correction of the offset frequency between LO and carrier frequencies
OPLL	Optical phase locked loop to ensure matching between channel lightwave carrier and that of the LO carrier (see the discussion on locking mechanism given in Chapter 5)

We thus can see that DSP algorithms are most critical in modern DSP-based optical receivers or transponders. This chapter introduces the fundamental aspects of these algorithms for optical transmission systems over dispersive channels, whether under linear or nonlinear distortion physical effects.

The functionalities of each processing block in Figure 6.1 can be categorized as shown in Table 6.1 [2].

6.2 General Algorithms for Optical Communications Systems

Indeed, in coherent optical communication systems incorporating DSP, the reception of the transmitted signals is implemented by mixing them with an LO whose frequency is identical or close to equality of the channel carrier, namely homodyne or intradyne techniques, respectively. This mixing can convert the optical signals back to the baseband electrical domain with preserved phase and amplitude states. One can thus apply a number of processing algorithms developed for high-speed modems [3] or algorithms for wireless communication systems [4].

Naturally, a number of steps of conversions from optical to electrical and vice versa at the receiver and transmitter have to be conducted with further optical amplifier noises accumulated along the optical transmission line and electronic noises.

In this chapter we assume that the signal level is quite high compared with the electronic noise level at the output of the electronic preamplifier, which follows the photodetector. Hence, the processing algorithms are conducted in the baseband electrical digital domain after the ADC stage.

6.2.1 Linear Equalization

Coherent optical communications systems can be categorized as synchronous serial systems. In transmitting optical signals over uncompensated optical fiber spans over a long-haul multi-span link or uncompensated metropolitan networks at very high bit rates, commonly now in the second decade of this century, at 25 GS/s or 28–32 GSy/s including forward error coding overloading, the received signals are distorted due to linear and nonlinear dispersive effects, hence leading to intersymbol interference (ISI) in the baseband after the coherent homodyne detection at the receiver. In Chapter 4 we describe in detail the processes of coherent detection, especially homodyne reception for optical signals, including noise processes. Over the non-dispersion compensating multi-span fiber links the most important signal distortions are the linear CD and PMD, especially when PDM is employed with two polarization orthogonal channels that are simultaneously transmitted. The ISI naturally occurs due to the band limiting of the signals over the limited bandwidth of the long fiber link as described in Chapter 2 on the transfer function of optical fibers. Other nonlinear distortion effects have also been included in the fiber transfer function whose passband is not even and narrower. These ISI effects will degrade the performance of the optical communications and need to be tackled with equalization processes discussed in this chapter.

The optical coherent detection processes and DSP of sampled digital signals may be classified into two separate groups.

The first of these, the received sampled digital signals, are fed through an equalizer that corrects the distortion introduced by the CD and PMD effects, and restores the received signals into a copy of the transmitted signals in the electrical domain. The received signals are then detected in the conventional manner as applied to any serial digital signals in the absence of ISI. In other words, the equalizer acts as the inverse of the channel so that together the equalizer and the channel introduce no signal distortion, and each data symbol is detected as it arrives, independently of each other. The equalizers can be either linear or nonlinear.

In the second group of detection processes, the decision process is modified to take into account the signal distortion that has been introduced by the channel, and no attempt need, in fact, be made here to reduce signal distortion prior to the actual decision process. Although no equalizer is now required, the decision process may be considerably complex, and hence complex algorithms to deal with this type of decision processing.

In addition to these two decision groups, there are additional issues to resolve involving the synchronization of the sampling at the receivers, the clock recovery and carrier phase estimation, so that a reference clocking instant can be established and the referenced phase of the carrier can be used to detect the phase states of the received samples. Concurrently, any skew or delay differences between the I- and Q-components as well as the polarized multiplexed channels must be detected and deskewed so that synchronous processes can be preserved in the coherent detection processes. These physical processes, although by software control remain in practical implementation, are not included in this book.

Furthermore, for any delay difference due to different propagation time of the polarized channels or electrical connection lines, the de-skewing process is to be conducted to eliminate this effect, as indicated in Figure 6.1.

Thus, in this section the LE process is introduced.

6.2.1.1 Basic Assumptions

In general, a simplified block diagram of the coherent optical communication system can be represented in the baseband, that is, after optical field mixing and detection by opto-electronic receivers and E-O conversion at the transmitter, as shown in Figure 6.2. The input to the transmitter is a sequence of regularly spaced impulses, assuming that the pulse width is much shorter than that of the symbol period, which can be represented by $\sum_i s_i \delta(t - iT)$, with the s_i of binary states taking both positive and negative values as the amplitude of the input signal sequence, thus we can write

$$s_i = \pm k; \quad k > 0 \tag{6.1}$$

The impulse sequence is a binary polar type. This type of signal representation is advantageous in practice, as it can be employed to modulate the optical modulators, especially the Mach–Zehnder interferometric modulator

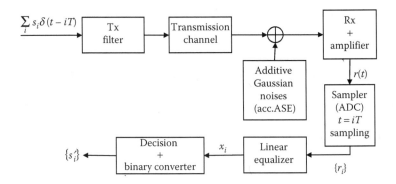

FIGURE 6.2
Simplified schematic of coherent optical communication system. Note equivalent baseband transmission systems due to baseband recovery of amplitude and phase of optically modulated signals.

(MZIM) or I–Q modulator, as described in Chapter 3, so that the phase of the carrier under the modulated envelope can take positive or negative phase angles, which can be represented on the phasor diagram. In practice, naturally the impulse can be replaced by a rectangular, rounded, or Gaussian waveform by modifying the transmitter filter, which can be a Nyquist-raised cosine type to tailor the transmitting pulse shape.

It is further noted that the phase of the optical waves and the amplitude modulated by the I- and Q-MZIM in the two paths of the I–Q modulator ensure the negative and positive position of the I- and Q-components on the real and imaginary axes of the constellation, as shown in Figure 3.42 (Figure 42 of Chapter 3, 4-ASK-1 and 4-ASK-2). Under coherent reception these negative amplitudes are recovered, as they are in the transmitter and with superimposed noises. We could see that the synchronization of the modulation signals fed to the two MZIM is very critical in order to avoid degradation of the sampling instant due to skew.

6.2.1.2 Zero-Forcing Linear Equalization (ZF-LE)

Consider a transmission system and receiver subsystem as shown in Figure 6.3 in which the transmitter transfer function can be represented by $H_{Tx}(f)$, the transfer function of the channel in linear domain $H_C(f)$, and the transfer function of the equalizer or filter $H_{Eq}(f)$, which must meet the condition for an overall transfer function of all subsystems equating to unity.

$$H(f) = H_{Tx}(f)H_C(f)H_{Eq}(f)e^{j\omega t_0} \tag{6.2}$$

Thus from Equation 6.2 we can obtain, if the desired overall transfer response follows that of a Nyquist raise-cosine filter, the equalizer frequency response of

$$H_{Eq}(f) = \frac{raise\_\cos\left(\dfrac{f}{1/T}, \rho\right)}{H_{Tx}(f)H_C(f)}e^{j\omega t_0} \tag{6.3}$$

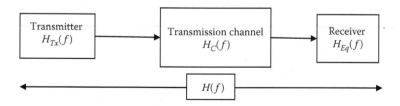

FIGURE 6.3
Schematic diagram of a transmission system including a zero-forcing equalizer/filter.

With the raised cosine function defined with β as the roll-off factor and the frequency response of the equalizer

$$H_{Eq}(f) = raise\_cos\left(\frac{f}{1/T}, \beta\right) = \begin{cases} 1 & 0 \le |x| \le \frac{1-\beta}{2} \\ \cos^2\left(\frac{\pi}{2}\frac{|x| - \frac{1-\beta}{2}}{\beta}\right) & \frac{1-\beta}{2} \le |x| \le \frac{1+\beta}{2} \\ 0 & |x| > \frac{1+\beta}{2} \end{cases}$$

(6.4)

assuming that the filter is a Nyquist filter of raised cosine shape, ω is signal baseband frequency, and t_0 is the time at the observable instant. In the case of two channels being observed simultaneously, the right-hand side of Equation 6.2 is a unity matrix and the unitary condition must be satisfied, that is, the conservation of energy.

From Equation 6.3 the magnitude and phase response of the equalizer can be derived. This condition guarantees a suppression of the ISI, thus the equalizer can be named as a linear equalizer zero-forcing (LE-ZF) filter.

6.2.1.3 ZF-LE for Fiber as a Transmission Channel

The ideal zero-forcing equalizer is simply the inverse of the channel transfer function given in Equation 6.2. Thus, the required equalizer transfer function $H_{Eq}(f)$ is given by

$$H_{Eq}(f) = H_{SSMF}^{-1}(f) = e^{j\alpha\omega^2}$$

(6.5)

where ω is the radial frequency of the signal envelope and α is the parameter dependent on dispersion, wavelength, and the velocity of light in a vacuum as described in Chapter 4. However, this transfer function does not maintain the conjugate symmetry, that is

$$H_{Eq}(f) = e^{j\alpha\omega^2} \ne H_{Eq}^{-1}(f)$$

(6.6)

This is due to the square-law dependence of the dispersion parameter of the fiber on frequency.

Thus, the impulse response of the equalizing filter is complex. Consequently, this filter cannot be realized by a baseband equalizer using only one baseband received signal, which explains the limited capability of LEs employed in direct-detection receivers to mitigate the CD. On the other hand, it also explains why fractionally spaced LEs that are used within a coherent receiver can potentially extend the system reach to distances that are only limited by the number of equalizer taps [5]. Under coherent detection and particularly the quadrature amplitude modulation (QAM), both real and imaginary parts are extracted separately and processed digitally. The schematic of a fractionally spaced finite impulse response (FIR) filter is shown in Figure 6.4, in which the delay time between taps is only a fraction of the bit period. The weighting coefficients $W_0(n)$ ... $W_{N-1}(n)$ are determined and then multiplied with the signal paths $S(n)$ at the outputs of the taps. These outputs are then summed up and subtracted by the expected filter coefficients $d(n)$ to obtain the errors of the estimation. This process is then iteratively operated until the final output sequence $Y(n)$ is close to the desired response.

Under ZFE, at the sampling instant the noise may have been increased due to errors in the delay of the zero crossing point of adjacent pulses, hence ISI effects. Thus the signal-to-noise ratio (SNR) may not be optimal.

The number of taps must cover the whole length of the dispersive pulses; for example, for a 1000 km SSMF transmission line and BW channel of 25 GHz (or ~0.2 nm at 1550 nm), the total pulse spreading of the pulse is 17 ps/nm/km × 1000 km × 0.2 nm = 3400 ps. The sampling rate would be 50 GSa/s (2 samples per period at 25 Gb/s, assuming modulation that results in the BW equal to symbol rate), and one would have 3400/40 period or ~85 periods or taps with 170 samples altogether.

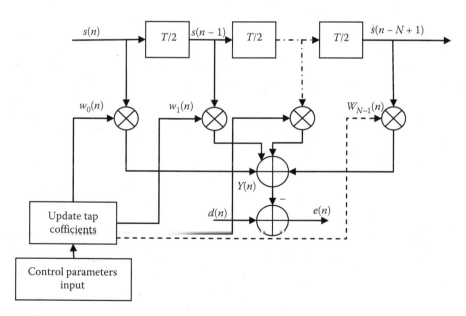

FIGURE 6.4
Schematic of a fractionally spaced FIR filter.

6.2.1.4 Feedback Transversal Filter

Transversal equalizers are a type of linear feedforward (FF) transversal filters commonly employed in practice over several decades since the invention of g = digital communications. Equalization may alternatively be achieved by a feedback transversal filter, provided that certain conditions can be met by the channel responses, in this case the linear transfer function of the fiber transmission line.

The z-transform of the sampled impulse response of the channel, the fiber field propagation, and coherently detected, are represented by

$$Y(z) = y_0 + y_1 z^{-1} + y_2 z^{-2} + \cdots\cdots + y_g z^{-g} \tag{6.7}$$

where $y_0 \neq 0$ and $V(z)$ is the z-transform derived from the scaling of $Y(z)$ so that the first term becomes unified. Thus, we have

$$V(z) = \frac{1}{y_0} Y(z) = 1 + \frac{y_1}{y_0} z^{-1} + \frac{y_2}{y_0} z^{-2} + \cdots\cdots + \frac{y_g}{y_0} z^{-g} \tag{6.8}$$

or written in cascade factorized form,

$$V(z) = \left(1 + \beta_1 z^{-1}\right)\left(1 + \beta_2 z^{-1}\right)\cdots\cdots\left(1 + \beta_g z^{-1}\right) \tag{6.9}$$

with β_i taking either real or complex value. The feedback transversal filter can then be configured as shown in Figure 6.5, which depicts the operation just described. Obviously the received sequence is multiplied by a factor $1/y_0$ in order to normalize the sequence, is delayed by one period with a multiplication ratio then summed up, and with the result is subtracted by the normalized incoming received sequence. The output is taped from the feedback path as shown in the diagram, at the point where the transversal operation is started.

6.2.1.5 Tolerance of Additive Gaussian Noises

Consider a linear FFE with $(m + 1)$ taps as shown in Figure 6.6. Assuming that the equalizer has equalized the input signals in the baseband, the output of the equalizer is given as

$$x_{i+h} \simeq s_i + u_{i+h} \quad \text{with} \begin{cases} s_i = \pm k; \text{ binary level signal amplitude} \\ u_{i+h} = \text{noise component} \end{cases} \tag{6.10}$$

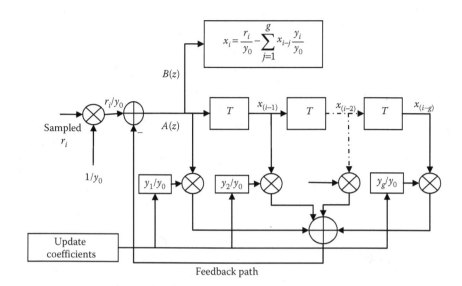

FIGURE 6.5
Schematic structure of a linear feedback transversal equalizer for a channel with an impulse response in Z-transform of $Y(z)$.

Note that noise always adds to the signal level so that for positive k the noise u_{i+h} follows the same sign and vice versa for a negative value.

The output noise signals of the FFE can thus be written as

$$u_{i+h} = \sum_{j=0}^{m} w_{i+h-j} c_j \qquad (6.11)$$

The noise components $\{w_i\}$ are statistically independent Gaussian random variables with a zero mean and variances σ^2 so that the total accumulated

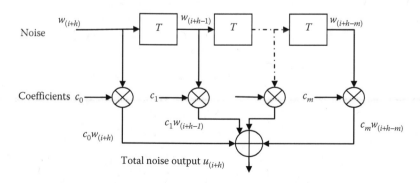

FIGURE 6.6
Noisy signals in a $(m + 1)$ tap linear FFE.

noise output given by Equation 6.11 is also an uncorrelated random Gaussian variable with a zero mean and a variance depicted as

$$\eta^2 = \sum_{j=0}^{m} \sigma^2 c_j^2 = \sigma^2 CC^T = \sigma^2 |C|^2 \qquad (6.12)$$

where $|C|$ is the Euclidean length of the $(m+1)$ component vector $[C] = [c_0 \ c_1 \ \dots \ c_m]$, the sampled channel impulse response. Thus η^2 is the mean square noise of the process and is also the mean square error of the output sample values x^{i+h}.

Thus, the probability noise density function of this Gaussian noise is given by

$$p(u) = \frac{1}{\sqrt{2\pi\eta^2}} e^{-\frac{u^2}{2\eta^2}} \qquad (6.13)$$

With the average value of the magnitude of the received signal value of k (equalized without any distortion), the probability of error of the noisy detection process is given by

$$P_e = Q\left(\frac{k}{\eta}\right) = Q\left(\frac{k}{\sigma|C|}\right) \qquad (6.14)$$

and the electrical SNR is given as

$$\text{SNR} = 20\text{Log}_{10}(k/\sigma) \text{ dB} \qquad (6.15)$$

with $Q(x)$ as the quality function of x, that is the complementary error function as commonly defined. In practice this SNR can be identified by measuring the standard deviation of the constellation point and the geometrical distance from the center of the constellation to the center of the constellation point. In the next section one can see that the equalized constellation would offer a much higher SNR as compared with that of the nonequalized constellation. Thus, under Gaussian random distribution, the noises contributed to the equalization process are also Gaussian and the probability density function would follow a straightforward, Euclidean distance from the channel impulse response.

6.2.1.6 Equalization with Minimizing MSE in Equalized Signals

It has been recognized that a linear transversal filter that equalizes and minimizes the mean square error in its output signals [6–8] generally gives a more effective degree of equalization than an equalizer that minimizes the peak distortion. Thus, an equalizer that minimizes the mean square difference between the actual and ideal sampled sequence at its output for a given number of taps would be advantageous when there is noise.

Now revisiting the transmission, in which the sampled impulse response of the channel; the sampled impulse response of $(m + 1)$ tap linear FFE; and the combined sampled impulse response of the channel and the equalizer, $[Y]$, $[C]$, and $[E]$ respectively, are given by

$$[Y] = \begin{bmatrix} y_0 & y_1 & \cdots & y_g \end{bmatrix}$$
$$[C] = \begin{bmatrix} c_0 & c_1 & \cdots & c_m \end{bmatrix} \tag{6.16}$$
$$[E] = \begin{bmatrix} e_0 & e_1 & \cdots & e_{m+g} \end{bmatrix}$$

Under the condition that the combined impulse response of the channel and the equalizer can be recovered, then the output sequence is only a pure delay of h sampling period of the input sampled sequence we have

$$x_{h+i} \approx s_i + u_{h+i} \tag{6.17}$$

or it is a pure delay superimposed by the Gaussian noises of the transmission line. In the case of optical transmission lines, these are the accumulated ASEs over many spans of the dispersive optically amplified transmission spans. Thus the resultant *ideal* vector at the output of the equalizer can be written as

$$[E] = \begin{bmatrix} \underbrace{0 \quad \ldots 0}_{h} & 1 & 0 & 0 & 0\ldots & 0 \end{bmatrix}; \quad h = \text{integer} \tag{6.18}$$

but in fact, the equalized output sequence would be (as seen from Equation 6.16)

$$x_{h+i} \approx \sum_{j=0}^{m+g} s_{i+h} - je_j + u_{h+i}; \quad j = 0,1\ldots,(m+g) \tag{6.19}$$

The input sampled values are statistically independent and have equal probability of taking values of $\pm k$, which indicates that the expected value (denoted by symbol $\xi[\cdot]$) is given by

$$\xi\left[s_{i+h-j}s_{i+h-l}\right] = 0; \quad \forall j \neq l \quad \text{and} \quad \xi\left[s_{i+h-j}^2\right] = k^2 \tag{6.20}$$

The noises superimposed on the sampled values are also statistically independent of the sampled signals with a zero mean, thus the sampled signals and noises are orthogonal so we have the expected value $\xi[s_{i+h-j}u_{i+h}] = 0$.

Thus an LE that minimizes the mean square error in its output signals would also minimize the mean square value of $[x_{i+h} - s_i]$, which can be written as

$$\xi\left[(x_{i+h} - s_i)^2\right] = \left\{ \begin{array}{c} \xi\left[\left(\sum_{j=0}^{m+g} s_{i+h-j}e_j + u_{i+h} - s_i\right)^2\right] \\ = k^2\left|E - E_h\right|^2 + \sigma^2\left|C\right|^2 \end{array} \right. \tag{6.21}$$

with

$$\left|E - E_h\right|^2 = \sum_{\substack{j=0 \\ j \neq h}}^{m+g} e_j^2 + k^2(e_h - 1)^2$$

where $|E - E_h|$ and $|C|$ are the Euclidean distances of the vectors $[E - E_h]$ and $C = [C]$, respectively. We can now observe that the first and second terms of Equation 6.21 are the mean square error in x_{i+h} due to ISI and the mean square error from the Gaussian noises. Thus we can see that the MSE process minimizes not only the distortion in terms of the mean square error due to ISI, but also minimizes that of the superimposing noises.

6.2.1.7 Constant Modulus Algorithm for Blind Equalization and Carrier Phase Recovery

6.2.1.7.1 Constant Modulus Algorithm

Suppose the ISI, additive noise, and carrier FO are considered. Then a received signal before processing, $x(k)$, can be written as

$$x(k) = \sum_{i=0}^{M-1} h(i)a(k - i)e^{j\phi(k)} + n(k) \tag{6.22}$$

where $h(k)$ is the overall complex baseband equivalent impulse response of the overall transfer function, including the transmitter, unknown channel, and receiver filter. Note that all E/O and O/E with coherent reception steps are removed and assume only noise contribution but no distortion in both phase and amplitude. $n(k)$ is the overall noise which is assumed to be Gaussian, mainly due to ASE noises over cascaded amplifiers with zero mean and standard deviation amount. The input data sequence $a(k - i)$ is assumed to be independent and identically distributed, and $\phi(k)$ is the carrier phase difference between the signal carrier and the LO laser.

Now let the impulse response of the equalizer be $\mathbf{W}(k)$ as denoted in the schematic given in Figure 6.7. The output of the equalized signals can be obtained by

$$y(k) = \mathbf{X}^T(k)\mathbf{W}(k) \qquad (6.23)$$

where $\mathbf{W}(k) = [w_0, w_1, \ldots, w_k]^T$ is the tap weight vector of the equalizer as is described in previous sections of this chapter, and $\mathbf{X}(k) = [x(k), x(k-11), \ldots, x(k-N+1)]^T$ is the vector input to the equalizer, where N is the length of the weight vector.

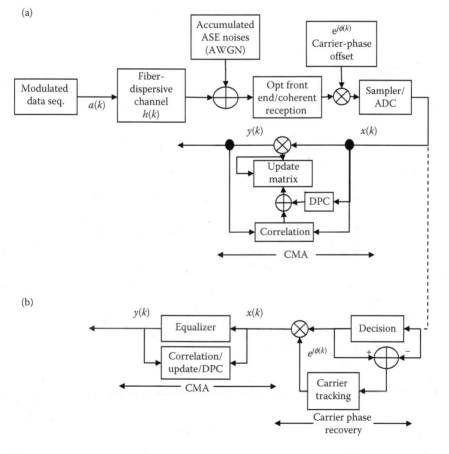

FIGURE 6.7
Equivalent model for baseband equalization. (a) Blind CMA, and (b) modified CMA by cascading CMA with carrier phase recovery. Signal inputs come from the digitalized received samples after coherent reception. DPC = differential phase compensation.

In the fiber transmission there is a pure phase distortion due to the CD effects in the linear region and the constellation is rotating in the phase plane. In addition, the carrier phase also creates an additional spinning rotation of the constellation, as observable in Equation 6.22. To equalize the linear phase rotation effects one must cancel the spinning effects due to this carrier phase difference created by the mixing between the LO and the signal carrier. Thus we can see that the CMA must be associated with the carrier phase recovery so that an addition phase rotation in reverse would be superimposed on the phase rotation due to channel phase distortion to fully equalize the constellation.

The constant modulus (CM) criterion may be expressed by the nonnegative cost function $J_{cma,p,q}$ parameterized by positive integers p and q such that

$$J_{cma,p,q} = \frac{1}{pq} E\left\{ \left| y_n \right|^p - \gamma \left|{}^q\right| \right\}; \quad \text{with } \gamma = \text{constant} \tag{6.24}$$

where $E(\cdot)$ indicates the expected statistical value. The CM criterion is usually implemented as CMA 2–2 where $p, q = 2$. Using this cost function the weight vectors can be updated by writing

$$\begin{aligned} \mathbf{W}(k+1) &= \mathbf{W}(k) - \mu \cdot \nabla \mathbf{J}(k) \\ &= \mathbf{W}(k) - \mu \cdot \left(y(k) \left| y(k) \right|^p \left| y(k) \right|^{p-2} \left(\left| y(k) \right|^p - R_p \right) \right) \cdot \mathbf{X}^*(k) \end{aligned} \tag{6.25}$$

where R_p is the constant depending on the type of constellation. Since the final output of the equalizer system would converge with the original input states, one can rewrite Equation 6.25 as

$$\left(a(k) \left| y(k) \right|^p \left| a(k) \right|^{p-2} \left(\left| a(k) \right|^p - R_p \right) \right) = 0 \tag{6.26}$$

Assuming that the inphase and quadrature phase components of $a(k)$ will be decorrelated with each other, and using the convolution of the input sequence $a(k)$ and the convolution with the impulse response of the channel $h(k)$ we have $x(k) = h(k) * a(k)$, the constant R_p can be determined as

$$R_p = \frac{E\left[\left| a(k) \right|^{2p} \right]}{E\left[\left| a(k) \right|^p \right]} \tag{6.27}$$

The flow of update procedure for the weight coefficients of the equalizer filter is shown in Figure 6.8.

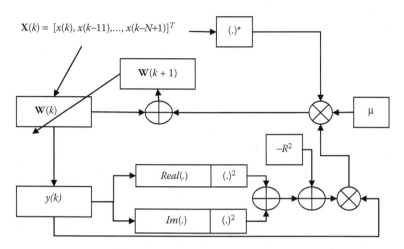

FIGURE 6.8
Block diagram of a tap update with CMA cost function.

6.2.1.7.2 Modified CMA: Carrier Phase Recovery Plus CMA

As we know from the cost function (6.24) used in the CMA, since the cost function is phase blind, the CMA can converge even in the presence of the phase error. Although it is the merit of the CMA, at convergence the equalizer output will have a constant phase rotation. This phase rotation is commonly due to the difference of the frequency of the LO and the carrier or LOFO. Furthermore, the constellation will be spinning at the carrier FO rate (considered as the phasor rotation) by lack of carrier frequency locking. While this phase-blind nature of the CMA is not a serious problem for the constant phase rotation, for some parameters such as the random PMD and the fluctuation of the phase of the carrier, for example, the 100 MHz oscillation of the feedback mirror in the external cavity of the LO, the performance of the CMA is severely degraded by randomly rotating phase distortion. Thus the carrier phase recovery is essential to determining the LOFO, the offset of the LO, and the carrier of the signal channel. The block diagram of combined CMA blind equalization and carrier tracking and locking is illustrated in Figure 6.7. The carrier recovery (CR) loop uses the error between the output of the equalizer and the corresponding decision. The phase updating rule is given by $\phi(k + 1) = \phi(k) - \mu_\phi I[z(k)e^*(k)]$, where μ_ϕ is the step-size parameter, and $e(k)$ is the error signal, given by $e(k) = z(k) - \hat{a}(k)$, where $z(k) = y(k)e^{-j\phi(k)}$ is the equalized output with phase error correction and $\hat{a}(k)$ is the estimation of $z(k)$ by a decision device. The carrier tracking loop described above gives the estimate of the phase error as shown in Figure 6.7b. Then a differential phase compensation (DPC) can be implemented.

The modified CMA is implemented by cascading the carrier phase recovery first and then cascaded with the standard CMA. Typical CMA operation with carrier phase recovery is shown in Figure 6.9 for 16-QAM constellation with optical coherent transmission and reception, as described in Chapter 4, in which 100 MHz oscillation of the LO laser is observed. We can observe that the constellation of the nonequalized processing is noisy and certainly cannot be used to determine the bit-error-rate of the transmission system. However, in Figure 6.9b, if only the CMA is employed then the phase rotation due to the phase offset between the LO and the signal channel carrier can be recovered by a reverse rotating of the phase. When carrier phase recovery is applied the constellation can then be recovered, as shown in Figures 6.9c and d, with and without a decision-directed (DD) processing stage, which will offer similar constellations. The DD processing will be described in a section dealing with NLEs or decision feedback equalizers (DFEs) of this chapter.

Ref. [9] has reported simulation of coherent receivers under polarization multiplexed QPSK employing CMA algorithms under standard and modified decision-directed algorithms under effects of finite linewidth of the laser (carrier source). There are certain penalty within 0.5 dB on the receiver sensitivity under coherent detection and DSP-based processing in the electronic domain with a 1e–3 FEC level for the linewidth—symbol rate product of 1e–4 to 1e–3.

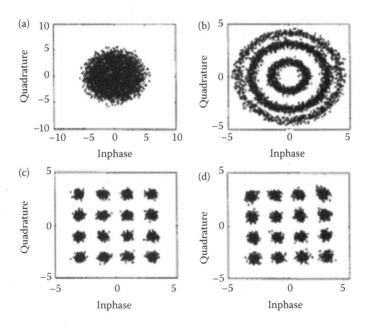

FIGURE 6.9
Constellations of 16-QAM modulated signals with a carrier FO with respect to that of the LO laser. (a) Unequalized output; (b) output equalized by CMA; (c) output by joint CMA and carrier phase DD recovery; and (d) output by modified CMA, carrier phase recovery, and CMA is cascade.

6.2.2 Nonlinear Equalizer or DFEs

6.2.2.1 DD Cancellation of ISI

It has been demonstrated in both wireless and optical communications systems over recent years that nonlinear (decision feedback) equalizers achieve better performance than their linear counterparts. The method of operation for these DD equalizers for cancellation of ISI is as follows.

Consider the transmission system given in Figure 6.10a, which is similar to Figure 6.2 except that the LE is replaced with the nonlinear one. The linear FF equalizer or transversal filter in this NLE is given in Figure 6.10c. The arrow directing from the decision block across the FF transversal filter indicates the updating of the filter coefficients.

Similar to the LE case, the input sampled data sequence also follows the notations given in Section 6.2 of this chapter. A multi-level data symbol can be treated the same way without unduly affecting any of important results in this analysis.

The pure NLE uses the detected data symbols $\{s_i'\}$ to synthesize the ISI component in a received signal $\{r_i\}$ and then it removes the ISI by subtraction. This is the process of DD cancellation of ISI.

Mathematically setting, by following the notation used from Equation 6.10 for the section just before the decision/detection block of Figure 6.10a or the sampled sequence at the input of Figure 6.10b, the signals entering the detector at the instant $t = iT$ are given as

$$x_i = \frac{r_i}{y_0} - \sum_{j=1}^{g} s_{i-j}' v_j = s_i + \sum_{j=1}^{g} s_{i-j}' v_j + \frac{w_i}{y_0} - \sum_{j=1}^{g} s_{i-j}' v_j \qquad (6.28)$$

so that with corrected detection of each sampled value s_{i-j} such that

$$s_{i-j}' = s_{i-j} \quad \text{for } j = 1, 2 \ldots g \rightarrow x_i = s_i + \frac{w_i}{y_0} \qquad (6.29)$$

In the correct detection and equalization we have recovered the sampled sequence $\{x_i\}$. Thus it is clear that so long as the data symbols $\{s_i\}$ are correctly detected, their ISIs are removed from the equalized signals followed by the NLE, and the channel continues to be accurately equalized.

The tolerance of this pure NLE operation with a channel-sampled impulse response whose z-transform is given by $\{Y(z)\}$ can be defined in Equation 6.5. The equalized sampled signals at the detector input are given by

$$x_i = s_i + \frac{w_i}{y_0} \qquad (6.30)$$

$$s_i = \pm k; \quad \text{binary amplitude of input seq.}$$

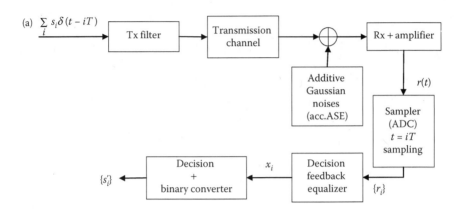

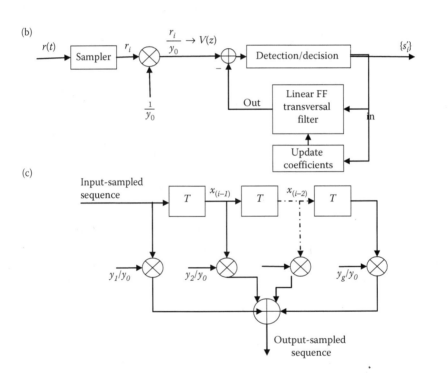

FIGURE 6.10
(a) Simplified schematic of a coherent optical communication system incorporating an NLE. Note equivalent baseband transmission systems due to baseband recovery of amplitude and phase of optically modulated signals. (b) Structure of the DFE employing linear FFE and a decision device. (c) Structure of the FF transversal filter.

It is assumed that the sampled sequence is identified as $+k$ or $-k$ depending whether x_i takes negative or positive values. Error would occur if the noise levels corrupt the magnitude, that is, if the noise magnitude is greater than k. As in a previous section related to the Gaussian noise (Section 6.2.1.5 of this chapter), we can obtain the probability density function of the random variable noise, w, as

$$p(w) = \frac{1}{\sqrt{2\pi y_0^{-2}\sigma^2}} e^{-\frac{w^2}{2y_0^{-2}\sigma^2}} \tag{6.31}$$

and the probability of error can be found as

$$P_e = \int_k^\infty p(w)dw = \int_k^\infty \frac{1}{\sqrt{2\pi y_0^{-2}\sigma^2}} e^{-\frac{w^2}{2y_0^{-2}\sigma^2}} dw = \int_{k\frac{|y_0|}{\sigma}}^\infty \frac{1}{\sqrt{2\pi}} e^{-\frac{w^2}{2}} dw = Q\left(k\frac{|y_0|}{\sigma}\right) \tag{6.32}$$

The most important difference between the nonlinear and linear equalizers is that the NLE can handle the equalization process when there are poles on the unit circle of the Z-plane of the transmission channel Z-transform impulse response.

A relative comparison of the LE and NLE is that (i) when all zeroes of the channel transfer function in the Z-plane lie inside the unit circle in the Z-plane, the NLE with samples, including the initial instant, would gain an advantage to the tolerance of the Gaussian additive noise over that of LE. The NLE would now be preferred in contrast to the LE, as the number of taps of the NLE would be smaller than that required for LE. However, (ii) when the transmission is purely phase distortion, that means that the poles outside the unit circle of the Z-plane and the zeroes are reciprocal conjugates of the poles (inside the unit circles), in which case the LE is a matched filter and can offer better performance than that of NLE, and close to a maximum likelihood (ML) detector. The complexity of NLE and LE are nearly the same in terms of taps and noise tolerances. (iii) Sometimes there would be about 2–3 dB better performance between NLE and LE processing.

6.2.2.2 Zero-Forcing Nonlinear Equalization

When an NLE is used in lieu of the LE of the zero-forcing nonlinear equalization (ZE-FE) illustrated in Figure 6.3, then the equalizer is called ZF-NLE, as illustrated in the transmission system shown in Figure 6.11. ZF-NLE consists

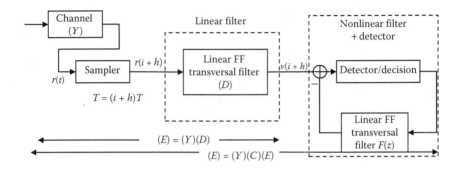

FIGURE 6.11
DFE incorporating a linear and nonlinear filter with overall transfer impulse responses.

of a linear FF transversal filter, a nonlinear equalizer with a linear feedback transversal, filter and a decision/detector operation.

The DFE performs the equalization by zero forcing, which is similar to ZF-LE (described in Section 6.2.1.2). The operational principles for this type of NLE are as follows:

- The linear filter of Figure 6.11 partially equalizes the channel by zero setting all components of the channel-sampled impulse response, preceding that of the largest magnitude without affecting the relative values of the remaining components.
- The NLE section then completes the equalization process by the operations described in Section 6.2.2.1.

Similar to the use of the notations assigned for the input sequence, the channel impulse response of g-sampling instants in the z-transform $Y(z)$ given in Equations 6.10 through 6.12, we can determine that the required equalizer impulse response in the z-domain $C(z)$ of m-tap would be

$$Y(z)C(z) \simeq z^{-h} \quad \text{with} \quad 0 < h < m + g \tag{6.33}$$

where h is an integer in the range of 0 to $(m + g)$ with

$$[Y] = \begin{bmatrix} y_0 & y_1 & \cdots & y_g \end{bmatrix}; \quad \text{Channel impulse response}$$

$$[C] = \begin{bmatrix} c_0 & c_1 & \cdots & c_m \end{bmatrix}; \quad \text{LE impulse response}$$

so that

$$Y(z) = y_0 + y_1 z^{-1} + \cdots + y_g z^{-g}; \quad \text{and } C(z) = c_0 + c_1 z^{-1} + \cdots + c_m z^{-m} \tag{6.34}$$

Let the LE have $[D]$ as a $(n + 1)$-tap gain filter with the impulse response $D(z)$ of $(n + 1)$-taps expressed as

$$[D] = \begin{bmatrix} d_0 & d_1 & \cdots & d_n \end{bmatrix}; \quad \text{impulse response linear transversal filter}$$

$$\rightarrow \quad D(z) = c_0 + c_1 z^{-1} + \cdots + c_m z^{-m} \tag{6.35}$$

Thus, the overall transfer function of the channel and the equalizer can be written when setting $[D] = [C][E]$, as

$$Y(z)C(z)E(z) \simeq z^{-h}E(z) \quad \text{with } n = m + g - 1 \tag{6.36}$$

with $[E]$ as the required sampled impulse response so that the NLE can equalize the input pulse sequence. This is to satisfy the condition that the product of the two transfer functions must be unitary, or

$$[E] = \begin{bmatrix} 1 & \dfrac{y_{l+1}}{y_l} & \dfrac{y_{l+2}}{y_l} & \dfrac{y_g}{y_l} \end{bmatrix} \rightarrow z\text{-transform impulse response}$$

$$\tag{6.37}$$

$$E(z) = 1 + \frac{y_{l+1}}{y_l} z^{-1} + \frac{y_{l+2}}{y_l} z^{-2} + \cdots + \frac{y_g}{y_l} z^{-g+l}$$

where y_l is the largest component of the channel-sampled impulse response. When this condition is satisfied, the nonlinear filter ZF-NLE acts the same as ZF-LE but with the tap gain order reduced to $(g - l)$ taps, with coefficients of $y_{l+1}/y_l, y_{l+2}/y_l, \ldots, y_g/y_l$ instead of g-taps with gains $v_1, v_2, \ldots, v_g$. The probability of error can be evaluated in a similar manner as described above, for additive noises of Gaussian distribution of the probability density function with zero mean and a standard deviation σ, to give

$$P_e = Q\left(\frac{k}{\sigma|D|}\right); \quad |D| = \text{Euclidean length of impulse response } [D] \tag{6.38}$$

6.2.2.3 Linear and Nonlinear Equalization of a Factorized Channel Response

In a normal case the z-domain transfer function of the channel $Y(z)$ can be factorized into a cascade of two sub-channels, given as

$$Y(z) = Y_1(z)Y_2(z) \tag{6.39}$$

So that we can employ the first linear equalizer to equalize the sub-channel $Y_1(z)$ and the NLE to tackle the second sub-channel $Y_2(z)$. The procedure is combined by cascading the two processing stages which have been described above.

6.2.2.4 *Equalization with Minimizing MSE in Equalized Signals*

We have discussed equalization using nonlinear equalizers in the previous section. The decision or detection will enhance the SNR performance of the system. However, under which criteria can the NLE make the decision block so that an optimum performance can be achieved? The mean square error can be employed as described above to minimize the errors in the decision stage of the NLE.

6.3 MLSD and Viterbi

Consider the detection of the first data symbol s_0 of the sequence $\{s_i\}$; $i = 0...m$ by an arrangement by delay of the h-sampling intervals increased to $m + g + 1$ such that all signals $\{v_i\}$ at the output of the filter D that are dependent on any of the m received data symbols $\{s_0 \quad s_1 \quad ... \quad s_{m-1}\}$ (see also Figure 6.11) are involved in the detection process at once. Since s_0 is the first detected value of the sequence of data symbols, no DD cancellation of the ISI can be carried out and the detection process operates on the $m + g$ signals $\{v_0 \quad v_1 \quad ... \quad v_{m+g-1}\}$. The detected value s_0' of the initial data symbol is therefore taken as its possible value, for which

$$\sum_{i=0}^{m+g=1} \left(v_i - \sum_{j=0}^{g} s_{i-j}' e_j \right)^2 \rightarrow \text{minimum for } \forall \text{ combinations} \qquad (6.40)$$

of all possible values of $\{s_0' \quad s_1' \quad ... \quad s_{m-1}'\}$ with the note that $s_j' = s_i = 0$; for $-i, 0$ and $i > m - 1$. This detection process is the maximum likelihood detection (MLSD) or estimation (MLSE), when the estimation is used in place of detection of the data symbols $\{s_0 \quad s_1 \quad ... \quad s_{m-1}\}$ from the corresponding received sequence of signals $\{v_0 \quad v_1 \quad ... \quad v_{m+g-1}\}$ the output of the filter D.

Under the assumed conditions, the detection process minimizes the probability of error in the detection of the complete sequence of data symbols. Thus all the data symbols are detected simultaneously in an optimal single detection process. Unfortunately, when m is large, especially in DSP, operating at an ultra-high processing speed would not allow such operation. Thus, successive processing of the data symbols is necessary and hence requirement of memory storage so that the detection process can be considered to be equivalent to simultaneous process. This sequential detection can be implemented using is the Viterbi algorithm detector [10].

The Viterbi algorithm operates as follows. Consider the data transmission as shown in Figure 6.11, in which the detector (by thresholding) and the transversal filter F are replaced by a more complex sub-system, assuming that the l-level data symbols $\{s_i\}$ are transmitted. After the optical to electronic detection and preamplification, the received signals are sampled and the stored components of $l^g h + 1$ component vectors of sequences $\{Q_{i-1}\}$ can be represented as

$$Q_{i-1} = [x_{i-h-1} \; x_{i-h} \quad \dots \quad x_{i-1}] \tag{6.41}$$

The symbol x_j for any j in the range of $\{0 \dots m - 1\}$ may take on any possible values of s_j and $h > g$. Each vector Q_{i-1} is formed by the last $(h + 1)$ components of the corresponding i-component vector $X_{i-1} = [x_0 \quad x_1 \quad \dots \quad x_{i-1}]$, which represents a possible received sequence of data symbols $\{s_j\}$. Associated with each stored vector $\{Q_{i-1}\}$ is its cost

$$c_{i-1} = \sum_{j=0}^{i-1} \left(v_j - z_j\right)^2; \quad \text{with } v_j = z_j + u_j'$$

$$z_j = \sum_{k=0}^{g} x_{j-k} e_k \quad x_j = 0 \quad \text{for } j < 0 \text{ and } j > m - 1 \tag{6.42}$$

The symbol z_j takes the value of v_j, which is received in the absence of noises. The cost c_{i-1} is the Euclidean distance between vectors $V_{i-1} = [v_0 \quad v_1 \quad \dots \quad v_{i-1}]$ and $Z_{i-1} = [z_0 \quad z_1 \quad \dots \quad z_{i-1}]$. Since Z_{i-1} is the sequence up to the time instant $t = (i - 1)T$ resulting from the transmitted sequence X_{i-1} in the absence of noises, the cost c_{i-1} is the measure of the probability that this sequence is correct. Thus the smaller value of this cost the more likely the corresponding sequence X_{i-1} is to be correct.

Thus, in summary, in MLSE decision is performed by calculating the transmitted bit stream with the highest probability. This means that the probability of all bit combinations for the whole sequence is taken into account and the combination of highest probability is assumed as the transmitted bit sequence.

6.3.1 Nonlinear MLSE

MLSE is a well-known technique in communications for equalization and detection of the transmitted digital signals. The concept of MLSE is discussed briefly. An MLSE receiver determines a sequence b as the most likely transmitted sequence when the conditional probability $\Pr(y|b)$ is maximized and where y is the received output sequence. If the received signal y is corrupted by a noise vector n, which is modeled as a Gaussian source (i.e., $y = b + n$), it is shown that the above maximization operation can be equivalent to the process of minimization of the Euclidean distance d [11]:

$$d = \sum_k | y_k - b_k |^2 \tag{6.43}$$

MLSE can be carried out effectively with the implementation of the Viterbi algorithm based on state trellis structure.

6.3.2 Trellis Structure and Viterbi Algorithm

6.3.2.1 Trellis Structure

Information signal b is mapped to a state trellis structure by a finite state machine (FSM), giving the mapping output signal c as shown in Figure 6.12. The FSM can be a convolutional encoder, a trellis-based detector or, as presented in more detail later on, the fiber medium for optical communications.

A state trellis structure created by the FSM is illustrated in Figure 6.13. At the epoch nth, the current state B has two possible output branches connecting to states E and F. In this case, these two branches correspond to the two possible transmitted symbols of "0" or "1," respectively. This predefined state trellis applies to the 1-bit-per-symbol modulation format such as binary ASK or DPSK formats. In cases of multi-bit modulation per symbol, the phase trellis has to be modified. For example, in the case of QPSK, there are four possible branches leaving the current state B and connecting the next states, corresponding to the possible input symbols of "00," "01," "10," and "11."

Furthermore, for a simplified explanation, several assumptions are made. These assumptions and according notations can be described as follows: (i) current state B is the starting state of only two branches, which connect to states E and F, denoted as: b_{BE} and b_{BF}. In general, b_{B*} represents all the possible branches in the trellis, which starts from state B. In this case, these two branches correspond to the possible binary transmitted information symbols of "0" or "1," respectively. The notation $c(B, a = 0)$ represents the encoded symbol (at the FSM output) for the branch BE, b_{BE}, which starts from state B and corresponds to a transmitted symbol $a = 0$. Similar conventions

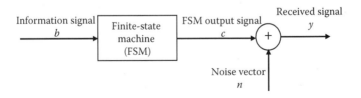

FIGURE 6.12
Schematic of the MLSE equalizer as an FSM.

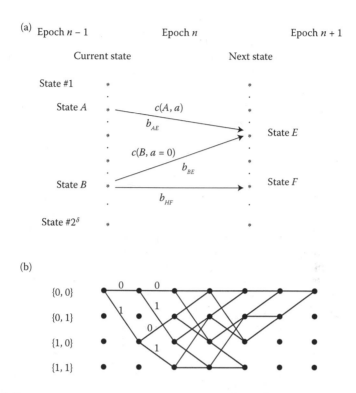

FIGURE 6.13
State trellis of an FSM. (a) Trellis path; (b) tracing of possible maximum likelihood estimation with assigned bit pattern.

are applied for the case of branch BF, b_{BF}, that is, $c(B,a=1)$ and the all the branches in the trellis. (ii) There are only two possible output branches ending at state E (from state A and state B). They are denoted as: b_{AE} and b_{BF}. In general, b_{*E} represents all the possible branches in the trellis that end at state E. The number of states 2^δ in the trellis structure are determined by the number of memory bits, δ, also known as the constraint length. Figure 6.13b also shows a possible trace of the trellis with the assigned states of the bit pattern.

6.3.2.2 Viterbi Algorithm

In principle, the Viterbi algorithm follows two main phases:

a. *Phase 1: Calculation of state and branch metrics*
   At a sampling index *n*, epoch *n*, a *branch metric,* is calculated for each of the branches in the trellis. For example, considering the connecting branch between the current state *B* and the next state *E*,

which corresponds to the case when "0" symbol is transmitted, the *branch metric* is calculated as

$$BM_{BE}(n) = |y_{(n)} - c_{BE}|^2$$

(6.44)

The state metric $SM_E(n)$ of the state E is calculated as the sum between the metric of the previous state, that is, state B and the branch metric obtained above. The calculation of state metrics $SM_E(n)$ of state E is given by

$$SM_E(n) = \min_{\forall B \to E}\left\{SM_B(n-1) + BM_{BE}(n)\right\}$$

(6.45)

The branch giving the minimum state metric $SM_E(n)$ for state E is called the preferred path.

b. *Phase 2: Trace-back process*

The process of calculating state and branch metrics continues along the state trellis before terminating at epoch N_{trace}, which is usually referred to as the trace-back length. A rule of thumb as referred to in Refs. [12,13] is that the value of N_{trace} is normally taken to be 5 times that of the sequence length. This value comes from the results showing that the solution for the Viterbi algorithm converges, giving a unique path from epoch 1 to epoch N_{trace} for the MLSE detection. (i) At N_{trace}, the terminating state with the minimum state metric and its connecting preferred path are identified. (ii) The previous state is then identified as well as its previous preferred path. (iii) The trace-back process continues until reaching the epoch 1.

6.3.3 Optical Fiber as a Finite State Machine

A structure of the FSM shown in Figure 6.13 is now replaced with the optical channel based on optical fiber. It is significant to understand how the state trellis structure is formed for an optical transmission system. In this case, the optical fiber involves all the ISI sources causing the waveform distortion of the optical signals, which includes CD, PMD, and filtering effects. Thus, the optical fiber FSM excludes the signal corruption by noise as well as the random-process nonlinearities. The ISI is caused by dispersion effects, and nonlinear phase and ASE and nonlinear effects [14].

6.3.4 Construction of State Trellis Structure

At epoch n, it is assumed that the effect of ISI on an output symbol of the FSM c_n is caused by both executive δ pre-cursor and δ post-cursor symbols on each side. In this case, c_n is the middle symbol of a sequence, which can be represented as

$$c = (c_{n-\delta},...,c_n,...,c_{n+\delta}) \quad b \in \{0,1\}; \quad n > 2\delta + 1 \qquad (6.46)$$

Unlike the conventional FSM used in wireless communications, the output symbol c_n of the optical fiber FSM is selected as the middle symbol of the sequence c. This is due to the fact that the middle symbol gives the most reliable picture about the effects of ISI in an optical fiber channel.

A state trellis is defined with a total number of $2^{2\delta}$ possible data sequences. At epoch n, the current state B is determined by the data sequence

$$(b_{n-2\delta+1},...,b_{n-\delta},...b_n) \quad b \in \{0,1\}; \quad n \geq 2\delta + 1 \qquad (6.47)$$

where b_n is the current input symbol into the optical fiber FSM. Therefore, the state trellis structure can be constructed from the optical fiber FSM and ready to be integrated with the Viterbi algorithm, especially for the calculation of a branch metric.

The MLSE nonlinear equalizer can effectively combat all the above ISI effects and it is expected that the tolerance of both CD and PMD of optical MSK systems with OFDR receiver can be improved significantly.

6.3.5 Shared Equalization between Transmitter and Receivers

Figures 6.14a and b depict the arrangement of equalizers at the transmitter only and combined at both the transmitter and receiver sides, respectively. There are advantages to when equalization can be conducted at both sides of the transmitter and receiver in the digital domain. At the transmitter, the equalizer performs the pre-distortion to compensate for the chromatic dispersion by chirping the phase. However, the dispersion of installed fiber links would never be determined exactly, and so it is always necessary to have an equalizer at the reception subsystem so that not only residual CD effects by others such as PMD, but also nonlinear impairment can be equalized.

6.3.5.1 Equalizers at the Transmitter

Consider the arrangement so that the equalizer acting as a pre-distortion is placed at the transmitter, as shown in Figure 6.14a. An ultra-high speed ADC would be best and flexible in generating pre-emphasis digital signals for compensation of the linear CD distortion effects. The nonlinear equalizer in this case converts the sequence of data symbols $\{s_i\}$ into a nonlinear channel $\{f_i\}$ at the output of the transmitter, normally an optical modulator driven by the output of the nonlinear electrical equalizer as shown in Figure 6.14b and remodeled as shown in Figure 6.15. These symbols are sampled

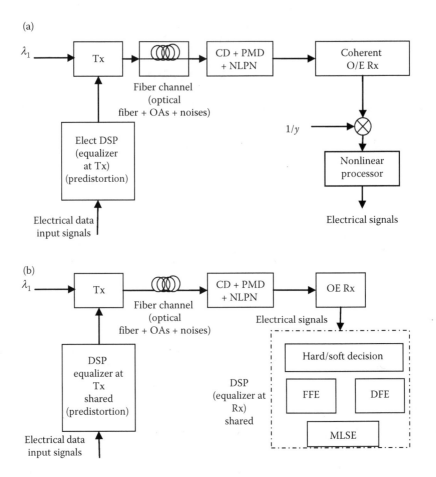

FIGURE 6.14
Schematic arrangement of electronic DSP for equalization in a digital processing unit incorporated with an optoelectronic receiver. Electronic DSP for pre-distortion, electronic DSP at the Rx is for post-equalization and shared pre-distortion and post-equalization at the Tx and Rx. (a) Equalization at the transmitter-side, (b) equalization at the receiver side, and (c) equalization shared between Tx and Rx sides.

and transmitted as sampled impulses $\{f_i\delta(t - iT)\}$. The SNR in this case is given by $E[f_i^2]/\sigma^2$, where the energy is the expected average energy of the pre-distorted signal sequence and must be equal to k^2.

A nonlinear equalizer is necessary at the output before the final recovery of the system. This processor performs a modulo-m operation on the received sequence $x[\text{modulo}] - m = x - jm$ with j an appropriate integer. The z-transform $F(z)$ at the output of the nonlinear distortion·equalizer is given by

$$F(z) = M\Big[S(z) - F(z)(y_0^{-1}Y(z) - 1)\Big] \qquad (6.48)$$

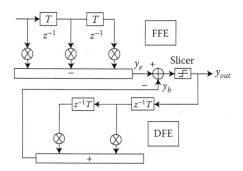

FIGURE 6.15
Schematic of a nonlinear feedback equalizer at the transmitter.

where $S(z)$ is the z-transform of the sampled input signals and $Y(z)$ is the z-transform of the channel transfer function. Then the z-transform at the output of the nonlinear processor at the receiver can be written as

$$X(z) = M[F(z)Y(z) + W(z)] = M[F(z)(y_0^{-1}Y(z) - y_0^{-1}W(z)] \quad \text{or}$$

$$X(z) = M[S(z) - y_0^{-1}W(z)] \tag{6.49}$$

that is

$$x_i = M[y_0^{-1}r_i] = M[s_i + y_0^{-1}w_i]$$

with x_i denoting the signal at the output of the nonlinear processor, and clearly all ISI has been eliminated except the noise contribution. The nonlinear processor M operates as a modulo-m as

$$M[q] = [(q + 2k) \text{ modulo} - 4k] - 2k = q - 4jk \quad -2k \le M[q] \le 2k \tag{6.50}$$

Thus an error would happen when

$$(4j - 3)k \le y_0^{-1}w_i \le (4j - 1)k \quad \forall j \tag{6.51}$$

The probability of error is then given as

$$P_e = 2\int_k^\infty \frac{1}{\sqrt{2\pi y_0^{-2}\sigma^2}} e^{-\frac{w^2}{2y_0^2\sigma^2}} dw = 2Q\left(\frac{k|y_0|}{\sigma}\right) \tag{6.52}$$

6.3.5.2 Shared Equalization

Long-haul transmission requires the equalization so that the extension of the transmission is as long as possible. Equalization at the receiver can be supplemented with that at the transmitter to further increase the reach.

The equalizer at the transmitter of the above section can be optimized by inserting another filter at the receiver, as shown in Figure 6.14. Consider the channel transfer function $Y(z)$ that can be written as a cascade of two linear transfer functions, $Y_1(z)$ and $Y_2(z)$, as

$$Y(z) = Y_1(z)Y_2(z) \quad \text{with} \quad \begin{cases} Y_1(z) = 1 + p_1 z^{-1} + \cdots + p_{g-f} z^{-g+f} \\ Y_2(z) = q_0 + q_1 z^{-1} + \cdots + q_f z^{-f} \end{cases} \quad (6.53)$$

$Y_1(z)$ has all the zeros inside the unit circle and $Y_2(z)$ has all zeros outside the unit circle. Then we could find a third system with a transfer function that has all the coefficients in reverse of those of $Y_2(z)$, given as

$$Y_3(z) = q_f + q_{f-1} z^{-1} + \cdots + q_0 z^{-f} \quad (6.54)$$

With the reverse coefficients, the zeroes of the $Y_3(z)$ now lie inside the unit circle.

The linear FFE filter D of the nonlinear filter formed by cascading an FFE and a DFE as shown above can minimize the length of the vector of D, and minimizing e_0 can be satisfied provided that

$$D(z) = z^{-h} Y_2^{-1}(z) Y_3(z) \quad (6.55)$$

which represents an orthogonal transformation without attenuation or gain. The linear equalizer with a z-transform $Y_2(z)Y_3^{-1}(z)$ is an all-pass pure phase distortion.

This transfer function of the all-pass pure phase distortion is indeed a cascade of optical interferometers, known as half-band filters [15,16]. This function can also be implemented in the electronic domain. The equalizer can then be followed by an electronic DFE. Hence, the feedback linear filter $F(z)$ in Figure 6.16 can be written as

$$F(z) = Y_1(z)Y_3(z) \quad (6.56)$$

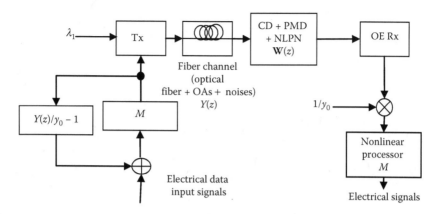

FIGURE 6.16
Schematic of a nonlinear feedback equalizer combining FFE and DFE equalization-slicer or decision/detector.

The probability of error can be similarly evaluated and given by

$$P_e = \int_k^\infty \frac{1}{\sqrt{2\pi y_0^2 \sigma^2}} e^{-\frac{w^2}{2y_0^2\sigma^2}} dw = Q\left(\frac{k|q_f|}{\sigma}\right) \tag{6.57}$$

with q_f taken as the first coefficient of $Y_3(z)$.

6.4 Maximum a Posteriori Technique for Phase Estimation

6.4.1 Method

Assuming that we want to estimate an unobserved population parameter θ on the basis of observation variable x, let f be the sampling distribution of x, so that $f(x|\theta)$ is the probability of x when the underlying population parameter is θ. Then the function that transforms $\theta \to f(x|\theta)$ is known as the likelihood function, and the estimate $\hat{\theta}_{ML}(x) = \max_\theta f(x|\theta)$ is the ML estimate of θ.

Now assume that a prior distribution g over θ exists. This allows us to treat θ as a random variable as in Bayesian statistics. Then the posterior distribution of θ can be determined as

$$\theta \to f(\theta|x) = \frac{f(x|\theta)g(\theta)}{\int_{v\in\Theta} f(x|v)g(v)dv} \tag{6.58}$$

where g density function of is θ, Θ is the domain of g. This is a straight-forward application of Bayes' theorem. The method of MAP estimation then estimates θ as the mode of the posterior distribution of this random variable

$$\hat{\theta}_{MAP} = \arg\_\max \frac{f(x|\theta)g(\theta)}{\int_v f(x|v)g(v)dv} = \arg\_\max_{\theta} f(x|\theta)g(\theta) \qquad (6.59)$$

The denominator of the posterior distribution (the so-called partition function) does not depend on θ, and therefore plays no role in the optimization. Observe that the MAP estimate of θ coincides with the ML estimate when the prior g is uniform (i.e., a constant function). The MAP estimate is a limit of Bayes estimators under a sequence of 0–1 loss functions, but generally not a Bayes estimator *per se*, unless θ is discrete.

6.4.2 Estimates

MAP estimates can be computed in several ways:

1. Analytically, the mode(s) of the posterior distribution can be obtained in closed form. This is the case when the conjugate priors are used; then

2. Either via numerical optimization such as the conjugate gradient method or Newton's method, which may usually require first or second derivatives, they are to be evaluated analytically or numerically; or

3. Via a modification of an expectation–maximization algorithm. This does not require derivatives of the posterior density.

One of the possible processing algorithms can be the ML phase estimation for coherent optical communications [17]. The differential QPSK (DQPSK) optical system can be considered as an example. In this system, as described in Chapter 5, the phase-diversity coherent reception technique with a mixing of an LO can be employed to recover the in-phase (I) and quadrature-phase (Q) components of the signals. Such a receiver is composed of a 90° optical hybrid coupler to mix the incoming signal with the four quadruple states associated with the LO inputs in the complex-field space. A $\pi/2$ phase shifter can be used to extract the quadrature components in the optical domain. The optical hybrid can then provide four lightwave signals to two pairs of balanced photodetectors to reconstruct I and Q information of the transmitted signal. In the case that the LO is the signal itself, then the reception is equivalent to a self-homodyne detection. For this type of reception subsystem, the balanced receiver would integrate a one-bit delay optical interferometer so as to compare the optical phases of two consecutive bits. We can assume a complete matching of the polarization between the signal and the LO fields so that only the impact of phase noise should be considered. Indeed, the ML algorithm is the modern MLSE technique [18].

The output signal reconstructed from the photocurrents can be represented by

$$r(k) = E_0 \exp[j(\theta_s(k) + \theta_n(k))] + \tilde{n}(k) \tag{6.60}$$

where k denotes the kth sample over the time interval $[kT, (k+1)T]$ (T is the symbol period); $E_0 = \Re\sqrt{P_{LO}p_s}$, R is the photodiode responsivity; P_{LO} and p_s are the powers of the LO and received signals, respectively; $\theta_s(k) \in \{0, \pi/2, \pi, -\pi/2\}$ is the modulated phase, the phase difference carrying the data information; $\theta_n(k)$ is the phase noise during the transmission; and $\{\tilde{n}(k)\}$ is the complex white Gaussian random variables with $E[\tilde{n}(k)] = 0$ and $E[|\tilde{n}(k)|^2] = N_0$.

To retrieve information from the phase modulation $\theta_s(k$ at time $t = kT$, the carrier phase noise $\theta_n(k)$ is estimated based on the received signal over the immediate past L symbol intervals, that is, based on $\{r(l), \ k - L \le l \le k - 1\}$. In the decision-feedback strategy, a complex phase reference vector $\mathbf{v}(k)$ is computed by the correlation of L, received signal samples $r(l)$, and decisions on L symbols $\{\hat{m}(l), \ k - L \le l \le k - 1\}$, where $\hat{m}(l)$ is the receiver's decision on $\exp(j\theta_s(l))$:

$$v(k) = \sum_{l=k-L}^{k-1} r(l)\hat{m}^*(l) \tag{6.61}$$

Here, * denotes the complex conjugation. An initial L-symbol long-known data sequence is sent to initiate the processor/receiver. Alternatively, the ML decision processor can be trained prior to the transmission of the data sequence. Of importance is that not only the $\pi/4$-radian phase ambiguity can be resolved, but also that the decision-aided method is now totally linear and efficient to implement based on Equation 6.61. To some extent, the reference $v(k)$ assists the receiver to acquire the channel characteristic. With the assumption that $\theta_n(k)$ varies slowly compared to the symbol rate, the computed complex reference $v(k)$ from the past L symbols forms a good approximation to the phase noise phasor $\exp(j\theta_n(k))$ at time $t = kT$.

Using the phase reference $\mathbf{v}(k)$ from (6.61), the decision statistic of the ML receiving processor is given as

$$q_i(k) = r(k)C_i^* \bullet v(k), \ i = 0, 1, 2, 3 \tag{6.62}$$

where $C_i \in \{\pm 1, \pm j\}$ is the DQPSK signal constellation, and $\bullet$ denotes the inner product of two vectors. The detector declares the decision $\hat{m}(k) = C_k$ if $q_k = \max q_i$. This receiver/processor has been shown to achieve coherent detection performance if the carrier phase is a constant and the reference length L is sufficiently long.

The performance of ML, the receiving processor, can be evaluated by simulation using Monte Carlo simulations in two cases: a linear and a nonlinear phase noise system. For comparison, the Mth power phase estimator and differential demodulation can also be employed.

Nonlinearity in an optical fiber can be ignored when the launch optical power is below the threshold, so the fiber optic can be modeled as a linear channel. The phase noise difference in two adjacent symbol intervals, that is, $\theta_n(k) - \theta_n(k-1)$, obeys a Gaussian distribution with the variance σ^2 determined by the linewidth of the transmitter laser and the LO, given by the Lorentzian linewidth formula $\sigma^2 = 2\pi\Delta vT$, where Δv is the total linewidth of transmitter laser and the LO [19]. In the simulations, σ can be set to 0.05, corresponding to the overall linewidth $\Delta v = 8$ MHz when the baud rate is 40 Symbols/s.

As observed in Figure 6.17, the ML processor outperforms the phase estimator and differential demodulation by about 0.25 and 1 dB, respectively. Although the performance gap between ML processor and Mth power phase estimation not very large, in this case of the MPSK system, an Mth power phase estimator requires more nonlinear computations, such as an arctan(•), which incurs a large latency in the system and leads to phase ambiguity in estimating the block phase noise. On the contrary, the ML processor is a linear and efficient algorithm, and there is no need to deal with the $\pi/4$-radian phase ambiguity, thus it is more feasible for on-line processing for the real systems. We also extend our ML phase estimation technique to the QAM

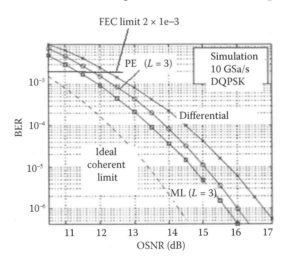

FIGURE 6.17
Simulated BER performances of a 10-Gsymbols/s DQPSK signal in linear optical channel with laser linewidth 2 MHz ($\sigma = 0.05$) and decision-aided length $L = 3$. The FEC limit can be set at $2 \times 1e\text{-}3$ to determine the required OSNR. (Modified from S. Zhang et al., *A Comparison of Phase Estimation in Coherent Optical PSK System*, Photonics Global '08, Paper C3–4A-03. Singapore, December 2008.)

system. The conventional Mth-power phase estimation scheme suffers from the performance degradation in QAM systems since only a subgroup of symbols with phase modulation $\pi/4 + n \bullet \pi/2$ ($n = 0, 1, 2, 3$) can be used to estimate the phase reference. The maximum tolerance of linewidth per laser in *Square_16QAM* can be improved 10 times compared to the Mth-power phase estimator scheme.

Let us now consider the case with nonlinear phase noise. In a multi-span optical communication system with erbium-doped fiber amplifiers (EDFAs), the performance of optical DQPSK is severely limited by the nonlinear phase noise converted by amplified ASE noise through the fiber nonlinearity [21]. The laser linewidth is excluded to only consider the impact of the nonlinear phase noise on the receiver. The transmission system can comprise a single channel DQPSK-modulated optical signal transmitted at 10–40 GSymbols/s over N 100-km equally spaced amplified spans, which are fully compensated for the CD and gain equalized optical power along the fiber as shown in Figure 6.18. The following parameters are assumed for the transmission system: $N = 20$, fiber nonlinear coefficient $\gamma = 2$ W^{-1} km^{-1}, fiber loss $\alpha = 0.2$ dB/km, the amplifier gain $G = 20$ dB with an NF of 6 dB, the optical wavelength $\lambda = 1553$ nm, and the bandwidth of optical filter and electrical filter of $\Delta\lambda = 0.1$ nm and 7 GHz, respectively. The OSNRis defined at the location just in front of the optical receiver as the ratio between the signal power and the noise power in two polarization states contained within a 0.1-nm

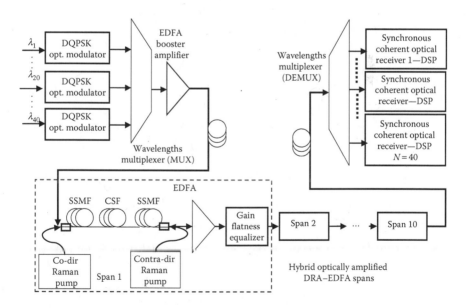

FIGURE 6.18
DWDM Optical DQPSK system with N-span dispersion-compensated optic fiber transmission link using a synchronous self-homodyne coherent receiver to compensate for nonlinear phase noise under both Raman amplification and EDF amplification.

spectral width region. As shown in Figure 6.19a, at the BER level of 10^{-4} the receiver sensitivity of the ML processor has improved by about 0.5 and 3 dB over phase estimation and differential demodulation, respectively. Besides, it is noteworthy that differential demodulation has exhibited an error floor due to severe nonlinear phase noise with high launch optical power at the transmitter.

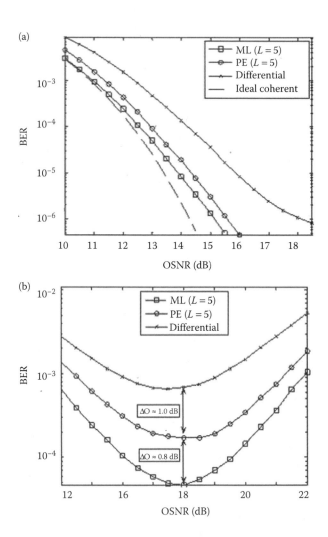

FIGURE 6.19
Monte Carlo Simulated BER of a 10-Gsymbol/s DQPSK signal in an *N*-span optical fiber system with a template length of *L* = 5: (a) *N* = 20; (b) *N* = 30 under different detection schemes of ML estimator, phase estimator, differential demodulation, and ideal coherent detection. ML = maximum likelihood, PE = phase estimation. (Extracted from D.-S. Ly-Gagnon et al., *J. Lightwave Technol.*, 24, 12–21, 2006.)

To obtain the impact of the nonlinear phase on the demodulation methods and the noise loading effects, the number of amplifiers can be increased to $N = 30$. The Q-factor is used to show a numerical improvement of ML processor receiver. As shown in Figure 6.19b, with the increase of launch power, the ML receiver/processor approximates the optimum performance because of the reducing variance of the phase noise. At the optimum point, the ML receiver/processor outperforms the Mth power phase estimator by about 0.8 dB, while having about 1.8 dB receiver sensitivity improvement compared to differential detection. It is also found that the optimum performance occurs at an OSNR of 16 dB, when the nonlinear phase shift is almost 1 rad. With the launch power exceeding the optimum level, nonlinear phase noise becomes severe again. Only the ML phase estimator can offer BER beyond 10^{-4}, while both the Mth power phase estimator and differential detection exhibit error floor before BER reaches 10^{-4}. From the numerical results, the ML processor carrier phase estimation shows a significantly better receiver sensitivity improvement than the Mth power phase estimator, and conventional differential demodulation for optical DQPSK signals in systems dominated by nonlinear phase noise. Again, we want to emphasize that the ML estimator is a linear and efficient algorithm, and there is no need to deal with the $\pi/4$-radian phase ambiguity; thus, it is feasible for on-line processing for the real systems.

In summary, the performance of synchronous coherent detection with ML processor carrier phase estimation may offer a linear phase noise system and a nonlinear optical noise system separately. The ML DSP phase estimation can replace the optical PLL, and the receiver sensitivity is improved compared to conventional differential detection and Mth power phase estimator. The receiver sensitivity is improved by approximately 1 and 3 dB in these two channels (20 spans for the nonlinear channel), respectively, compared to differential detection. For the 30-span nonlinear channel, only the ML processor can offer BER beyond 10^{-4} (using Monte Carlo simulation), while both the Mth power phase estimator and differential detection exhibit error floor before BER reaching 10^{-4}. At the optimum point of the power, the Q-factor of the ML receiver/processor outperforms the Mth power phase estimation by about 0.8 dB. In addition, an important feature is the linear and efficient computation of the ML phase estimation algorithm, which enables the possibility of real-time, on-line DSP processing.

6.5 Carrier Phase Estimation

6.5.1 Remarks

Under the homodyne or intradyne reception, the matching between the LO laser and that of the signal carrier is very critical in order to minimize the deviation of the reception performance, hence enhancement of the receiver sensitivity so as to maximize the transmission system performance. This difficulty remains a considerable obstacle for the first generation of coherent systems developed in the mid-1980s, employing analog or complete hardware circuitry.

The first few parts of this chapter deal with the recovery of carrier phase. Then later parts introduce advanced processing algorithms to deal with the carrier phase recovery under the scenario that the frequency of the LO laser is an oscillator, an operational condition to get the stabilization of an external cavity laser, which is essential for homodyne reception with high sensitivity.

In DSP-based optical reception sub-systems, this hurdle can be overcome by processing algorithms that are installed in a real-time memory processing system, the application-specific integrated circuit (ASIC), which may consist of an ADC, a digital signal processor, and high-speed fetch memory.

The FO between the LO and the signal lightwave carrier is commonly due to the oscillation of the LO, commonly used is an external cavity incorporating an external grating that is oscillating by a low frequency of about 300 MHz so as to stabilize the feedback of the reflected specific frequency line to the laser cavity. This oscillation, however, degrades the sensitivity of the homodyne reception systems. A later part of this section illustrates the application of DSP algorithms described in the above sections of this chapter to demonstrate the effectiveness of these algorithms in real-time experimental set up.

It is noted that the carrier phase recovery must be implemented before any equalization process can be implemented, as explained in Sections 6.2.1 and 6.2.2.

Also noted here is that the optical phase locking (OPL) technique presented in Chapter 5 is another way to reduce the FO between the LO and the signal carrier with operation in the optical domain, while the technique described in this section is at the receiver output in the digital processing domain.

6.5.2 Correction of Phase Noise and Nonlinear Effects

Kikuchi's group [222] at the University of Tokyo uses electronic signal processing based on the Mth power phase estimation to estimate the carrier phase. However, DSP circuits for the Mth power phase estimator need nonlinear computations, thus impeding the potential possibility of real-time processing in the future. Furthermore, the Mth power phase estimation method requires dealing with the $\pi/4$-radian phase ambiguity when estimating the phase noise in adjacent symbol blocks. While the electronic DSP is based on the ML processing for carrier phase estimation to approximate the ideal synchronous coherent detection in an optical phase modulation system, which requires only linear computations, it eliminates the optical PLL and is more feasible for online processing of the real systems. Some initial simulation results show that the ML receiver/processor outperforms the Mth power phase estimator, especially when the nonlinear phase noise is dominant, thus significantly improving the receiver sensitivity and tolerance to the nonlinear phase noise.

Liu's group [23,24] at Alcatel-Lucent Bell Labs uses optical delay differential detection with DSP to detect differential BPSK (DPSK) and DQPSK signals. Since direct detection only detects the intensity of the light, the improvement of DSP is limited after the direct detection. While after the synchronous coherent detection of our technique, the DSP, such as electronic equalization of CD and PMD, offer better performance since the phase information is retrieved. As the level of phase modulation increases, it becomes more and more difficult to apply the optical delay differential detection because of the rather complex implementation of the receiver and the degraded SNR of the demodulated signal, while the ML estimator/processor can still demodulate other advanced modulation formats, such as 8-PSK, 16-PSK, 16-QAM, and so on.

6.5.3 Forward Phase Estimation QPSK Optical Coherent Receivers

Recent progress in DSP [25,26] with the availability of ultra-high sampling rates allows the possibility of DSP-based phase estimation and polarization management techniques that make the coherent reception robust and practical. This section is dedicated to the new emerging technology that will significantly influence the optical transmission and detection of optical signals at ultra-high speed.

Recalling the schematic of a coherent receiver in Figures 6.20 and 6.21, it shows a coherent detection scheme for QPSK optical signals in which they are mixed with the LO field. In the case of modulation format, QPSK or DQPSK, a $\pi/2$ optical phase shifter is needed to extract the quadrature component, thus the real and imaginary parts of a phase-shift keyed signal can be deduced at the output of a balanced receiver from a pair of identical photodiodes connected back to back. Note that for a balanced receiver the quantum shot noise

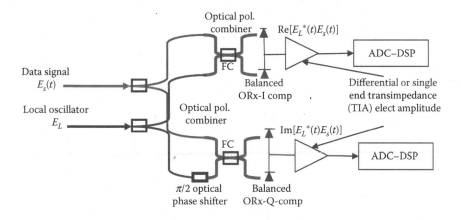

FIGURE 6.20
Schematic of a coherent receiver using a balanced detection technique for I-Q component phase and amplitude recovery, incorporating DSP and ADC.

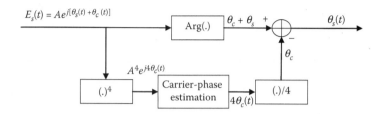

FIGURE 6.21
Schematic of a QPSK coherent receiver.

contributed by the photodetectors is doubled, as noise is always represented by the noise power and no current direction must be applied.

When the received signal is raised to the quadratic power, the phase of the signal disappears because $e^{j4\theta_s(t)} = 1$ for DQPSK, for which states are $0, \pi/2, \pi, 3\pi/2, 2\pi$. The data phase is excluded and the carrier phase can be recovered with its fluctuation with time. This estimation is computed using the DSP algorithm. The estimated carrier phase is then subtracted by the detected phase of the signal to give only the phase states of the signals as indicated in the diagram. This is an FF phase estimation and most suitable for DSP implementation either offline or in real time.

6.5.4 CR in Polarization Division Multiplexed Receivers: A Case Study

Advanced modulation formats in combination with coherent receivers incorporating DSP subsystems enable high capacity and spectral efficiency [27,28]. Polarization division multiplexing (PDM), QAM, and coherent detection are dominating the next-generation, high-capacity optical networks since they allow information encoding in all the available degrees of freedom [29] with the same requirement of the optical SNR (OSNR). The coherent technique is homodyne or intradyne mixing of the signal carrier and the LO. However, as just mentioned, a major problem in homodyne coherent receivers is the CR so that matching between the channel carrier and the LO can be achieved to maximize the signal amplitude and phase strength, hence the SNR in both amplitude and phase plane. This section illustrates an experimental demonstration of the CR of a PDM–QPSK transmission scheme carrying 100 Gbits/s per wavelength channel, including two polarized modes, H (horizontal) and V (vertical) polarized modes in standard single-mode optical fiber (SSMF). That is, 28 GSy/s × 2 (2 bits/symbol) × 2 (polarized channels) using some advanced DSP algorithms, especially the Viterbi-Viterbi (V-V) algorithm with MLSE nonlinear decision feedback procedures [30].

Feed-forward carrier phase estimation (FFCPE) [31] has been commonly considered the solution for this problem. The fundamental operation principle is to assume a time invariant of the phase offset between carrier and local laser during N ($N > 1$) consecutive symbol periods.

In coherent optical systems using tunable lasers, the maximum absolute FO can vary by 5 GHz. To accommodate such a large FO, coarse digital frequency estimation (FDE) and recovery (FDR) techniques [32,33] can be employed to limit the FO to within allowable range so that phase recovery can be implemented/managed using fine FFCPE and feedfoward carrier phase recovery (FFCPR) techniques [34]. The algorithm presented in Ref. [23], the so-called "Differential Viterbi" (DiffV), estimates phase difference between two successive complex symbols. This method enables the estimation of a large FO of multiple phase shift keying (M-PSK) modulated signal up to $\pm f_s/(2M)$, where f_s stands for the symbol rate frequency. Larger FO is possible to estimate by the technique presented in Ref. [6], which generates insignificant estimation error that can be handled by the FFCPE and FFCPR. However, the FFCPE or V-V would not perform well when the absolute residual FO at the input of the estimation circuit is larger than $\pm f_s/(2MN)$. This problem becomes more serious when the laser frequency oscillates and the coarse FFE can be realized by a slow-speed processor to decrease complexity of integrated digital electronic circuits, and release valuable time for processing the signals at an ultra-high sampling rate, which is a very important factor in the cost and power consumption in an ultra-high-speed coherent DSP-based optical receiver.

An algorithm for coupling carrier phase estimations in polarization division multiplexed PDM systems is presented in Ref. [35]. The algorithm requires two loop filters and the coupling factor should be carefully selected. In this section, we have described an enhanced CR concept covering large FO and enabling almost zero residual FO for the V-V CPE. Furthermore, a novel polarization coupling algorithm with reduced complexity is briefly given based on the work of Ref. [22].

6.5.4.1 FO Oscillations and Q-Penalties

Figure 6.22 shows the typical DSP of medium complexity for a coherent PDM optical transmission system. CD and polarization effects are compensated in

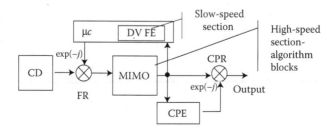

FIGURE 6.22
Typical digital signal process structure in PDM-coherent receivers, including a low-speed processor and high-speed DSP system. CD = chromatic dispersion; CPE = carrier phase estimation; CPR = carrier phase recovery; FR = frequency recovery; MIMO = multiple input multiple output.

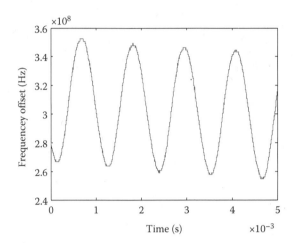

FIGURE 6.23
Laser frequency oscillation of EMCORE external cavity laser as measured from an external cavity laser.

CD and MIMO blocks, respectively. The DV FE is realized in the μc. Several thousands of data are periodically loaded to the micro controller that calculates FO. A small bandwidth of the estimation loop delivers only averaged FO value to the CMOS ASIC part. However, long-time experimental tests with commercially available lasers, for example, the EMCORE laser, show that laser frequency oscillations are sinusoidal with a frequency of 888 Hz in order to stabilize central frequency by the use of feedback control, and amplitude of more than 40 MHz (see Figure 6.23 for measured oscillation of the EMCORE external cavity laser as LO). This oscillation is created by the vibration of the external reflector—a grating surface—so that the cavity can be stabilized. The mean offset is close to 300 MHz. So, the DiffV enables the averaging of the FO of 300 MHz while the residual offset of ±50 MHz be compensated by the V-V algorithm.

Penalties caused by residual offset of 50 MHz have been investigated for the case of V-V CPE. Figure 6.24 shows the simulation values of the Q penalties versus V-V average window length (WL) for two scenarios under off-line data processing. In the first case, we multiplexed the 112 G PDM-QPSK channel with twelve 10 G on–off keying (OOK) neighbors (10 G and 100 G over 1200 km link), while the second scenario includes the transmission of eight 112 G PDM-QPSK channels (100 G WDM – 1500 km link). Channel spacing was 50 GHz (ITU-grid). The FO was partly compensated with the residual offset of 50 MHz to check the V&V CPR performance. Three values of launch power have been checked: optimum launch power (LP_{opt}), $LP_{opt} - 1$ dB (slightly linear regime), and $LP_{opt} + 1$ dB (slightly nonlinear regime). Maximum Q penalties of 0.9 dB can be observed that strongly depend on the average WL parameter.

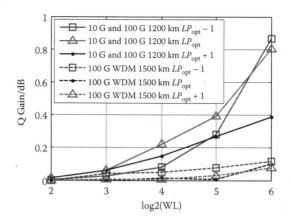

FIGURE 6.24
Penalties of V&V CPR with residual FO of 50 MHz.

6.5.4.2 Algorithm and Demonstration of Carrier Phase Recovery

A robust FFCPE and FFCPR with polarization coupling circuit is shown in Figure 6.25. The FFCFE and FFCPR are developed for the recourses of the V-V, and by simple recourses reallocation of an enhanced CR is obtained. The proposed method is designed to maximize CR performance with a moderate realization complexity. The method consists of four main parts: coarse carrier frequency estimation by processing in slow-speed DSP sub-systems (indicated by μc) and recovery (CCFE-R), feedback carrier frequency estimation and recovery (FBCFE-R, red line in Figure 6.4), XY phase offset estimation and compensation (H-V [or XY]-POE-C, gray line in Figure 6.25), and joint H-V FF carrier phase estimation and recovery (JFF-CPE-R).

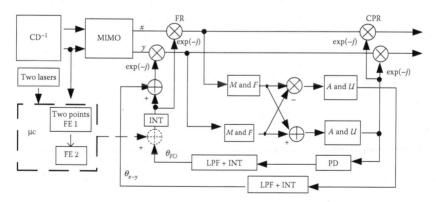

FIGURE 6.25
Modified structure of algorithms for CR. Legend: A&U = arctan and unwrapping; INT = integrator; LPF = low-pass filter; M&F = Mth power operation and averaging filter; PD = phase difference calculation. (Extracted from J.P. Gordon and L.F. Mollenauer, *Opt. Lett.* 15, 1351–1353, 1990.)

The M-F block conducts the Mth power operation and averaging in the case of the M-PSK modulation format [6]. For the K-QAM signal, the M constellation points on the ring with specific radii may be selected and used in Mth power block. More rings can be selected to improve the estimation procedure (e.g., inner and outer rings in 16-QAM can be used). So, the use of the method is not limited to PSK schemes.

Large FO (FO > $\pm f_s/(2MN)$) can be estimated by the method presented in Ref. [36] (denoted by FE1 in Figure 6.25), which can estimate FO within ±10 GHz FO (linear curve in case of 112 G PDM-QPSK—polarization division multiplexed QPSK). A small number of data is transferred into the μc for calculating coarse range of the FO. Depending on FO estimation sign, one can superimpose certain FOs (e.g., 4 GHz) of opposite sign. Using a linear interpolation (FE2), the current FO can be estimated to an error within a few hundreds of MHz. Similarly, depending on available electrical bandwidth and receiver structure, the compensation of such a large FO can be conducted either in CMOS ASIC or by controlling the frequency of LO, commonly an external cavity laser.

The residual FO after CCFE and CCFER may have values of several hundred MHz, plus 50 MHz generated by low processing speed and laser frequency oscillations. The FBCFE estimates the residual frequency that is compensated by the CFR feedback loop. The frequency estimation is in the range of FBCFE-R is $\pm f_s/(2MN)$ due to its implementation based on the DiffV algorithm.

The FBCFE-R compensates almost the total frequency offset. The residual FO and finite laser line width influence can be compensated by the JFF-CPE-R. This operation can be supported by ideal coupling of the carrier estimations from both polarizations. Prior to the coupling of estimations, both X and Y constellations are aligned using the XY-POE-C; the phase difference between adjacent outputs of the lower unwrapping block can then be used for the alignment. It is sufficient to rotate the Y polarization for the constellation angle mismatch.

The algorithm employing MIMO and V-V CPE is compared with the CR as presented in Figure 6.24, in which the uncompensated FO after the MIMO block is set to 50 MHz to simulate a realistic scenario. The measured FO after the MIMO is around 300 MHz. All data processing algorithms are done block-wise with parallelization of 64 symbols per block and a realistic processing delay is added [22].

Figure 6.26 shows the gain in the quality factor Q for the two measurement cases. Since the FO is recovered by the feedback CR, the penalties shown in Figure 6.24 are completely compensated, with an additional gain coming from the estimations of coupling.

The 10 G and 100 G hybrid offline optical transmission with homodyne coherent reception experiments are used to demonstrate the convergence of the feedback loops of the modified algorithm shown in Figure 6.25. As shown in Figure 6.27, the FO of 327 MHz can be acquired after 3.4 μs or 1500

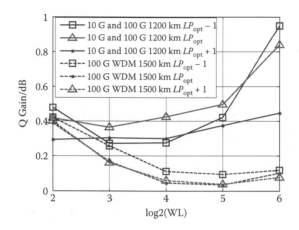

FIGURE 6.26
CR Q gain.

blocks of processing [22], the steady state of the plateau region. The XY constellation offset is acquired after 3000 blocks due to the demand for correct FO compensation. The estimated XY phase offset is 0.8 rad. The CR can cope with a laser frequency oscillation of 36 kHz. Acquisition and tracking times are shown in Figure 6.28. The CR is able to track the laser frequency changes after 100 ms. Note that a feedback loop can cause delay between real oscillations and estimation (delay between two lines in Figure 6.28), hence resulting in some penalties.

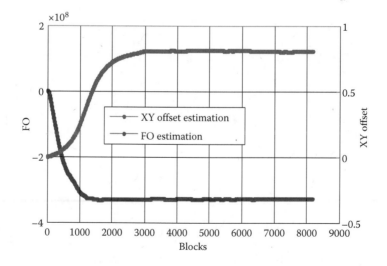

FIGURE 6.27
10 G hybrid off-line experiment with LOFO and X–Y phase offset estimation by FO variation with respect to the number of blocks. FO = frequency offset.

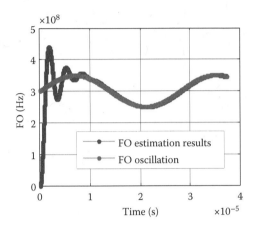

FIGURE 6.28
36 kHz oscillation variation of the laser frequency.

High-precision CR is achieved by the use of the V-V circuits under appropriate allocation and connections. Using feedback and FF CR, significant improvements can be achieved. Laser FO can be compensated by complex circuit design and appropriate parameters selection. The ideal coupling of estimations coming from two polarizations further enhance CR performance with less complexity.

6.6 Systems Performance of MLSE Equalizer–MSK Optical Transmission Systems

6.6.1 MLSE Equalizer for Optical MSK Systems

6.6.1.1 Configuration of MLSE Equalizer in Optical Frequency Discrimination Receiver

Figure 6.29 shows the block diagram of a narrowband filter receiver integrated with nonlinear equalizers for the detection of 40 Gb/s optical MSK signals. The MSK transmitter was described in Chapter 3 on modulation technique, and it has the 2 bits/symbol modulation as in the case of QPSK, except that the phase change from one state to the other is continuous rather than a discrete step change. Continuous phase change implies frequency step change. So when the frequency difference between states is equal to a quarter of the bit rate, the minimum distance between them so that the channel states can be recovered with minimum ISI, then the two frequency states are orthogonal and then the modulation is termed as the MSK.

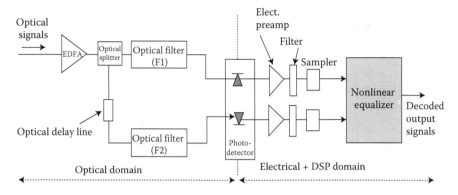

FIGURE 6.29
Block diagram of a narrowband filter receiver integrated with nonlinear equalizers for the detection of 40 Gb/s optical MSK signals.

In the reception of MSK channels, two narrowband filters can be used to discriminate the upper sideband (USB) channel and the lower sideband (LSB) channel frequencies that correspond to logic "1" and "0" transmitted symbols, respectively. A constant optical delay line, easily implemented in integrated optics, is introduced on one branch to compensate any differential group delay (DGD) $t_d = 2\pi f_d \beta_2 L$ between f_1 and f_2, where $f_d = f_1 - f_2 = R/2$, β_2 represents group velocity delay (GVD) parameter of the fiber, and L is the fiber length as described in Chapter 2. Alternatively, de-skewing can be implemented in the digital domain to compensate for this delay difference. If the DGD is fully compensated, the optical lightwaves in two paths arrive at the photodiodes simultaneously. The outputs of the filters are then converted to electrical domain through the photodiodes. These two separately detected electrical signals are sampled before being fed as the inputs to the non-linear equalizer.

It is noted that the MSK reception system described here can be a self-homodyne detection or coherent with the signal channel employed as the LO. Alternatively, the reception sub-system is coherent one, where the LO oscillator has the power level as that of the average power of the signal.

6.6.1.2 MLSE Equalizer with Viterbi Algorithm

At epoch k, it is assumed that the effect of ISI on an output symbol of the FSM c_k is caused by both executive δ pre-cursor and δ post-cursor symbols on each side. First, a state trellis is constructed with 2δ states for both detection branches of the OFDR. A lookup table per branch corresponding to symbols "0" and "1" transmitted and containing all the possible $2^{2\delta}$ states of all 11 symbol-length possible sequences is constructed by sending all the training sequences incremented from 1 to $2^{2\delta}$.

The output sequence $c(k) = f(b_k, b_{k-1}, \ldots, b_{k-2\delta-1}) = (c_1, c_2, \ldots, c_{2\delta})$ is the non-linear function representing the ISI caused by the δ adjacent pre-cursor and post-cursor symbols of the optical fiber FSM. This sequence is obtained by selecting the middle symbols c_k of 2δ as possible sequences, with a length of $2\delta + 1$ symbol intervals.

The samples of the two filter outputs at epoch k y_k^i, $i = 1, 2$ can be represented as $y^i(k) = c^i(k) + n_{ASE}^i(k) + n_{Elec}^i(k)$, $i = 1, 2$. Here, n_{ASE}^i and n_{Elec}^i represent the amplified spontaneous emission (ASE) noise and the electrical noise, respectively.

In linear transmission of an optical system, the received sequence y_n is corrupted by ASE noise of the optical amplifiers n_{ASE}, and the electronic noise of the receiver, n_E. It has been proven that the calculation of branch metric and hence state metric is optimum when the distribution of noise follows the normal/Gaussian distribution, that is, the ASE accumulated noise and the electronic noise of the reception sub-system are collectively modeled as samples from Gaussian distributions. If noise distribution departs from the Gaussian distribution, the minimization process is sub-optimum.

The Viterbi algorithm subsystem is implemented on each detection branch of the OFDR. However, the MLSE with Viterbi algorithm may be too computationally complex to be implemented at 40 Gb/s with the current integration technology. However, there have been commercial products available for 10 Gb/s optical systems. Thus a second MLSE equalizer using the technique of reduced-state template matching is presented in the next section.

In an optical MSK transmission system, narrowband optical filtering plays the main role in shaping the noise distribution back to the Gaussian profile. The Gaussian-profile noise distribution is verified in Figure 6.30. Thus, branch metric calculation in the Viterbi algorithm, which is based on minimum Euclidean distance over the trellis, can achieve optimum performance.

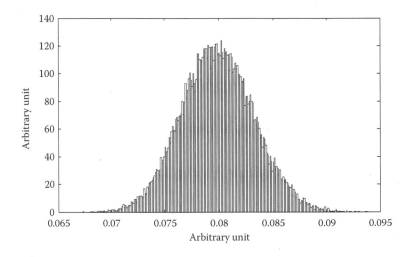

FIGURE 6.30
Noise distribution following Gaussian shape due to narrowband optical filtering.

Also, the computational effort is less complex than ASK or DPSK systems due to the issue of non-Gaussian noise distribution.

6.6.1.3 MLSE Equalizer with Reduced-State Template Matching

The modified MLSE is a single-shot template matching algorithm. First, a table of $2^{2\delta+1}$ templates, $\mathbf{g}^k$, $k = 1,2, \ldots, 2^{2\delta+1}$, corresponding to the $2^{2\delta+1}$ possible information sequences, I^k, of length $2\delta + 1$ is constructed. Each template is also a vector of size $2\delta + 1$, which is obtained by transmitting the corresponding information sequence through the optical channel and obtaining the $2\delta + 1$ consecutive received samples. At each symbol period in the sequence, $\hat{\mathbf{I}}_n$, with the minimum metric is selected as $c(k) = \arg\min_{b(k)}\{m(b(k), y(k))\}$. The middle element of the selected information sequence is then output as the nth decision, $\hat{I}_n$, that is, $\hat{I}_n = \hat{\mathbf{I}}_n(\delta + 1)$. The minimization is performed over the information sequences $\mathbf{I}^k$, which satisfy the condition that the δ-2 elements are equal to the previously decoded symbols $\hat{I}_{n-\delta}, \hat{I}_{n-(\delta-1)}, \ldots \hat{I}_{n-2}$, and the metric, $m(\mathbf{I}^k, \mathbf{r}_n)$, is given by $m(\mathbf{I}^k, \mathbf{r}_n) = \{\mathbf{w} \bullet (\mathbf{g}^k - \mathbf{r}_n)\}^T \{\mathbf{w} \bullet (\mathbf{g}^k - \mathbf{r}_n)\}$, where $\mathbf{w}$ is a weighting vector that is chosen carefully to improve the reliability of the metric. The weighting vector is selected so that when the template is compared with the received samples, less weighting is given to the samples further away from the middle sample. For example, we found through numerical results that a weighting vector with elements $\mathbf{w}(i) = 2^{-|i-(\delta+1)|}$, $i = 1,2,\ldots,2\delta + 1$ gives good results. Here, $\bullet$ represents Hadamard multiplication of two vectors, $(.)^T$ represents the transpose of a vector, and $|.|$ is the modulus operation.

6.6.2 MLSE Scheme Performance

6.6.2.1 Performance of MLSE Schemes in 40 Gb/s Transmission Systems

Figure 6.31 shows the simulation system configuration used for the investigation of the performance of both the above schemes when used with the narrowband optical Gaussian filter receiver for the detection of non-coherent

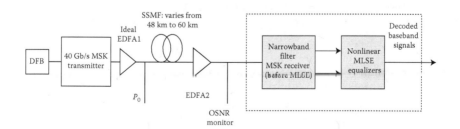

FIGURE 6.31
Simulation set-up for performance evaluation of MLSE and modified MLSE schemes for detection of 40 Gb/s optical MSK systems.

TABLE 6.2

Key Simulation Parameters Used in the MLSE MSK Modulation Format

Input power: $P_0 = -3$ dBm	Narrowband Gaussian filter: $B = 5.2$ GHz or $BT = 0.13$
Operating wavelength: $\lambda = 1550$ nm	Constant delay: $t_d = \|2\pi f_D \beta_2 L\|$
Bit rate: $R = 40$ Gb/s	Pre-amp EDFA of the OFDR: $G = 15$ dB and $NF = 5$ dB
SSMF fiber: $\|\beta_2\| = 2.68\text{e}{-}26$ or $\|D\| = 17$ ps/nm/km	$i_d = 10$ nA
Attenuation: $\alpha = 0.2$ dB/km	$N_{eq} = 20$ pA/(Hz)$^{1/2}$

40 Gb/s optical MSK systems. The input power into fiber (P_0) is −3 dBm, which is much lower than the nonlinear threshold power. The EDFA2 provides 23 dB gain to maintain the receiver sensitivity of −23.2 dBm at BER = 10^{-9}. As shown in Figure 6.31, the optical receiver power (P_{Rx}) is measured at the input of the narrowband MSK receiver and the OSNR is monitored to obtain the BER curves for different fiber lengths. Length of SSMF is varied from 48 to 60 km in a step of 4 km to investigate the performance of the equalizers to the degradation caused by fiber-cumulative dispersion. The narrowband Gaussian filter with the time bandwidth product of 0.13 is used for the detection filters.

Electronic noise of the receiver can be modelled with an equivalent noise current density of the electrical amplifier of 20 pA/$\sqrt{\text{Hz}}$ and dark current of 10 nA. The key parameters of the transmission system are given in Table 6.2.

The Viterbi algorithm used with the MLSE equalizer has a constraint length of 6 (that is 2^5 number of states) and a trace back length of 30. Figure 6.32 shows the bit error rate (BER) performance of both nonlinear equalizers plotted against the required OSNR. The BER performance of the optical MSK receiver without any equalizers for 25 km SSMF transmission is also shown in Figure 6.32 for quantitative comparison. The numerical results are obtained via Monte Carlo simulation (triangular markers as shown in Figure 6.32) with the low BER tail of the curve linearly extrapolated. The OSNR penalty (at BER = 10^{-9}) versus residual dispersion corresponding to 48, 52, 56, and 60 km SSMF are presented in Figure 6.33. The MLSE scheme outperforms the modified MLSE schemes, especially at low OSNR. In the case of 60 km SSMF, the improvement at BER = 10^{-9} is approximately 1 dB, and in the case of 48 km SSMF, the improvement is about 5 dB. In case of transmission of 52 and 56 km SSMF, MLSE with Viterbi algorithm has 4 dB gain in OSNR compared to the modified scheme. With residual dispersion at 816, 884, 952, and 1020 ps/nm, at BER of 10^{-9}, the MLSE scheme requires 12, 16, 19, and 26 dB OSNR penalty, respectively, while the modified scheme requires 17, 20, 23, and 27 dB OSNR, respectively.

6.6.2.2 Transmission of 10 Gb/s Optical MSK Signals over 1472 km SSMF Uncompensated Optical Link

Figure 6.34 shows the simulation set-up for 10 Gb/s transmission of optical MSK signals over 1472 km SSMF. The receiver employs an optical narrowband

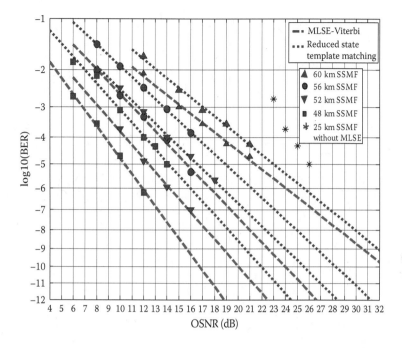

FIGURE 6.32
Performance of modified Verterbi-MLSE and template matching schemes, BER versus OSNR.

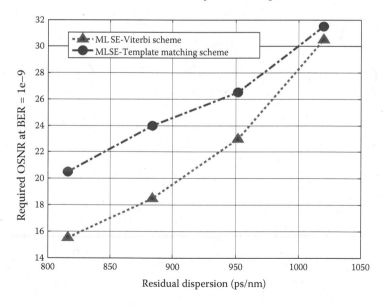

FIGURE 6.33
Required OSNR (at BER = 1e–9) versus residual dispersion in Viterbi-MLSE and template-matching MLSE schemes.

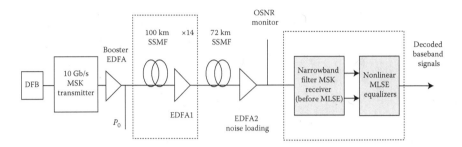

FIGURE 6.34
Simulation set-up for transmission of 10 Gb/s optical MSK signals over 1472 km SSMF with MLSE-Viterbi equalizer integrated with the narrowband optical filter receiver.

frequency discrimination receiver integrated with an 1024-state Viterbi-MLSE post-equalizer. The input power into fiber (P_0) is –3 dBm lower than the fiber non-linear threshold power. The optical amplifier EDFA1 provides an optical gain to compensate the attenuation of each span completely. The EDFA2 is used as a noise loading source to vary the required OSNR. The electronic noise receiver is modeled with equivalent noise current density of the electrical amplifier of 20 pA$\sqrt{\text{Hz}}$ and dark current of 10 nA for each photodiode. A narrowband optical Gaussian filter with two-sided bandwidth of 2.6 GHz (one-sided $BT = 0.13$) is optimized for detection. A back-to-back OSNR = 8 dB is required for BER at 10^{-3} for each branch. The corresponding received power is –25 dBm. This low OSNR is possible due to large suppression of noise after being filtered by narrowband optical filters. A trace back length of 70 is used in the Viterbi algorithm. Figure 6.34 shows the simulation results of BER versus the required OSNR for 10 Gb/s optical MSK transmission over 1472 and 1520 km SSMF uncompensated optical links with 1, 2, and 4 samples per bit, respectively. An SSMF of 1520 km, with one sample per bit, is seen as the limit for 1024-state Viterbi algorithm due to the slow roll-off and the error floor. However, 2 and 4 samples per bit can obtain error values lower than the FEC limit of 10^{-3}.

Thus, 1520 km SSMF transmission of 10 Gb/s optical MSK signals can reach the error-free detection with use of a high-performance FEC. In the case of 1472 km SSFM transmission, the error events follow a linear trend without sign of error floor and, therefore, error-free detection can be comfortably accomplished.

Figure 6.35 also shows the significant improvement in OSNR of 2 and 4 samples per bit over a 1 sample per bit counterpart, with values of approximately 5 dB and 6 dB, respectively. In terms of OSNR penalty at a BER of 1e–3 from back-to-back setup, 4 samples per bit for 1472 km and 1520 km transmission distance suffers 2 and 5 dB penalty, respectively.

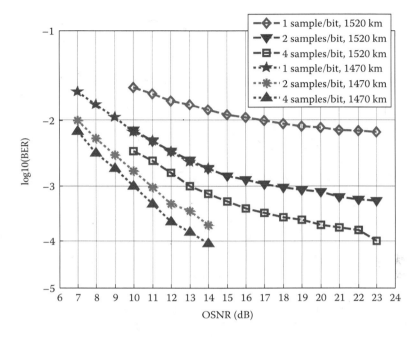

FIGURE 6.35

BER vs required OSNR for transmission of 10 Gb/s optical MSK signals over 1472 and 1520 km SSMF uncompensated optical links.

6.6.2.3 Performance Limits of Viterbi–MLSE Equalizers

The performance limits of the Viterbi-MLSE equalizer to combat ISI effects are investigated against various SSMF lengths of the optical link. The number of states used in the equalizer is incremented accordingly to this increase and varied from 2^6 to 2^{10}. This range was chosen as reflecting the current feasibility and the future advance of electronic technologies that can support high-speed processing of the Viterbi algorithm in the MLSE equalizer. In addition, these numbers of states also provide feasible time for simulation.

In addition, one possible solution to ease the requirement of improving the performance of the equalizer without increasing the complexity is by multi-sampling within one bit period. This technique can be done by interleaving the samplers at different times. Although a greater number of electronic samplers are required, they only need to operate at the same bit rate as the received MSK electrical signals. Moreover, it will be shown later on that there is no noticeable improvement with more than 2 samples per bit period. Hence, the complexity of the MLSE equalizer can be affordable while improving the performance significantly.

Figure 6.36 shows the simulation set-up for 10 Gb/s optical MSK transmission systems with lengths of uncompensated optical links varying up to

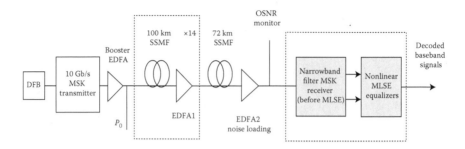

FIGURE 6.36
Simulation set-up of OFDR-based 10 Gb/s MSK optical transmission for study of performance limits of the MLSE-Viterbi equalizer.

1472 km SSMF. In this set-up, the input power into fiber P_0 is set to be −3 dBm, thus avoiding the effects of fiber nonlinearities. The optical amplifier EDFA1 provides an optical gain to compensate the attenuation of each span completely. The EDFA2 is used as a noise-loading source to vary the required OSNR values. Moreover, a Gaussian filter with two-sided 3-dB bandwidth of 9 GHz (one-sided BT = 0.45) is utilized as the optical discrimination filter because this BT product gives the maximized detection eye openings. The receiver electronic noise is modeled with equivalent noise current density of the electrical amplifier of $20\,pA/\sqrt{Hz}$ and a dark current of 10 nA for each photodiode. A back-to-back OSNR of 15 dB is required for a BER of 10^{-4} on each branch and a correspondent receiver sensitivity of −25 dBm. A trace back length of 70 is used for the Viterbi algorithm in the MLSE equalizer. Figures 6.36, 6.37, and 6.38 show the BER performance curves of 10 Gb/s OFDR-based MSK optical transmission systems over 928, 960, 1280, and 1470 km SSMF uncompensated optical links for a different number of states used in the Viterbi-MLSE equalizer. In these figures, the performance of Viterbi-MLSE equalizers is given by the plot of the BER versus the required OSNR for several detection configurations: balanced receiver (without incorporating the equalizer), the conventional single-sample per bit sampling technique, and the multi-sample per bit sampling techniques (2 and 4 samples per bit slot). The simulation results are obtained by the Monte-Carlo method.

The significance of multi-samples per bit-slot in improving the performance for MLSE equalizer in cases of uncompensated long distances is shown. It is found that the tolerance limits to the ISI effects induced from the residual CD of a 10 Gb/s MLSE equalizer using 2^6, 2^8, 2^{10} states are approximately equivalent to lengths of 928, 1280, and 1440 km SSMF, respectively. The equivalent numerical figures in the case of 40 Gb/s transmission are corresponding to lengths of 62, 80, and 90 km SSMF respectively.

In the case of 64 states over a 960 km SSMF uncompensated optical link, the BER curve encounters an error floor that cannot be overcome even by using high-performance FEC schemes. However, at 928 Km, the linear BER curve indicates the possibility of recovering the transmitted data with the

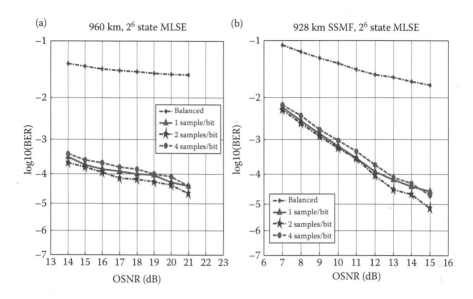

FIGURE 6.37

Performance of 64-state Viterbi-MLSE equalizers for 10 Gb/s OFDR-based MSK optical systems over (a) 960 km and (b) 928 km SSMF uncompensated link.

use of high-performance FEC. Thus, for 10 Gb/s of OFDR-based MSK optical systems, a length of 928 km SSMF can be considered the transmission limit for the 64-state Viterbi-MLSE equalizer. Results shown in Figure 6.38 suggest the length of an 1280 km SSMF uncompensated optical link is the transmission limit for the 256-state Viterbi-MLSE equalizer when incorporating with OFDR optical front-end. It should be noted that this is achieved when using the multi-sample per bit sampling schemes.

It is observed from Figure 6.39 that the transmission length of 1520 km SSMF with one sample per bit is seen as the limit for 1024-state Viterbi algorithm due to the slow roll-off and the error floor. However, 2 and 4 samples per bit can obtain error values lower than the FEC limit of 10^{-3}. Thus, 1520 km SSMF transmission of 10 Gb/s optical MSK signals can reach the error-free detection with use of a high-performance FEC. In the case of 1472 km SSFM transmission, the error events follow a linear trend without sign of error floor and, therefore, error-free detection can be achieved. Figure 6.39 also shows the significant OSNR improvement of the sampling techniques with 2 and 4 samples per bit, compared to the single-sample per bit counterpart. In terms of OSNR penalty at a BER of 1e–3 from back-to-back setup, 4 samples per bit for 1472 and 1520 km transmission distance suffers a 2 and 5 dB penalty, respectively.

It is found that for incoherent detection of optical MSK signal based on the OFDR, an MLSE equalizer using 2^4 states does not offer better performance than the balanced configuration of the OFDR itself, that is, the

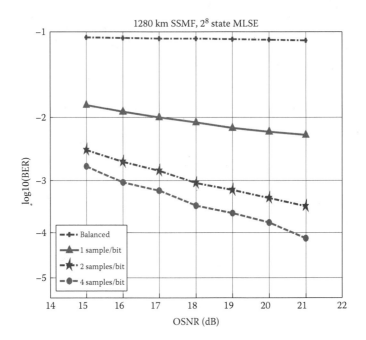

FIGURE 6.38
Performance of a 256-state Viterbi-MLSE equalizer for 10 Gb/s OFDR-based MSK optical systems over a 1280 km SSMF uncompensated link.

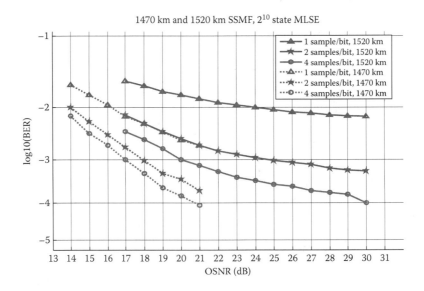

FIGURE 6.39
Performance of a 1024-state MLSE equalizer for 10 Gb/s optical MSK signals vs required OSNR over 1472 and 1520 km SSMF uncompensated optical links.

uncompensated distance is not over 35 km SSMF for 40 Gb/s or 560 km SSMF for 10 Gb/s transmission systems, respectively. It is most likely that in this case, the severe ISI effect caused by the optical fiber channel has spread beyond the time window of 5-bit slots (2 pre-cursor and 2 post-cursor bits), the window of which a 16-state MLSE equalizer can handle. Thus it is necessary that the number of taps or states of MLSE covers the full length of the dispersive pulse.

A trace-back length of 70 is used in the investigation, which guarantees the convergence of the Viterbi algorithm. The longer the trace-back length, the larger memory is required. With state-of-the-art technology for high storage capacity nowadays, memory is no longer a big issue. Very fast processing speed at 40 Gb/s operations hinders the implementation of 40 Gb/s Viterbi-MLSE equalizers at the mean time. Multi-sample sampling schemes offer an exciting solution for implementing fast signal processing. This challenge may also be overcome in the near future, along with the advance of the semiconductor industry. At present, the realization of Viterbi-MLSE equalizers operating at 10 Gb/s has been commercially demonstrated [37].

6.6.2.4 Viterbi–MLSE Equalizers for PMD Mitigation

Figure 6.40 shows the simulation test-bed for the investigation of MLSE equalization of the PMD effect. The transmission link consists of a number of spans comprised of 100 km SSMF (with $D = +17$ ps/nm · km, $\alpha = 0.2$ dB/km) and 10 km DCF (with D = -170 ps/nm · km, $\alpha = 0.9$ dB/km). Launched power into each span (P_0) is -3 dBm. The EDFA1 has a gain of 19 dB, hence providing input power into the DCF to be -4 dBm, which is lower than the nonlinear threshold of the DCF. The 10 dB gain of EDFA2 guarantees that input power into the next span goes unchanged, with an average power of -3 dBm. An OSNR of 10 dB is required for receiver sensitivity at BER of 10^{-4} in the case of back-to-back configuration. Considering the practical aspect and complexity a Viterbi-MLSE equalizer for PMD equalization, a small

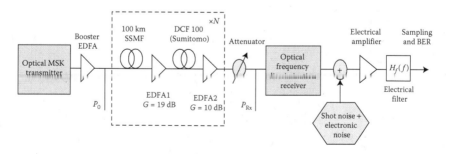

FIGURE 6.40
Simulation test-bed for investigation of effectiveness of an MLSE equalizer to PMD.

number of 4-state bits, or effectively 16 states, were chosen for the Viterbi algorithm in the simulation study.

The performance of MLSE against the PMD dynamic of optical fiber is investigated. BER versus required OSNR for different values of normalized average DGD <Δt> are shown in Figure 6.41. The mean DGD factor is normalized over one bit period of 100 ps and 25 ps for 10 and 40 Gb/s bit rate, respectively. The numerical studies are conducted for a range of values from 0 to 1 of normalized DGD <Δt>, which equivalently corresponds to the instant delays of up to 25 ps or 100 ps in terms of 40 and 10 Gb/s transmission bit rate, respectively.

The advantages of using multiple samples per bit over the conventional single sample per bit in MLSE equalizer are also numerically studied, and the Monte Carlo results for normalized DGD values of 0.38 and 0.56 are shown in Figures 6.42 and 6.43. However, the increase from 2 to 4 samples per bit does not offer any gain in the performance of the Viterbi-MLSE equalizer. Thus, 2 samples per symbol are preferred to reduce the complexity of the equalizer. Here and for the rest of this chapter, multi samples per bit implies the implementation of 2 samples per symbol. Performance of MLSE equalizer for different normalized DGD values with 2 samples per bit is shown in Figure 6.44. From Figure 6.44, the important remark is that a 16-state MLSE equalizer implementing 2 samples per symbol enables the optical MSK transmission systems achieving a PMD tolerance of up to one bit period at BER = 10^{-4} with a required OSNR of 8 dB. This delay value starts introducing a BER floor, indicating the limit of the 16-state MLSE equalizer. This problem

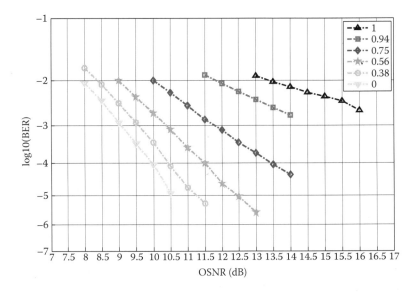

FIGURE 6.41
Performance of an MLSE equalizer versus normalized <$\Delta\tau$> values for 1 sample per bit.

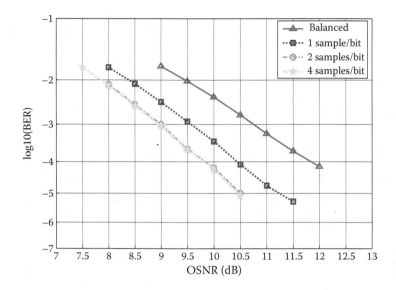

FIGURE 6.42
Comparison of MLSE performance for configurations of 1, 2, and 4 samples per bit with normalized <Δτ>; value of 0.38.

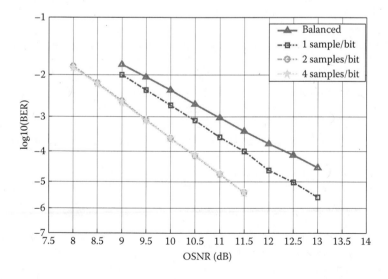

FIGURE 6.43
Performance of a Viterbi-MLSE equalizer for configurations of 1, 2, and 4 samples per bit with a normalized <Δt>; value of 0.56.

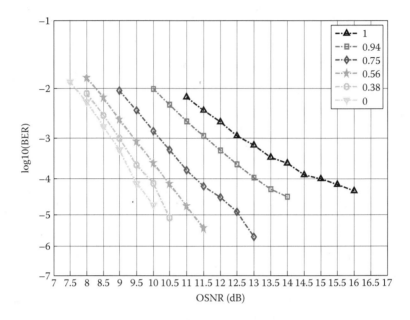

FIGURE 6.44
Performance of a Viterbi-MLSE equalizer for different normalized <Δt>; values with 2 and 4 samples per bit.

can be overcome with a high-performance FEC. However, the DGD mean value of 0.94 can be considered the limit for an acceptable performance of the proposed MLSE equalizer without aid of high-performance FEC. Figure 6.45 shows the required OSNR for MLSE performance at 10^{-4} versus normalized <Δt> values in configurations of a balanced receiver, 1 sampler per bit and 2 samples per bit.

A balanced receiver with no MLSE equalizer incorporated requires an OSNR = 5 dB to obtain a BER = 10^{-4} when there are no effects of the PMD compared to the required OSNR of 3 dB, and 1 dB in cases of 1 sample per bit and 2 samples per bit, respectively. The OSNR penalty for various normalized <Δt> values of the above three transmission system configurations are shown in Figure 6.46.

It can be observed that the OSNR penalties (back-to-back) of approximately 3 and 1 dB apply to the cases of balanced receiver and 1 sample per bit, respectively, with reference to the 2 samples per bit configuration. Another important remark is that a 16-state Viterbi-MLSE equalizer, which implements 2 samples per bit, enables the optical MSK transmission systems to achieve a PMD tolerance of up to one bit period at a BER of 10^{-4} with a power penalty of about 6 dB. Moreover, the best 2-dB penalty occurs at 0.75 for the value of normalized <Δt>. This result shows that the combination of OFDR-based MSK optical systems and Viterbi-MLSE equalizers, particularly with the use of multi-sample sampling schemes, was found to be highly effective

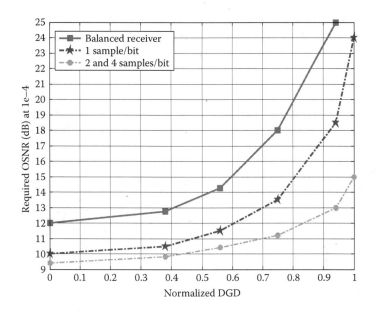

FIGURE 6.45
Required OSNR for MLSE performance at 10^{-4} versus normalized $<\Delta t>$; values in configurations of balanced receivers and 1 and 2 samples per bit.

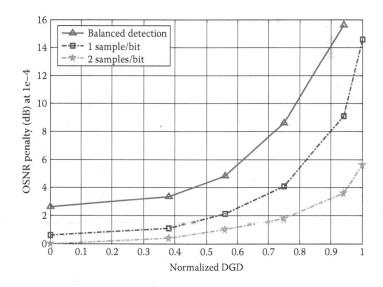

FIGURE 6.46
OSNR penalties of MLSE performance at 10^{-4} for various normalized $<\Delta t>$ values in configurations of balanced receiver, single-sample, and 2 samples per bit sampling schemes.

in combating the fiber PMD dynamic impairment and better than recently reported PMD performance for OOK and DPSK modulation formats [38].

6.6.2.5 On the Uncertainty and Transmission Limitation of Equalization Process

The fundamental limitations of a quadratic phase media on the transmission speed are formulated in signal space for significant applications in signal equalization. These limitations are quantitatively derived for both coherent and incoherent optical systems. This section investigates the dispersive effect of optical fiber on the transmission speed of the media. The main effect of dispersion is pulse broadening (and frequency chirping) as the signal pulses propagate through the quadratic phase media. This causes ISI and thereby limits the speed of transmission. There are various ways of mitigating the harmful effect of ISI. One way is to use optical fiber with antidispersive properties to equalize the dispersion. In practice, this optical scheme reduces signal level and thus requires amplification. The amplifier introduces noise, which in turn limits the channel speed (Shannon's information capacity theorem). Another way of combating ISI is by digital electronic equalization means, such as MLSE, which is applicable for both coherent and incoherent detection schemes [39–43]. These reported results show clearly that there is a fundamental limit on the achievable transmission speed for a given fiber length.

This section studies the limitations of quadratic phase channel, the single-mode optical fiber, in terms of signal space. The signal space is chosen so that it consists of binary signals in 8-bit block code. As the 8-bit symbols propagate down the quadratic phase channel, naturally the waveform patterns of the symbols would become less and less distinctive, hence more and more difficult to discriminate between symbols. How do we quantify the detrimental effect of the quadratic phase channel? Although it is obvious that the more information we have at hand the more accurate one would be able to obtain the BER. This system performance is approached entirely from a digital communication perspective, and the results so derived are the fundamental limit imposed by the quadratic phase channel (pure phase distortion but quadratic or square dependence of frequency by the linear dispersion). It is virtually independent of the detection scheme used. Two mechanisms that limit the transmission speed are explained. One is brought about by the finite time window available for detection. In all practical schemes, the decoder must decode each symbol within a finite time. This requires the algorithm to be the least complex as possible. The other is brought about by not using the phase information for detection. Depending on the complexity of the detection scheme chosen, it is quite often that the phase information may be lost inadvertently when the optical signal is converted into the electrical signal. Another issue is the power consumption of the application-specific integrated circuit (ASIC) for implementation of the digital algorithm for real-time applications. The more complex the algorithms, the higher the number of digital circuits, and hence the more power consumption. It is expected that all ASICs must consume less than 70 W in real-time processing.

References

1. N. Benuveto and G. Cheubini, *Algorithms for Communications Systems and Their Applications*, J Wiley, Chichester, England, 2004.
2. S. Savoy, Digital coherent optical receivers, *IEEE J. Sel. Areas Quant. Elect*, 16(5), 1164–1178, 2010.
3. A.P. Clark, *Equalizers for High Speed Modem*, Pentech Press, London, 1985.
4. L. Hanzo, S.X. Ng, T. Keller, and W. Webb, *Quadrature Amplitude Modulation: Froom Basics to Adaptive Trellis Coded, Turbo-Equalized and Space-Time Coded OFDM CDMA and MC-CDMA Systems*, J. Wiley and Sons, 2nd Ed., Chichester, England, 2004.
5. J. Winters, Equalization in coherent lightwave systems using a fractionally spaced equalizer, *IEEE J. Lightw. Technol.*, 8(10), 1487–1491, Oct. 1990.
6. R.W. Lucky, J. Salz, and E.J. Weldon, *Principles of Data Communications*, McGraw-Hill, New York, pp. 93–165, 1968.
7. A. Papoulis, *Probability, Random Variables and Stochastic Processes*, McGraw-Hill, New York, 1965.
8. S. Bennetto and E. Biglieri, *Principles of Digital Transmission with Wireless Applications*, Kluver Academic, New York, 1999.
9. R. Noe, T. Pfau, M. El-Darawy, and S. Hoffmann, Electronic polarization control algorithms for coherent optical transmission, *IEEE J. Sel. Quant. Elect.*, 16(5), 1193–1199, Sept 2010.
10. A.P. Clark, Pseudobinary viterbi detector, *Proc IEE, Part F*, 131, 208–218, 1984.
11. J.G. Proakis and M. Salehi, *Communication Systems Engineering*, 2nd Ed., Prentice Hall, Inc, New Jersey, 2002.
12. J.G. Proakis and M. Salehi, *Communication Systems Engineering*, Prentice Hall, Inc, 2nd Ed., New Jersey, 2002.
13. J.G. Proakis, *Digital Communications*, McGraw-Hill, 4th Ed., New York, pp, 185–213, 2001.
14. L.N. Binh, *Digital Optical Communications*, CRC Press, Boca Raton, Chapter 5, 2009.
15. K. Jinguji and M. Oguma Optical half-band filters, *IEEE J. Lightw. Tech.*, 18(2), 252–259, Feb. 2000.
16. L. Binh and V.A.T. Tran, Design of Photonic Half-band Filters Using Multirate DSP Technique, Technical report MECSE-27–2004. http://www.ds.eng.monash.edu.au/techrep/reports/2004/MECSE-27–2004.pdf.
17. S. Zhang, P.Y. Kam, J. Chen, and C. Yu, *A Comparison of Phase Estimation in Coherent Optical PSK System*, Photonics Global '08, Paper C3–4A-03. Singapore, December 2008.
18. A.P. Clark, *Equalizers for Digital Modem*, Pentec Press, London, 1985.

19. G. Nicholson, Probability of error for optical heterodyne DPSK system with quantum phase noise, *Electron. Lett.*, 20(24), 1005–1007, Nov. 1984.

20. S. Zhang, P.Y. Kam, J. Chen, and C. Yu, *A Comparison of Phase Estimation in Coherent Optical PSK System*, Photonics Global '08, Paper C3-4A-03. Singapore, December 2008.

21. J.P. Gordon and L.F. Mollenauer, Phase noise in photonic communications systems using linear amplifiers, *Opt. Lett.*, 15, 1351–1353, 1990.

22. D.-S. Ly-Gagnon, S. Tsukamoto, K. Katoh, and K. Kikuchi, Coherent detection of optical quadrature phase-shift keying signals with carrier phase estimation, *J. Lightwave Technol.*, 24, 12–21, 2006.

23. X. Liu, X. Wei, R.E. Slusher, and C.J. McKinstrie, Improving transmission performance in differential phase-shift-keyed systems by use of lumped nonlinear phase-shift compensation, *Opt. Lett.*, 27(18), 1351–1353, Sep. 2002.

24. X. Wei, X. Liu, and C. Xu, Numerical simulation of the SPM penalty in a 10-Gb/s RZ-DPSK system, *IEEE Photon. Technol. Lett.*, 15(11), 1636–1638, Nov. 2003.

25. E. Ip and J.M. Kahn, Digital equalization of chromatic dispersion and polarization mode dispersion, *IEEE J. Lightw. Tech.*, 25, 2033–2043, 2007.

26. R. Noe', PLL-free synchronous QPSK polarization multiplex/diversity receiver concept with digital I&Q baseband processing, *IEEE Phton. Technology Lett.*, 17, 887–889, 2005.

27. L.N. Binh, *Digital Optical Communications*, CRC Press, Taylor & Francis Group, Boca Raton, FL, 2009.

28. J.G. Proakis, *Digital Communications*, 4th Ed., McGraw-Hill, New York, 2001.

29. Y. Han and G. Li, Coherent optical communication using polarization multiple-input-multiple output, *Optics Express*, 13, 7527–7534, 2005.

30. N. Stojanovic, Y. Zhao, B. Mao, C. Xie, F.N. Hauske, and M. Chen, *Robust Carrier Recovery in Polarization Division Multiplexed Receivers*, OFC 2013, Annaheim, March 2013.

31. H. Meyr et al., *Digital Communication Receivers*, John Wiley & Sons, 1998.

32. A. Leven et al., Frequency estimation in intradyne reception, *IEEE Photon. Technol. Lett.*, 19, 366–368, 2007.

33. Z. Tao et al., *Simple, Robust, and Wide-Range Frequency Offset Monitor for Automatic Frequency Control in Digital Coherent Receivers*, ECOC 2007, Berlin, Germany, paper Tu3.5.4.

34. A.J. Viterbi and A.M. Viterbi, Nonlinear estimation of PSK-modulated carrier phase with application to burst digital transmission, *IEEE Transactions on Information Theory*, **29**, 543–551, 1983.

35. M. Kuschnerov et al., DSP for coherent single-carrier receivers, *J. Lightw. Technol.*, 27, 3614–3622, 2009.

36. Z. Tao et al., *Simple, Robust, and Wide-Range Frequency Offset Monitor for Automatic Frequency Control in Digital Coherent Receivers*, ECOC 2007, Berlin, Germany, paper Tu3.5.4.

37. http://www.coreoptics.com/product/prod_ic.php, USA, March 2010.

38. T. Sivahumaran, T.L. Huynh, K.K. Pang, and L.N. Binh, Non-linear equalizers in narrowband filter receiver achieving 950 ps/nm residual dispersion tolerance for 40 Gb/s optical MSK transmission systems, *Proceedings of OFC'07*, paper OThK3, 2007, Annaheim, USA.

39. P. Poggiolini, G. Bosco, M. Visintin, S.J. Savory, Y. Benlachtar, P. Bayvel, and R.I. Killey, *MLSE-EDC versus optical dispersion compensation in a single-channel SPM-limited 800 km link at 10 Gbit/s*, paper 1.3, ECOC2007, Berlin 2007.
40. M.S.A.S. Al Fiad, D. van den Borne, F.N. Hauske, A. Napoli, A.M.J. Koonen, and H. de Waardt, Maximum-likelihood sequence estimation for optical phase-shift keyed modulation formats, *IEEE Journal of Lightwave Technology*, 27(20), 4583–4594, 2009.
41. T. Sivahumaran, T.L. Huynh, K.K. Pang, and L.N. Binh, Non-linear equalizers in narrowband filter receiver achieving 950 ps/nm residual dispersion tolerance for 40 Gb/s optical MSK transmission systems, *Proceedings of OFC'07*, paper OThK3, CA, 2007.
42. V. Curri, R. Gaudino, A. Napoli, and P. Poggiolini, Electronic equalization for advanced Modulation formats in dispersion-limited systems, *Photon. Tech. Lett.*, 16(11), 2004.
43. G. Katz, D. Sadot, and J. Tabrikian, Electrical dispersion compensation equalizers in optical long-haul coherent-detection system, *Proceedings of ICTON'05*, paper We.C1.5, 2005.

7

DSP-Based Coherent Optical Transmission Systems

7.1 Introduction

In Chapter 6, a generic flow chart of the digital signal process was introduced (see Figure 6.1). This diagram is now reintroduced here, where the clock/timing recovered signals are feedback into the sampling unit of the analog to digital converter (ADC) so as to obtain the best correct timing for sampling the incoming data sequence for digital signal processing (DSP). Any errors made at this stage of timing will result into high deviation of the bit error rate (BER) in the symbol decoder shown in Figure 7.1. It is also noted that the vertical polarized channel (V-pol) and horizontal polarized (H-pol) channels are detected, and their inphase (I) and quadrature (Q) components are produced in the electrical domain with signal voltage conditioned for the conversion to digital domain by the ADC.

The processing of a sampled sequence from the received optical data and photodetected electronic signals passing through the ADC relies on the timing recovery from the sampled events of the sequence. The flowing stages of the blocks given in Figure 7.1 may be changed or altered accordingly depending on the modulation formats and pulse shaping, for example, the Nyquist pulse shapes in Nyquist superchannel transmission systems.

This chapter attempts to illustrate the performance of the processing algorithms in optical transmission systems, employing coherent reception techniques over highly dispersive optical transmission lines, especially the multispan optically amplified non-dispersion compensating module (DCM) long-haul distance. First, the quadrature phase shift keying (QPSK) homodyne scheme is examined, and then the 16 QAM incorporating both polarized channels multiplexed in the optical domain; hence the term polarization division multiplexed channels (PDM)-QPSK, or PDM-16 QAM. We then expand the study for superchannel transmission systems, in which several subchannels are closely spaced in the spectral region to increase the spectral efficiency so that the total effective bit rate must reach at least 1 Tb/s. Owing to the over-

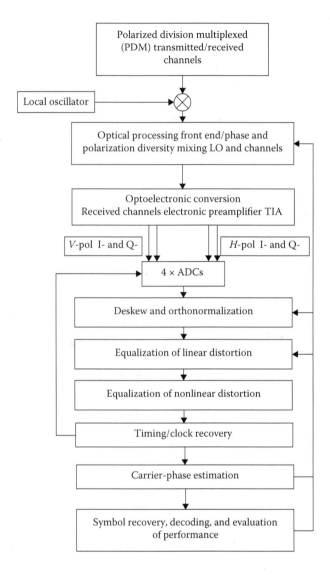

FIGURE 7.1

Flow of functionalities of DSP processing in a coherent optical receiver of a coherent transmission system. A modified diagram from Figure 6.1 with the feedback path diverted to ADC.

lapping of adjacent channels, there are possibilities that modifications of the processing algorithms are to be made.

Furthermore, the nonlinearity impairments on transmitted subchannels would be degrading the system performance and the application of back-propagation techniques described in Chapter 3 must be combined with linear and nonlinear equalization schemes so as to effectively combat performance degradation.

7.2 QPSK Systems

7.2.1 Carrier Phase Recovery

Homodyne coherent reception requires a perfect match of the frequency of the signal carrier and the LO. Any frequency difference will lead to phase noise of the detected signals. This is the largest hurdle for the first optical coherent system initiated in the mid-1980s. In DSP-based coherent reception systems, the recovery of the carrier phases and hence the frequency are critical to achieve the most sensitive reception with maximum performance in the BER, or evaluation of the probability of error. This section illustrates the recovery of the carrier phase for QPSK and 16 QAM optical transmission systems in constellations of single- or multilevel circular distribution. How would the DSP algorithms perform under the physical impairment effects on the recovery of the phase of the carrier?

7.2.2 112 G QPSK Coherent Transmission Systems

Currently, several equipment manufacturers are striving to provide advanced commercial optical transmission systems at 100 Gb/s employing coherent detection techniques for long-haul backbone networks and metro networks. Since the 1980s, it is well known that single-mode optical fibers can support such transmission due to the preservation of the guided modes and the polarized modes of the weakly guiding linearly polarized (LP) electromagnetic waves [1]. Naturally, both transmitters and receivers must satisfy the coherency conditions of narrow linewidth sources and coherent mixing with a local oscillator (LO), an external cavity laser (ECL), to recover both the phase and amplitude of the detected lightwaves. Thus far, both polarized modes of the LP modes can be stable over long distances in order to provide the PDM, even with polarization mode dispersion (PMD) effects. All linear distortion due to PMD and chromatic dispersion (CD) can be equalized in the DSP domain by employing algorithms in real-time processors, such as those provided in Chapter 6.

It is very important to ensure that these subsystems are performing the coherent detection and transmitting functions. This section thus presents a summary of the tests conducted with a back-to-back transmission of QPSK PDM channels. The symbol rate of the transmission system is 28 GSy/s under the modulation format PDM-CSRZ-QPSK. It is noted that the differential QPSK (DQPSK) encoder and the bit pattern generator are provided.

The transmission system is arranged as shown in Figure 7.2. The carrier-suppressed return-to-zero (CSRZ) QPSK transmitter consists of a CSRZ optical modulator, which is biased at the minimum transmission point of the transfer characteristics of the MZIM driven by sinusoidal signals with a frequency of half the symbol rate or 14 GHz for 28 GSy/s. A WDM multiplexer (mux) is employed to multiplex other wavelength channels located within the

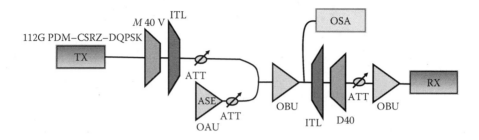

FIGURE 7.2
Setup of the PDM QPSK optical transmission system.

C-band (1530–1565 nm). An optical amplifier (EDFA type) is employed at the front end of the receiver so that noises can be superimposed on the optical signals to obtain the optical signal-to-noise ratio (OSNR). The DSP-processed signals in the digital domain are carried out offline and the BER is obtained.

The transmitter consists of an ECL, an encore type, a polarization splitter coupled with a 45° aligned ECL beam, two separate CS-RZ external LiNbO3 modulators, and then two I-Q optical modulators. The linewidth of the ECL is specified at about 100 kHz, and with external modulators, we can see that the spectrum of the output-modulated lightwaves is dominated by the spectrum of the baseband modulation signals. However, we observed that the laser frequency is oscillating about 300 MHz due to the integration of a vibrating grating so as to achieve stability of the optical frequency.

There are two types of receivers employed in this system. One is a commercialized type, Agilent N 4391 in association with an Agilent external LO ; the other one is a transimpedance amplifier (TIA) type, including a photodetector (PD) pair connected back to back (B2B) in a push-pull manner and then to a broadband TIA. In addition to the electronic reception part, a pi/2 hybrid coupler including polarization splitter, pi/2 phase shift, and polarization combiner that mixed the signal polarized beams and those of the LO (an ECL type identical to the one used in the transmitter) is employed as the optical mixing subsystem at the front end of the receiver. The mixed polarized beams (I-Q signals in the optical domain) are then detected by balanced receivers. I-Q signals in the electrical domain are then sampled and stored in the real-time oscilloscope (Tektronix 7200). The sampled I-Q signals are then processed offline using the algorithms provided in the sampling scope or externally loaded algorithms such as the evaluation of EVM described above for Q-factor and thence BER.

Both the transmitter and receivers are functioning with the required OSNR for the B2B of about 15 dB at a BER of 2e–3. It is noted that the estimation of the amplitude and phase of the received constellation is quite close to the received signal power and the noise contributed by the balanced receiver with a small difference, due to the contribution of the quantum shot noise contributed by the power of the LO. The estimation technique was described in Chapter 4. Figure 7.3 shows the BER versus OSNR for B2B QPSK

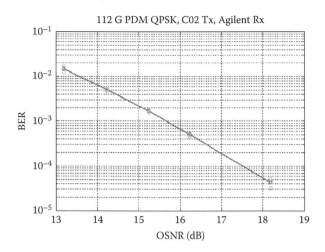

FIGURE 7.3
B2B OSNR versus BER performance with Agilent Rx in 112 Gb/s PDM-CSRZ-DQPSK system.

PDM channels. The variation of BER as a function of the signal energy over noise for different modulation formats is shown in Figure 7.4. For 4-QAM or QPSK coherent system then the SNR is expected at about 8 dB for BER of 1e–3. Experimental processing of such scheme in B2B configuration shows an OSNR of about 15.6 dB. This is due to 3 dB split by the polarized channels and then additional noises contributed by the receiver, hence about 15.6 dB OSNR required. The forward error coding (FEC) is set at 1e–3. The receiver is the Agilent type as mentioned above.

A brief analysis of the noises at the receiver can be as follows. The noise is dominated by the quantum shot noise generated by the power of the LO, which is at least 10 times greater than that of the signals. Thus, the quantum shot noises generated at the output of the PD are

$$i^2_{N-LO} = 2\,qI_{LO}B$$

$$I_{LO} = 1.8\,\text{mA} @ 0.9 - \text{PD\_quantum\_efficiency}$$

$$B = 31\,\text{GHz} - \text{BW\_U2t\_Agilent\_Rx}$$

$$i^2_{N-LO} = 2 \times 10^{-19} \times 1.8 \times 10^{-3} \times 30 \times 10^9$$

$$= 10.8 \times 10^{-12}$$

$$\longrightarrow i_{N-LO} = 3.286\,\mu\text{A}$$

The bandwidth of the electronic preamplifier of 31 GHz is taken into account. This shot noise current due to the LO imposed on the PD pair is compatible with that of the electronic noise of the electronic receiver given

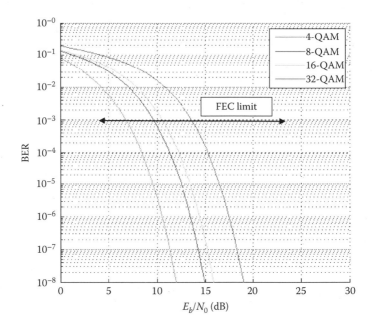

FIGURE 7.4
Theoretical BER versus SNR for different-level QAM schemes obtained by bertool.m of MATLAB.

that the noise spectral density equivalent at the input of the electronic amplifier of the U2t balanced receiver is specified at $80\,pA/\sqrt{Hz}$. For example,

$$(80 \times 10^{-12})^2 \times 30.10^9 = 19.2 \times 10^{-12} A^2 \xrightarrow[\text{noise\_current}]{} i_{Neq.} = 4.38\,\mu A$$

Thus, any variation in the LO would affect this shot noise in the receiver. It is thus noted that with the transimpedance of the electronic preamplifier estimated at $150\,\Omega$, a dBm difference in the LO would contribute to a change of the voltage noise level of about 0.9 mV in the signal constellation obtained at the output of the ADC. A further note is that the noise contributed by the electronic front end of the ADC has not been taken into account. We note that differential TIA offers at least 10 times higher transimpedance, around $3000\,\Omega$ over 30 GHz mid-band frequency response. However, these differential TIA does have narrow dynamic ranges due to the clipping of the output signals. These TIAs offer much higher sensitivities as compared to single-input TIA type [2,3].

7.2.3 I–Q Imbalance Estimation Results

Imbalance occurs due to the propagation of the polarized channels and the I and Q components. They must be compensated in order to minimize the error. The I-Q imbalance of the Agilent BalRx and U²t BalRx is less than 2

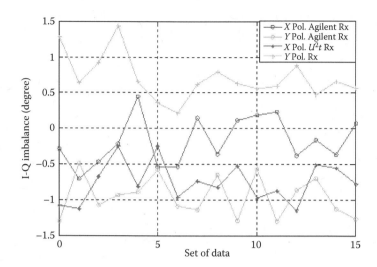

FIGURE 7.5
I-Q imbalance estimation results for each Rx. Note a maximum imbalance phase of ±1.5°.

degrees, which might be negligible for the system, as shown in Figure 7.5. This imbalance must be compensated for in the DSP domain.

7.2.4 Skew Estimation

In addition to the imbalance of the I and Q due to optical coupling and electronic propagation in high-frequency cables, there is also propagation delay time difference between these components that must be compensated. The skew estimation is shown in Figure 7.6, obtained over a number of data sets.

Abnormal skew variation from time to time was also observed, which should not happen if there is no modification on the hardware. Considering the skew variation happened with the Agilent receiver, which only has a very short RF cable and tight connection, there is a higher probability that the skew happened inside the Tektronix oscilloscope than at the optical or electrical connection outside.

Figure 7.7 shows the BER versus the OSNR when the skew and the imbalance between I and Q components of the QPSK transmitter are compensated. The OSNR of DQPSK is improved by about 0.3 ~ 0.4 dB at 1e–3BER, compared to the result without I-Q imbalance and skew compensation. In the time domain, compared to the result obtained in 2009, the required OSNR of DQPSK at a BER of 1e–3 is about 14.7 dB, is improved by 0.1 dB. The required OSNR at 1e–3BER of QPSK is about 14.7 dB, which is the common requirement for QPSK modulation format. For a BER = 1e–3, an imbalanced CMRR = –10 dB would create a penalty of 0.2 dB in the OSNR for the Agilent receiver and an improvement of 0.7 dB for a balanced commercial receiver employed in the Rx subsystems.

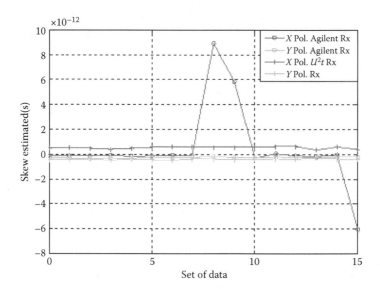

FIGURE 7.6
Skew estimation results for both types of Rx.

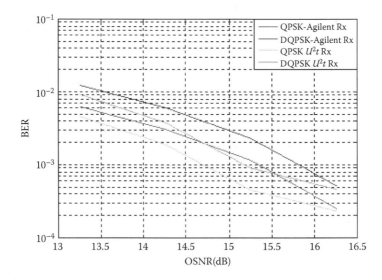

FIGURE 7.7
OSNR versus BER for two types of integrated coherent receivers after compensating for I-Q imbalance and skew.

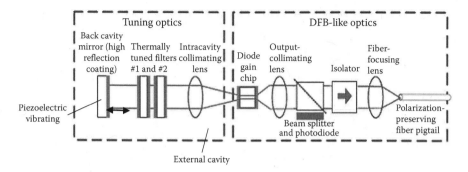

FIGURE 7.8
External cavity structure of the laser with the back mirror vibration in the external cavity structure.

Figure 7.8 shows the structure of the ECL incorporating a reflection mirror, which is vibrating at a slow frequency of around 300 MHz. A control circuit would be included to indicate the electronic control of the vibration and cooling of the laser so as to achieve stability and elimination of Brillouin scattering effects.

7.2.5 Fractionally Spaced Equalization of CD and PMD

Ip and Kahn [4] have employed the fractional-spaced equalization scheme with mean square error (MSE) to evaluate its effectiveness of PDM amplitude shift keying (ASK) with a nonreturn to zero (NRZ) or return to zero (RZ) pulse-shaping transmission system. Their simulation results are displayed in Figure 7.9 for the maximum allowable CD (normalized in ratio with respect to the dispersion parameter of the single-mode fiber as defined in Chapter 2), versus the number of equalizer taps N for a 2-dB power penalty at a launched OSNR of 20 dB per symbol for ASK, RZ, and NRZ pulse shapes using a Bessel antialiasing filter with sampling rate $1/T = M/KT_s$; T_s = symbol_period, with M/K as the fractional ratio. This is exhibited in Figure 7.10, with a fractional ratio of (a) $M/K = 1$, (b) 1.5, (c) 2 using Bessel filter type, and (d), (e), and (g) using a Butterworth antialiasing filter with the $M/K = 1$, 1.5, and 2, respectively. The sampling antialiasing filter is employed to ensure that artificial fold back to the spectrum is avoided. The filter structures of Bessel and Butterworth exhibit similar performance for fractional spaced equalizers but less tap for Bessel filtering cases when equally spaced equalizers are used (see Figure 7.10a and c).

7.2.6 Linear and Nonlinear Equalization and Back-Propagation Compensation of Linear and Nonlinear Phase Distortion

Ip and Kahn [5] first developed and applied the back propagation algorithm as given in Chapter 2 to equalize the distortion due to nonlinear impairment

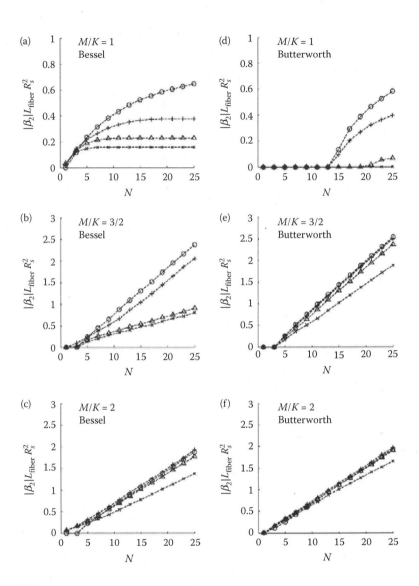

FIGURE 7.9

Maximum allowable CD versus the number of equalizer taps N for a 2-dB power penalty at an input SNR of 20 dB per symbol for RZ and NRZ pulse shapes, using a Bessel antialiasing filter. "o" denotes transmission using NRZ pulses, "x" denotes 33% RZ, "Δ" denotes 50% RZ, and "+" denotes 67% RZ, by fractional spaced equalizer (FSE) and sampling rate R_s. $1/T = M/KT_s$; T_s = symbol_period, M/K is the fractional ratio, (a) $M/K = 1$, (b) 1.5, (c) 2 using Bessel filter type and (d), (e), and (g) using a Butterworth antialiasing filter with $M/K = 1$, 1.5, and 2, respectively. The modulation scheme is ASK with NRZ and RZ pulse shaping. N, number of taps of the equalizer. (After E. Ip and J. M. Kahn, *IEEE J. Lightw. Tech.*, 25(8), 2033, August 2007. With permission.)

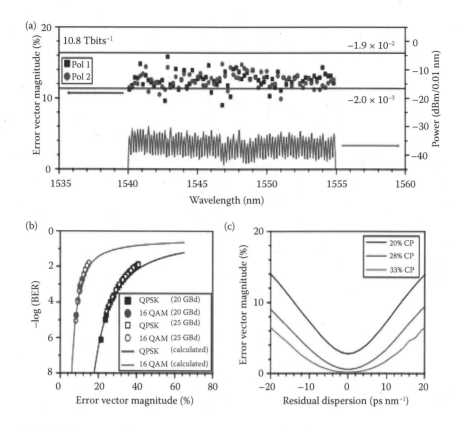

FIGURE 7.10
All-optical 10.8 Tbit/s OFDM results. (a) Measured error vector magnitude (EVM) for both polarizations (symbols) and for all subcarriers of the OFDM signal decoded with the all-optical FFT. The estimated BER for all subcarriers is below the third-generation FEC limit of 1.9e–2. The optical spectrum far left race is drawn beneath. (b) Relationship between BER and EVM. Measured points (symbols) and calculated BER as a function of EVM for QPSK and 16 QAM. (c) Tolerance of residual CD in the implemented system, decoded with the eight-point FFT for a cyclic prefix of 20, 28, and 33. (Extracted from I. Fatadin and S. J. Savory, *IEEE Photonic Tech. Lett.*, 23(17), 1246–1248, 2001.)

of optical channel transmission through the single-mode optical fibers. The back-propagation algorithm is simply a reverse-phase rotation at the end of each span of the multispan link. The rotating phase is equivalent to the phase exerted on the signals in frequency domain with a square of the frequency dependence. Thus, this back propagation is efficient in the aspect that the whole span can be compensated so as to minimize the numerical processes, hence, less processing time and central processing unit time of the digital signal processor.

Figure 7.11 shows the equalized constellations of a 21.4 GSy/s QPSK modulation scheme system after transmission through 25 × 80 km non-DCF spans

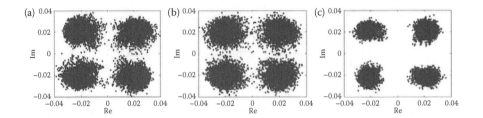

FIGURE 7.11
Constellation of QPSK scheme as monitored (a) with linear CD equalization only, (b) with nonlinear phase noise compensation, and (c) after back-propagation processing combining with linear equalization. (Simulation results extracted from E. Ip and J. M. Kahn, *IEEE J. Lightw. Tech.*, 26(20), 3416, October 15, 2008.)

under the equalization using (a) linear compensation only, (b) nonlinear equalization, and (c) using combined back propagation and linear equalization. Obviously the back propagation contributes to the improvement of the performance of the system.

Figure 7.12 shows the phase errors of the constellation states at the receiver versus launched power of 25 × 80 km multispan QPSK 21.4 GSy/s transmission system. The results extracted from Ref. [5] show the performance of back-propagation phase rotation per span for 21.4 Gb/s 50% RZ-QPSK transmitted

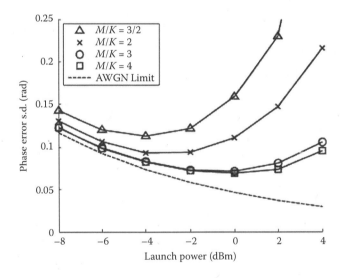

FIGURE 7.12
Phase errors at the receiver versus launched power or 25 × 80 km multispan QPSK at 21.4 GSy/s. Performance of back-propagation phase rotation per span for 21.4 Gb/s 50% RZ-QPSK transmitted over 25 × 80 km spans of SMF with 5 ROADMs, with 10% CD undercompensation using fractionally spaced equalizers with an M/K factor of 1.5 to 4. (Simulation results extracted from E. Ip and J. M. Kahn, *IEEE J. Lightw. Tech.*, 26(20), 3416, October 15, 2008.)

over 25×80 km spans of SMF with 5 reconfigurable optical add-drop module (ROADMs), with 10% CD undercompensation. The algorithm is processed offline with received sample data after 25×80 km SSMF propagation via the use of the nonlinear Schrödinger equation (NLSE) and coherent reception technique described in Chapter 4. It is desired that the higher the launched power, the better the OSNR that can be employed for longer-distance transmission. So, fractional space ratio of 3 and 4 offer higher launched power and thus are the preferred equalization schemes as compared with equal space or sampling rates equal to that of the symbol rate. The ROADM is used to equalize the power of the channel under consideration as compared to other DWDM channels.

7.3 16 QAM Systems

Consider the 16 QAM received symbol signal with a phase Φ denoting the phase offset. The symbol d_k denotes the magnitude of the QAM symbols and n_k the noises superimposed on the symbol at the sampled instant. The received symbols can be written as

$$r_k = d_k e^{j\phi} + n_k; \quad k = 1, 2, ..., L \tag{7.1}$$

Using the maximum likelihood sequence estimator (MLSE), the phase of the symbol can be estimated as

$$l(\phi) = \sum_{k=1}^{L} \ln \left\{ \sum_d e^{-\frac{1}{2\sigma^2}\left|r_k - de^{j\phi}\right|^2} \right\} \tag{7.2}$$

or effectively, one would take the summation of the contribution of all states of the 16 QAM on the considered symbol measured as the geometrical distance in natural logarithmic scale with the noise contribution of a standard deviation σ.

The frequency offset estimation for 16 QAM can be conducted by partitioning the 16 QAM constellation into a number of basic QPSK constellations, as shown in Figure 7.13. There are two QPSK constellations in the 16 QAM with symbols that can be extracted from the received sampled data set. They are then employed to estimate the phase of the carrier as described in the previous section on carrier phase estimation for QPSK modulated transmission systems. At first, a selection of the innermost QPSK constellation, classified as class I symbols, thence an estimation of the frequency offset of the 16 QAM transmitted symbols can be made using the constant amplitude

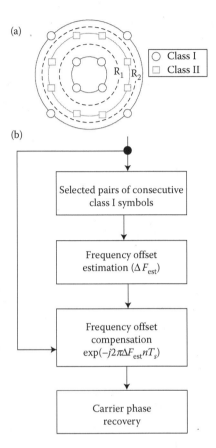

FIGURE 7.13

Processing of 16 QAM for carrier phase estimation [6]: (a) constellation of 16 QAM, and (b) processing for carrier phase recovery with classes I and II of circulator subconstellation.

modulus algorithm. Thence, an FO compensation algorithm is conducted and the phase recovery of all 16 QAM symbols can be derived. Further confirmation of the difference of carrier phase recovery or estimation can be conducted with the constellation of class I, as indicated in Figure 7.13a.

Carrier phase recovery based on the Viterbi–Viterbi algorithm on the class I QPSK subconstellation of the 16 QAM may not be sufficient, and so a modified scheme has been reported by Fatadin et al. in Ref. [7]. Further refining this estimation of the carrier phase for 16 QAM by partition and rotating is made so as to match certain symbol points to those of class I constellation of the 16 QAM. The procedures are as shown in the flow diagram of Figure 7.14.

Partition the constellation into different classes of constellation patterns and then rotate class 2 symbols with an angle either clockwise or anticlockwise of $\pm\theta_{rot} = \pi/4 - \tan^{-1}(1/3)$. To avoid opposite rotation with respect to the

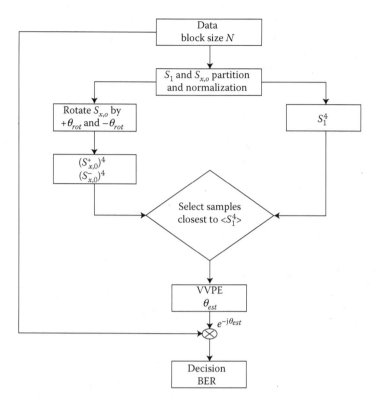

FIGURE 7.14
Refined carrier phase recovery of 16 QAM by rotation of class I and class II subconstellations.

real direction, the estimation of the error in the rate of changes of the phase variation or frequency estimation can be found by use of the fourth power of the argument of the angles of two consecutive symbols, given by

$$\Delta F_{est.} = \frac{1}{8\pi T_s} \arg\left(\sum_{k=0}^{N} S_{k+1} S_k\right)^4 \tag{7.3}$$

to check their quadratic mean, then select the closer symbol, then apply the standard Viterbi–Viterbi procedure. Louchet et al. did also employ a similar method [8] and confirmed the effectiveness of such a scheme. The effects on the constellation of the 16 QAM due to different physical phenomena are shown in Figure 7.15. Clearly, the FO would generate the phase noises in (a) and influence both the I and Q components by the CD of small amount (so as to see the constellation noises) and the delay of the polarized components on the I and Q components. These distortions of the constellation allow practical engineers to assess the validity of algorithms that are normally separate

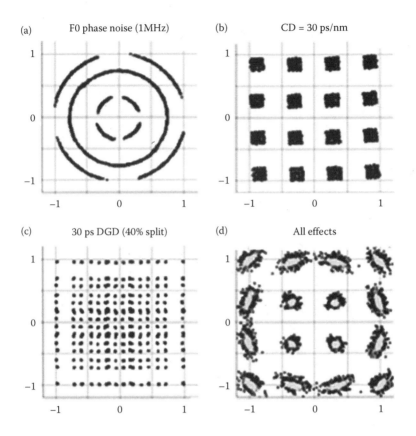

FIGURE 7.15
16 QAM constellation under influence: (a) phase rotation due to FO of 1 MHz and no amplitude distortion; (b) residual CD impairment; (c) DGD of PMD effect; and (d) total phase noises effect. (Adapted from I. Fatadin and S. J. Savory, *IEEE Photonic Tech. Lett.*, 23(17), 1246–1248, 2001.)

and independent and implemented in serial mode. This is in contrast to the constellations illustrated for QPSK as shown in Figure 7.16. Figure 7.17 also shows the real-time signals that result from the two sinusoidal waves of FO beating in a real-time oscilloscope.

Noe et al. [9] have simulated the carrier phase recovery for QPSK with polarization-division-multiplexed (PDM) channels under CMA and decision-directed (DD) with and without modification, in which the error detected in each stage would be updated. The transmission system under consideration is B2B with white Gaussian noises and phase noises superimposition on the signals. For an OSNR of 11 dB, the CMA with modification offers a BER of 1e–3 while it is 4e–2 for CMA without modification. This indicates that the updating of the matrix coefficients is very critical to recover the original data sequence. The modified CMA was also recognized to be valid for 16 QAM.

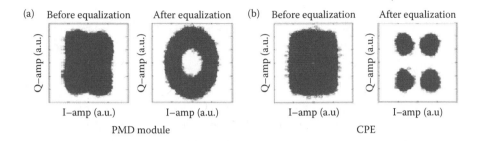

(a) Before equalization   After equalization (b) Before equalization   After equalization

PMD module   CPE

FIGURE 7.16
Constellation after PMD module (a) and after CPE algorithm module (b).

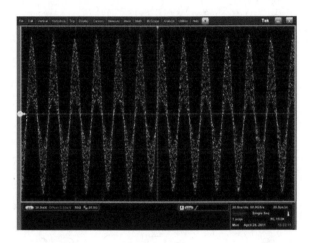

FIGURE 7.17
Beating signals of the two mixed lasers as observed by real-time sampling oscilloscope.

7.4 Tera-Bits/s Superchannel Transmission Systems

7.4.1 Overview

PDM-QPSK has been exploited as the 100 Gb/s long-haul transmission commercial system, as the optimum technologies for 400 GE/1TE transmission for next-generation optical networking have now attracted significant interests for deploying ultra-high-capacity information over the global Internet backbone networks. Further intense research on Tbps transmission systems have also attracted several research groups as the logical rate to increase from 100 Gb/s. The development of hardware platforms for 1 to N Tbps is critical for proving the design concept. The Tbps can be considered a superchannel, which is defined an optical channel comprising of a number of subrate subchannels whose spectra would be the narrowest allowable. Thus, to achieve efficient spectral efficiency, phase shaping is required, and one of the most efficient techniques is Nyquist pulse shaping. Thus, Nyquist QPSK can be considered the most effective format for the delivery of high spectral efficiency and coherent transmission and reception, as well as equalization at both transmitting and reception ends.

In this section, we describe a detailed design and experimental platform for the delivery of Tbps using Nyquist-QPSK at a symbol rate of 28–32 GSa/s and 10 subcarriers. The generation of subcarriers has been demonstrated using either recirculating frequency shifting (RFS) or nonlinear driving of an I-Q modulator to create five subcarriers per main carrier; thus, two main carriers are required. Nyquist pulse shaping is used to effectively pack multiplexed channels in which carriers are generated by a comb generation technique. A digital to analog converter (DAC) with a sampling rate varying from 56G to 64 GS/s is used for generating Nyquist pulse shape, including the equalization of the transfer functions of the DAC and optical modulators.

7.4.2 Nyquist Pulse and Spectra

The raised-cosine filter is an implementation of a low-pass Nyquist filter, that is, one that has the property of vestigial symmetry. This means that its spectrum exhibits odd symmetry, about $1/2T_s$, where T_s is the symbol-period. Its frequency-domain representation is a "brick-wall-like" function, given by

$$H(f) = \begin{cases} T_s & |f| \leq \dfrac{1-\beta}{2T_s} \\ \dfrac{T_s}{2}\left[1 + \cos\left(\dfrac{\pi T_s}{\beta}\left\{|f| - \dfrac{1-\beta}{2T_s}\right\}\right)\right] & \dfrac{1-\beta}{2T_s} < |f| \leq \dfrac{1+\beta}{2T_s} \\ 0 & \text{otherwise} \end{cases} \quad (7.4)$$

with $0 \leq \beta \leq 1$

This frequency response is characterized by two values: β, the *roll-off factor* (ROF), and T_s, the reciprocal of the symbol-rate in Sym/s, that is, $1/2T_s$ is the half bandwidth of the filter. The impulse response of such a filter can be obtained by analytically taking the inverse Fourier transformation of Equation 7.4, in terms of the normalized sinc function, as

$$h(t) = \text{sinc}\left(\dfrac{t}{T_s}\right)\dfrac{\cos\left(\dfrac{\pi\beta t}{T_s}\right)}{1 - \left(2\dfrac{\pi\beta t}{T_s}\right)^2} \quad (7.5)$$

where the ROF, β, is a measure of the *excess bandwidth* of the filter, that is, the bandwidth occupied beyond the Nyquist bandwidth, as from the amplitude

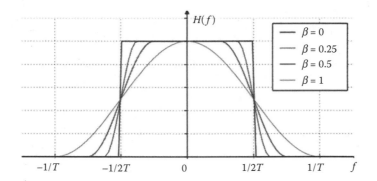

FIGURE 7.18
Frequency response of a raised cosine filter with various values of the ROF β.

at $1/2T$. Figure 7.18 depicts the frequency spectra of raised-cosine pulse with various ROFs. Their corresponding time domain pulse shapes are given in Figure 7.19.

When used to filter a symbol stream, a Nyquist filter has the property of eliminating intersymbol interference (ISI), as its impulse response is zero at all nT (where n is an integer), except when $n = 0$. Therefore, if the transmitted waveform is correctly sampled at the receiver, the original symbol values can be recovered completely. However, in many practical communications systems, a matched filter is used at the receiver, so as to minimize the effects of noise. For zero ISI, the net response of the product of the transmitting and receiving filters must equate to $H(f)$, thus, we can write

$$H_R(f)H_T(f) = H(f) \tag{7.6}$$

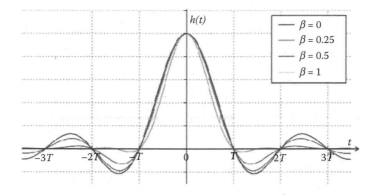

FIGURE 7.19
Impulse response of raised-cosine filter with the ROF β as a parameter.

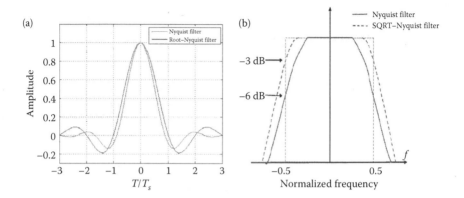

FIGURE 7.20
(a) Impulse and (b) corresponding frequency response of sinc. Nyquist pulse shape or root-raised-cosine (RRC) Nyquist filters.

Or alternatively we can rewrite that

$$|H_R(f)| = |H_T(f)| = \sqrt{|H(f)|} \tag{7.7}$$

The filters that can satisfy the conditions of Equation 7.7 are the root-raised-cosine filters. The main problem with root-raised-cosine filters is that they occupy larger frequency bands than those of the Nyquist sinc pulse sequence. Thus, for the transmission system, we can split the overall raised-cosine filter with the root-raised-cosine filter at both the transmitting and receiving ends, provided the system is linear. This linearity is to be specified accordingly. An optical fiber transmission system can be considered linear if the total power of all channels is under the nonlinear SPM threshold limit. When it is over this threshold, a weakly linear approximation can be used.

The design of a Nyquist filter influences the performance of the overall transmission system. Oversampling factor, selection of ROF for different modulation formats, and finite impulse response (FIR) Nyquist filter design are key parameters to be determined. If taking into account the transfer functions of the overall transmission channel, including fiber, WSS, and the cascade of the transfer functions of all O/E components, the total channel transfer function is more Gaussian-like. To compensate this effect in the Tx-DSP, one would thus need a special Nyquist filter to achieve the overall frequency response equivalent to that of the rectangular or raised cosine with ROF shown in Figures 7.20 and 7.21.

7.4.3 Superchannel System Requirements

Transmission distance: As the next generation of backbone transport, the transmission distance should be comparable to the previous generation, namely, 100 Gbps transmission system. As the most important requirement, we

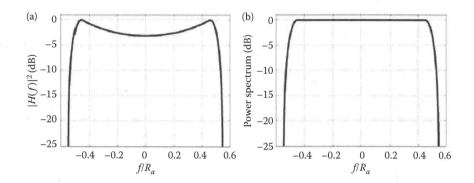

FIGURE 7.21
(a) Desired Nyquist filter for spectral equalization. (b) Output spectrum of the Nyquist filtered QPSK signal.

require that the 1 Tbps transmission for long-haul be 1500 ~ 2000 km, for metro application ~300 km.

CD tolerance: As the SSMF fiber CD factor/coefficient 16.8 ps/nm is the largest among the current deployed fibers, CD tolerance should be up to 30,000 ps/nm at the central channel, with a wavelength approximated at 1550 nm. At the edge of C-band, this factor is expected to increase by about 0.092 ps/(nm².km), or about 32, ps/nm at 1560 nm and 26,400 ps/nm at 1530 nm [10].

PMD tolerance: The worst case of deployed fiber with 2000 km would have a DGD of 75 ps or about three symbol period for 25 GSy/s per subchannel. So, the PMD (mean all-order DGD) tolerance is 25 ps.

SOP rotation speed: According to the 100 Gbps experiments, SOP rotation can be up to 10 kHz; we take the same spec as the 100G system.

Modulation format: PDM-QPSK for long-haul transmission; PDM-16QAM for metro application.

Spectral efficiency: Compared to the 100G system with an increasing of factor 2. Both Nyquist-WDM and CO-OFDM can fulfill this. However, it depends on technological and economical requirements that would determine the suitability of the technology for optical network deployment.

Table 7.1 tabulates the system specifications of various transmission structures with parameters of subsystems, especially when the comb generators employed are using either recirculating or nonlinear generation techniques. The DSP-reception and offline DSP is integrated in these systems.

7.4.4 System Structure

7.4.4.1 DSP-Based Coherent Receiver

A possible structure of a superchannel transmission system can be depicted in Figure 7.22. The received data streams can then be pulse-re-shaped, thence individual data streams can be formed. A DAC can be used to shape

TABLE 7.1

Tbps Offline System Specifications

Parameter	Superchannel		Some Specs	Remarks
	RCFS Comb Gen	Nonlinear Comb Gen		
Technique				
Bit rate	1,2,...,N Tbps (whole C-band)	1, 2,...,N Tbps	~1.28 Tbps @ 28–32 GB	20% OH for OTN, FEC
Number of ECLs	1	N × 2		
Nyquist roll-off α	0.1 or less	0.1 or less		DAC preequalization required
				Pending FEC coding allowance
Baud rate (GBauds)	28–32	28–32	28, 30, or 31.5 GBaud	
Transmission distance	2500	2500	1200 (16 span) ~ 2000 km (25 spans) 2500 km (30 spans)	20% FEC required for long-haul application
			500 km	Metro application
Modulation format	QPSK/16 QAM	QPSK/16 QAM	Multicarrier Nyquist WDM PDM-DQPSK/QAM	For long haul
				For long haul
			Multicarrier Nyquist WDM PDM-16QAM	For metro
Channel spacing			4 × 50 GHz	For long haul
			2 × 50 GHz	For long haul

Launch power	≪0 dBm if 20 Tbps is used		~−3 to 1 dBm lower if $N > 2$	Depending on QPSK/16 QAM and long haul/metro, can be different
B2B ROSNR @ 2e−2 (BOL) (dB)	14.5	14.5	15 dB for DQPSK 22 dB for 16 QAM	1 dB hardware penalty 1 dB narrow filtering penalty
Fiber type	SSMF G.652 Or G.655	SSMF G.652 (or 655)	G.652 SSMF	
Span loss	22	22	22 dB (80 km)	
Amplifier	EDFA (G > 22 dB); NF < 5 dB		EDFA (OAU or OBU)	
BER	2e−3	2e−3	Pre-FEC 2e−2 (20%) or 1e−3 classic FEC (7%)	
CD penalty (dB)			0 dB @ ±3000 ps/nm < 0.3 dB @ ±30,000 ps/nm	16.8 ps/nm/km and 0.092 ps/(nm²·km)
PMD penalty (DGD)			0.5 dB @ 75 ps, 2.5 symbol periods	
SOP rotation speed	10 kHz	10 kHz	10 kHz	OPLL may be required due to oscillation of the LO carrier
Filters cascaded penalty			<1 dB @ 12 pcs WSS	
Driver linearity	Required	Required	THD <3%	16 QAM even more strict

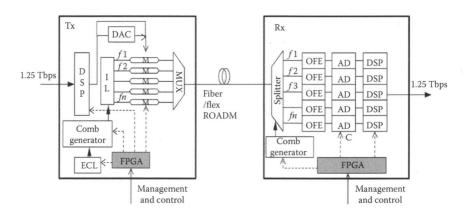

FIGURE 7.22
A possible structure of superchannel Tbps transmission system.

the pulse with a Nyquist-equivalent shape, hence a raised cosine shape pulse sequence can be formed whose spectrum follows that of a raised cosine function with roll-off factor β, varying from 0.1 to 0.5. If this off factor takes the value of 0.1, the spectra would follow an approximate shape of a rectangle. A comb generator can be used to generate equally spaced subcarriers for the superchannel from a single carrier laser source, commonly an ECL of a very narrow band of linewidth of about 100 kHz. These comb-generated subcarriers (see Figure 7.23) are then demultiplexed into subcarriers and fed into a bank of I-Q optical modulators as described in Chapter 3, and a Nyquist-shaped pulse sequence as output of the DAC is then employed to modulate these subcarriers to form the superchannels at the output of an optical multiplexer, shown in the block at the left side of Figure 7.22. More details of the transmitter for superchannels are shown in Figure 7.24. The superchannel of 1.25Tbps input data streams are fed into a bank of digital processors DSPs and are superimposed by the pulse shaping. They are then fed into microwave power amplifiers to drive the optical modulators to generate the inphase and quadrature optical waves to be launched into the transmission fiber link. It is noted that the generation of a comb source can be implemented by recirculating of the RF frequency-shifting of the original carrier around a close optical loop incorporating an optical modulator. The frequency shift is the spacing frequency between the subcarriers, so the Nth subcarrier would be the Nth time circulation of the original carrier. There would also be superimposing of noises due to the ASE incorporated in the loop, which would be minimized by inserting into the loop an optical filter with a bandwidth the same or wider than that of the superchannel.

The fiber transmission line can be optically amplified at each fiber span without incorporating any dispersion compensating fibers (DCF). The transmission is very dispersive. The broadening of a 40 ps width pulse would

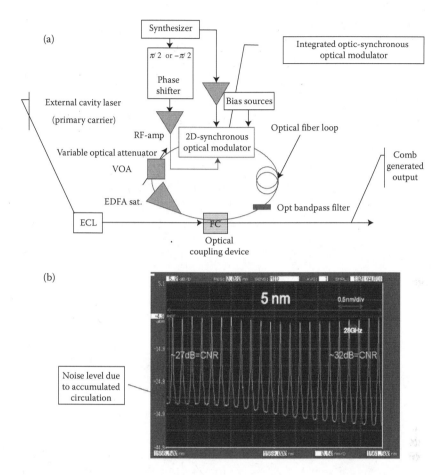

FIGURE 7.23

Block diagram of an RFS comb generator (a) and a typical generated spectrum of the comb generator (b) with 28 GHz spacing between channels over more than 5 nm in spectral region and about 30 dB carrier-to-noise ratio (CNR).

spread across at least 80–100 symbol period after propagating over 3000 km of SSMF. Thus, one can assume that the pulse launched into the fiber of the first span would be considered an impulse as compared to that after a 3000 km SSMF propagation.

After the propagation over the multispan non-DCF line, the transmitted subchannels are demuxed via a wavelength splitter into individual subchannels, with minimum crosstalk. Each subchannel is then coherently mixed with an LO, which is generated from another comb source incorporating an optical phase lock loop (OPLL) to lock the comb into that of the subcarriers of the superchannel. Thus, a comb generator is indicated in the right side, the reception system of Figure 7.22. The coherently mixed subchannels are then

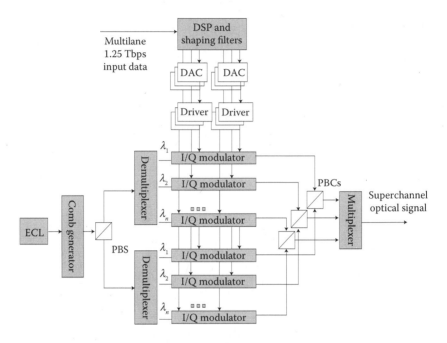

FIGURE 7.24
Generic detailed architecture of a superchannel transmitter. Generic schematic structure of optical recirculating and frequency shifting loop. DAC, digital to analog converter; ECL, external cavity laser; I/Q, inphase–quadrature phase; PBC, polarization beam combiner, PBS, polarization beam splitter.

detected by a balanced receiver (as described in Chapter 4 and Section 7.2 of this chapter), then electronically amplified and fed into the sampler and ADC. The digital signals are then processed in the DSP of each subchannel system or parallel and interconnected DSP system. In these DSPs, the sequence of processing algorithms is employed to recover the carrier phase and hence the clock recovery, compensating for the linear and nonlinear dispersion and the evaluation of the BER versus different parameters such as OSNR, and so on. Figure 7.25 shows the modulated spectra of five channels with subcarriers selected from the multiple subcarrier source of Figure 7.22. The modulation is QPSK with Nyquist pulse shaping.

7.4.4.2 Optical Fourier Transform-Based Structure

A superchannel transmission system can also be structured using optical fast Fourier transform (OFFT), as demonstrate in Ref. [11] and shown in Figure 7.26, in which MZDI components (see Figure 7.26a) act as spectral filters and splitters (see Figure 7.26b). The outputs of these MZDI are then fed into coherent receivers and processed digitally as in Figure 7.26c, with the electroabsorption modulator (EAM) performing the switching function so as to time demultiplex

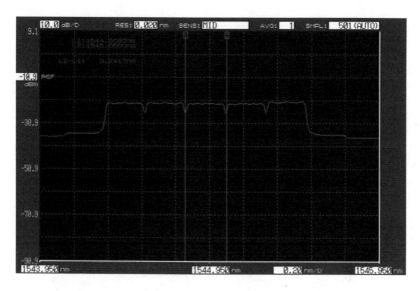

FIGURE 7.25
Selected five subcarriers with modulation.

the ultrafast signal speed to a lower-speed sequence so that the detection system can decode and convert to the digital domain for further processing.

The spectra of superchannels at different positions in the transmission system can be seen in Figure 7.27a through c. It is noted that the pulse shape follows that of a Nyquist function as given in Section 1.3.2.2 and the subchannels are placed close and satisfying the orthogonal condition; thus, the name optical orthogonal frequency division multiplexing (OFDM) is used to indicate this superchannel arrangement.

7.4.4.3 Processing

The processing of superchannels can be considered similar to the digital processing of individual subchannels, except when there may be crosstalk between subchannels due to overlapping of certain spectral regions between the considered channel and its adjacent channels.

Thus, for the Nyquist QPSK subchannel, the DSP processing would be much the same as for QPSK (dense wavelength division multiplexing (DWDM) for 112 G as described above, with care taken for the overlapping either at the transmitter or at the receiver. The error vector magnitude is a parameter that indicates the scattering of the vector formed by I and Q components departing from the center of the constellation point. The variance of this EVM in the constellation plane is used to evaluate the noises of the detected states, thence the Q-factor can be evaluated with ease and the BER evaluated by using the probability density function and the magnitude of the vector of a state on the constellation plane. The BER of the subchannels of the OFDM superchannel is shown

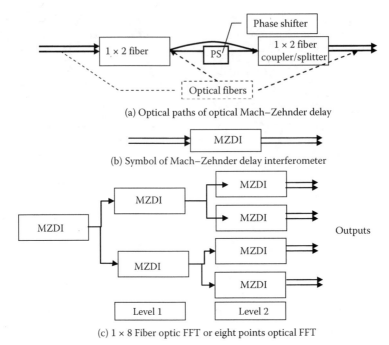

(a) Optical paths of optical Mach–Zehnder delay

(b) Symbol of Mach–Zehnder delay interferometer

(c) 1 × 8 Fiber optic FFT or eight points optical FFT

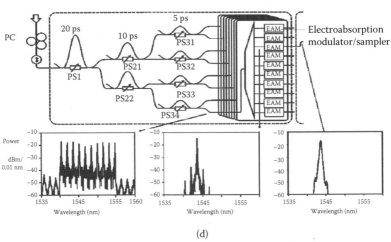

(d)

FIGURE 7.26
Operations by guided wave components using fiber optics (a) guided wave optical path of a Mach–Zehnder delay interferometer or asymmetric interferometer with phase delay, tunable by thermal or electrooptic effects; (b) block diagram representation; (c) implementation of optical FFT using cascade stages of fiber optical MZDI structure; and (d) spectra of optical signals at different locations of the optical FFT. Electroabsorption modulator (EAM) used for demultiplexing in time domain. Note the phase shifters employed in MZDIs between stages. Inserts are spectra of optical signals at different stages, as indicated by the OFFT (serial type). PC, polarization controller. (Extracted from D. Hillerkuss et al., *Nat. Photonics*, 5, 364, June 2011.)

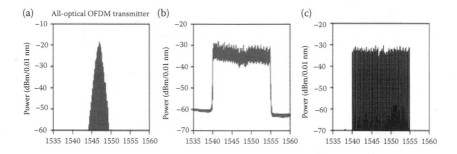

FIGURE 7.27
Spectrum of superchannels and demuxed channels. All subchannels are orthogonal and thus the name OFDM. (a) At output of transmitter, (b) after fiber propagation, and (c) after polarization demux. (Extracted from I. Fatadin and S. J. Savory, *IEEE Photonic Tech. Lett.*, 23(17), 1246–1248, 2001.)

in Figure 7.10b and c [9] for different percentage overloading due to FEC. The loading factor is important, as this will increase the speed or symbol rate of the subchannel one has to offer. The higher this percentage, the higher the increase in the symbol rate, thus requiring high-speed devices and components.

In the receiver of the optical OFDM superchannel system of Ref. [9], to judge the effectiveness of the optical FFT receiver, three alternative receiver concepts were tested for a QPSK signal. A QPSK signal is chosen because it was not possible to receive a 16 QAM signal with the alternative receivers, owing to their inferior performance [9]. First, a subcarrier with a narrow bandpass filter is used to extract a subchannel. The filter passband is then adjusted for best performance of the received signal (Figures 7.10 and 7.27a). The selected filter bandwidth is 25 GHz, and the constellation diagram shows severe distortion. When using narrow optical filtering, one has to accept a compromise between crosstalk from neighboring channels (as modulated OFDM subcarriers necessarily overlap) and ISI, owing to the increasing length of the impulse response when narrow filters are used. Narrow filters can be used, however, if the ringing from ISI is mitigated by additional time gating. The reception of a subcarrier using a coherent receiver is then performed. In the coherent receiver, the signal is down-converted in a hybrid coupler as described in Chapter 5, and detected using balanced detectors and sampled in a real-time oscilloscope. Using a combination of error low-pass filtering due to the limited electrical bandwidth of the oscilloscope and DSP, the subcarrier is extracted from the received signal. This receiver performs better than the filtering approach, but a larger electrical bandwidth and sampling rate of the ADC and additional DSP would be needed to eliminate the crosstalk from other subcarriers and then to achieve a performance similar to that of the optical FFT.

Thus, the optical OFDM may offer significant advantages for superchannel, but additional processing time would be required. The formation of Nyquist pulse-shaped QPSK superchannels offers better performance and reduces the complexity in the receiver DSP subsystem structure.

7.4.5 Timing Recovery in Nyquist QAM Channel

Nyquist pulse shaping is one of the efficient methods to pack adjacent sub-channels into a superchannel. The timing recovery of such Nyquist subchannels is critical for sampling the data received and improving the transmission performance. Timing recovery can be done either before or after the PMD compensator. The phase detector scheme is shown in Figure 7.28, a Godard type [12], which is a first-order linear scheme. After CD compensation (CD^{-1} blocks), the signal is sent to a state-of-polarization (SOP) modifier to improve the clock extraction. The clock performance of an NRZ QPSK signal in the presence of a first-order PMD characterized by a differential group delay (DGD) and azimuth is presented in Figure 7.3. The azimuth of 45° and DGD of a half symbol/unit interval (UI) completely destroys the clock tone. Therefore, the SOP modifier is required for enabling the clock extraction. In practical

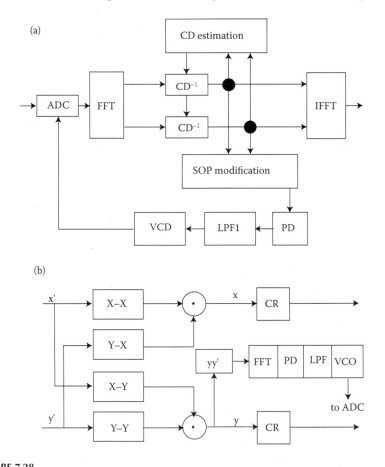

FIGURE 7.28
(a) Godard phase detector algorithm; (b) 4PPD. CD, chromatic dispersion; VCO, voltage control oscillator; FFT, fast Fourier transform; IFFT, inverse FFT; SOP, state of polarization.

systems, a raised-cosine filter is used to generate Nyquist pulses. A filter pulse response is defined by two parameters, the ROF β and the symbol period T_s, and described by taking the inverse Fourier transform of Equation 7.4. The Godard phase detector cannot recover the carrier phase, even with small β; thus, the channel spectra is close to rectangular. A higher-order phase detector must be used to effectively recover the timing clock period as shown in Ref. [13]. A fourth-power law PD (4PPD) with prefiltering presented in Ref. [14] and as shown in β values. Figure 7.28b can deal with small β values.

The 4PPD operates by first splitting and forming the combination of these components X- and Y-polarized channels then conducting the frequency domain detection and regenerating through the VCO the frequency shift required for the ADC to ensure the sampling timing is correct with the received sample for processing.

The use of such a phase detector in coherent optical receivers requires large hardware effort. Owing to PMD effects, the direct implementation of this method before PMD compensation is almost impossible. Therefore, the 4PPD implementation in the frequency domain after the PMD compensation is proven to be the most effective and well-performing even in the most extreme cases, with ROF equal to zero.

7.4.6 128 Gb/s 16 QAM Superchannel Transmission

An experimental setup by Dong et al. [15] for the generation and transmission of six channels carrying 128-Gbit/s under modulation format and a polarization multiplexing PDM-16QAM signal. The two 16-Gbaud electrical 16 QAM signals are generated from the two arbitrary waveform generators. Laser sources at 1550.10 nm and the second source with a frequency spacing of 0.384 nm (48 GHz) are generated from two ECLs, each with a line-width less than 100 kHz and the output power of 14.5 dBm. Two I/Q MODs are used to modulate the two optical carriers with I and Q components of the 64-Gb/s (16-Gbaud) electrical 16 QAM signals after the power amplification using four broadband electrical amplifiers/drivers. Two phase shifters with the bandwidth of 5 k–22.5 GHz provide two-symbol extra delay to decorrelate the identical patterns. For the operation to generate 16 QAM, the two parallel Mach–Zehnder modulators (MZMs) for I/Q modulation are both biased at the null point and driven at full swing to achieve zero-chirp and phase modulation. The phase difference between the upper and the lower branch of I/Q MZIM is also controlled at the null point. The data input is shaped so that about 0.99 roll-off factor of the raised-cosine pulse shape could be generated.

The power of the signal is boosted using polarization-maintaining EDFAs. The transmitted optical channels are then mixed with an LO and polarization demultiplexed via a pi/2 hybrid coupler, then I and Q components are detected by four pairs of balanced PDs. They are then transimpedance amplified and resampled.

The sampled data are then processed in the following sequence: CD compensation, clock recovery, then resampling and going through classical CMA, a three-stage CMA, frequency offset compensation, a feedforward phase equalization, LMS equalizer (LMSE), and then differentially detected to avoid cycle slip effects.

Furthermore, the detailed processing [16] for the electrical polarization recovery is achieved using a three-stage blind equalization scheme: (i) First, the clock is extracted using the "square and filter" method, then the digital signal is resampled at twice the baud rate based on the recovery clock. (ii) Second, a T/2-spaced time-domain finite impulse response (FIR) filter is used for the compensation of CD, where the filter coefficients are calculated from the known fiber CD transfer function using the frequency-domain truncation method. (iii) Third, adaptive filters employing two complex-valued, 13 tap coefficients and partial T/2-space are employed to retrieve the modulus of the 16 QAM signal.

The two adaptive FIR filters are based on the classic constant modulus algorithm (CMA) and followed by a three-stage CMA to realize multimodulus recovery and polarization demultiplexing. The carrier recovery is performed in the subsequent step, where the fourth power is used to estimate the frequency offset between the LO and the received optical signal. The phase recovery is obtained by feedforward and the least-mean-square (LMS) algorithms for LO frequency offset (LOFO) compensation. Finally, differential decoding is used for BER calculation after decision.

The spectra of the six Nyquist channels under BTB and 1200 km SSMF transmission are shown in Figure 7.29. It is noted that even with 0.01 roll-off of Nyquist channels, we do not observe the flatness of individual channels in the diagram. The constellation as expected would require a significant amount of equalization and processing.

On average, the achieved BER of 2e–3 with a launched power of –1 dBm over 1200 km SSMF nondispersion compensation link was demonstrated by

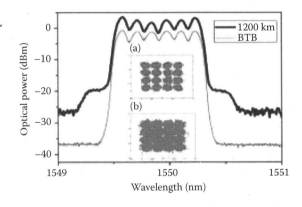

FIGURE 7.29
Spectra of Nyquist channels, 6 × 128G PDM 16 QAM back-to-back and 1200 km transmission.

the authors of Ref. [16]. The link is determined by a recirculating loop consisting of four spans, with each span equal to 80 km SSMF and an inline EDFA. Wavelength selective switch (WSS) is employed wherever necessary to equalize the average power of the subchannels. The optimum launched power is about –1 dBm for the six-subchannel superchannel transmission.

7.4.7 450 Gb/s 32 QAM Nyquist Transmission Systems

Further spectral packing of subchannels in a superchannel can be done with Nyquist pulse shaping and predistortion or preequalization at the transmitting side. Zhou et al. [17] have recently demonstrated the generation and transmission of 450 Gb/s wavelength-division multiplexed (WDM) channels over the standard 50 GHz ITU-T grid optical network at a net spectral efficiency of 8.4 b/s/Hz. This result is accomplished by the use of Nyquist-shaped PDM 32-QAM, or 5 bits/symbol × 2 (polarized modes) × 45GSy/s to give 450 Gb/s. Both pre- and posttransmission digital equalization techniques are employed to overcome the limitation of the DAC bandwidth. Nearly ideal Nyquist pulse shaping with ROF of 0.01 allows guard bands of only 200 MHz between subcarriers. To mitigate the narrow optical filtering effects from the 50 GHz-grid ROADM, a broadband optical pulse-shaping method is employed. By combining electrical and optical shaping techniques, the transmission of 5 × 450 Gb/s PDM-Nyquist 32 QAM on the 50 GHz grid over 800 km and one 50 GHz-grid ROADM was proven with soft DSP equalization and processing. The symbol rate is set at 28GSy/s.

It is noted that the transmission SSMF length is limited to 800 km due to the reduced Euclidean geometrical distance between constellation points of 32 QAM and by avoiding the accumulated ASE noises contributed by EDFA in each span. Raman optical amplifiers with distributed gain are used in a recirculating loop of 100 km ultralarge area fibers. The BER performance for all five subchannels is shown in Figure 7.30a with insertion of the spectra of all subchannels. Note the near-flat spectrum of each subchannel that indicates the near-Nyquist pulse shaping. Figure 7.30b and c shows the spectra of a single subchannel that performs as spectra shaping before and after WSS with and without optical filtering. The original pulse shape can be compared with that displayed in Figure 1.15 of Chapter 1. In addition to this published work, Zhou et al. have also demonstrated the transmission of time-multiplexed 64 QAM over 1200 km [19], that is, three circulating times around the ring perimeter of a 400 km which is formed by 4 × 100 km span of SSMF and Raman-pumped amplifiers. About 5 dB penalty between 32 QAM and 64 QAM in receiver sensitivity at the same FEC BER of 6e–3 is obtained. Note that electrical time division multiplexing of digital sequences from the arbitrary waveform generators was implemented by interleaving so that higher symbol rates can be achieved. Furthermore, 20% soft decision FEC using quasicyclic low density parity code (LDPC) is employed to achieve a BER threshold of 2.4e–2; thus, the 20% extra overhead is required on the symbol rate.

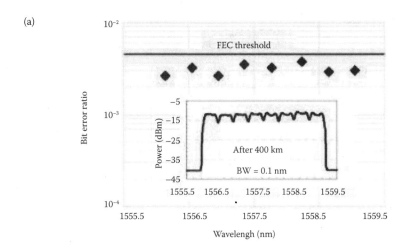

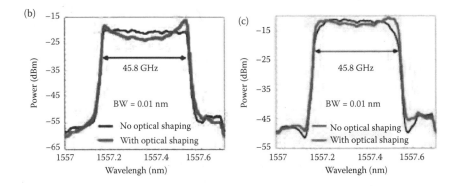

FIGURE 7.30

(a) Spectra of five subchannels of 450 Gb/s channel, that is, 5 × 450 Gb/s superchannels after 400 km (one loop circulating) of 5 × 80 km plus Raman amplification. BER versus wavelength and FEC threshold at 2.3e–3. (b) Spectrum of a single subchannel before and (c) after WSS with and without optical shaping by optical filters. Dark line for original spectra, light line for spectra after spectral shaping. (Extracted from X. Zhou et al., *OFC 2012*, CA, USA with permission.)

Simulated results by Bosco et al. [19] show that without using soft FEC, the variation of the maximum reach distance with a BER of 2e–3 for channels in the C-band (bottom axis) and spectral efficiency for PM-BPSK (polarization multiplexed binary phase shift keying), PM-QPSK (polarization multiplexed quadrature phase shift keying), PM-8QAM (polarization multiplexed quadrature amplitude modulation 8-level), and PM-6QAM (polarization multiplexed quadrature amplitude modulation 16-level) as shown in Figure 7.31 for SSMF non-DCF optical transmission lines as well as nonzero dispersion shifted fibers (NZ-DSF) indicated by dashed lines. The design of NZ-DSF was described in Chapter 2.

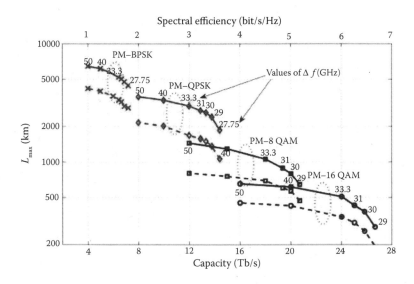

FIGURE 7.31

Transmission reach distance variation with respect to spectral efficiency and total capacity of superchannels. Different QAM schemes are indicated. PM, polarization multiplexing. The numbers are for bandwidth of the subchannel. (Extracted from G. Bosco et al., *IEEE J. Lightw. Tech.*, 29(1), 53, January 2011.)

7.4.8 DSP-Based Heterodyne Coherent Reception Systems

We have so far described optical transmission systems under homodyne coherent reception, that is, when the frequencies of the lightwave carriers and that of the LO are equal. The original motivation of using homodyne detection is to eliminate the 3 dB degradation as compared to heterodyne technique, and this is possible via the DSP-based reception algorithm to avoid the difficulties of locking of the LO and the channel carrier. Under the classical heterodyne reception, the "at least" 3 dB loss comes from the splitting of the received signals in the electrical domain and then multiplying them by the sinusoidal *cosine* and *sine* RF oscillator to extract the inphase and quadrature components.

So far we have discussed homodyne reception DSP-based optical transmission systems, which are considered for extensive deployment in commercial coherent communication systems for 100G, 400G, 1T, or beyond. However, with the development of large-bandwidth and high-speed electronic ADCs and PDs, once again, coherent detection with DSP has been allowed the mitigation of impairments in optical transmission and can be compensated by equalization in the electrical domain. As we have seen in the previous sections of this chapter and in Chapter 3, for homodyne detection in PDM systems, the I/Q components of each polarization state should be separated in optical domain with full information. Thus, four balanced

PD pairs incorporating a photonic dual-hybrid structure and four-channel time-delay synchronized ADCs are required.

By upconverting I and Q components to the intermediate frequency (IF) at the same time, not only can heterodyne coherent detect half the number of the balanced PDs and ADCs of the coherent receiver, but there is also no need to consider the delays between I and Q components in the PDM signal. Therefore, the four output ports of the optical hybrid can also be halved accordingly. However, this heterodyne technique can possibly be restricted by the bandwidth of the PDs. Furthermore, the external down conversion of the IF signals may enhance the complexity of the reception system. Currently, the tremendous progresses in increasing the sampling rate and bandwidth of ADCs and PDs give a high possibility to exploit a simplified heterodyne detection. With large-bandwidth PDs and ADCs, down conversion of the IF, I/Q separation of quadrature signal, and equalization for the PDM and nonlinear effect can all be realized in the digital domain. The time domain signals are then sampled by an ADC and then processed in the DSP. A heterodyne detection in a transmission system is a limited 5-Gb/s 4-ary quadrature amplitude modulation (4 QAM) signal over 20 km in Ref. [20] and limited 20-Mbaud 64 and 128 QAM over 525 km in Ref. [21], then reaching Tbps by Dong et al. [22]. High-order modulation formats, such as PDM-16QAM and PMD-64QAM, taking advantage of high capacity, can offer spectral effectiveness for 100 G or beyond by adopting heterodyne detection.

For the 100 G or beyond coherent system with required transmission distance shorter than 1000 km, the inferiority of the SNR sensitivity in heterodyne detection is not so obvious. Conversely, less number of ADCs and easy implementation of the DSP for IF down conversion make heterodyne detection a potential candidate for a transmission system of 100 G or beyond. Figure 7.32a shows the schematic diagram of the heterodyne reception with digital processing. The quadrature amplitude-modulated signals are transmitted and imposed onto the $\pi/2$ hybrid coupler, which is now simplified, and there is no $\pi/2$ phase shifter as compared to the hybrid coupler discussed in Chapter 4 and the previous sections of this chapter. The coherent mixing of the LO and the modulated channel would result in time domain signals, which are RF envelopes covering the lightwaves and limited within the symbol period. Thus, both the real and imaginary parts appear in the electronic signals produced after the balanced detection in the PDPs, where the beating happens (refer to Figure 7.32b). The electronic currents produced after the PDPs are then amplified via the TIA to produce voltage-level signals that are conditioned to appropriate levels so that they can be sampled by the ADCs. Thence, doing the FFT will produce the two-sided spectrum that will exhibit the frequency shifting of the baseband to the RF or IF frequency. Both the inphase and quadrature components are embedded in the two-sided spectrum. One sideband of the spectrum can be used to extract the I and Q parts of the QAM signals for further processing. Ideally, the compensation of the CD should be done in the first stage, followed by the carrier phase recovery, and then resampling with correct timing.

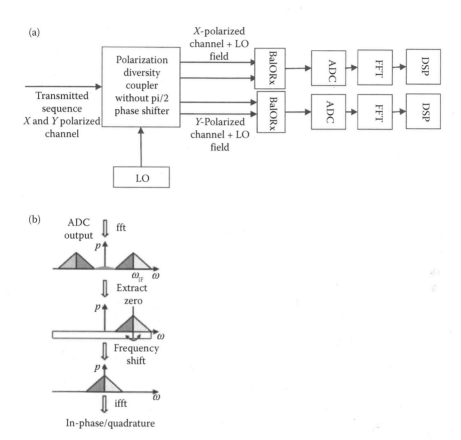

FIGURE 7.32

The principles of heterodyne coherent reception. (a) Principal blocks. (b) Spectra to recover baseband signals. (Adapted from Z. Dong et al., *Opt. Express*, 21(2), 1773, January 28, 2013.)

The following sequence of processing in the digital domain was conducted: (i) sampling in ADC and conversion from analog voltage level to digital sampled states; (ii) FFT to obtain frequency domain spectrum; then (iii) extract one-sided spectrum and do a frequency shifting to obtain the baseband samples of the spectrum; (iv) resampling with 2 times the sampling rate; hence (v) compensation of CD and (vi) carrier phase and clock recovery; (vii) using the recovered clock to resample the data sequence; then (viii) conducting normal CMA and three-stage CMA to obtain the initial constellation; (ix) equalization of frequency difference of the LO; (x) feedforward phase equalization and LMS equalization for PMD compensation; then finally (xi) differential decoding of the samples and symbol to determine the transmission performance BER with respect to certain OSNR measured in the optical domain and launched power at the input of the transmission line.

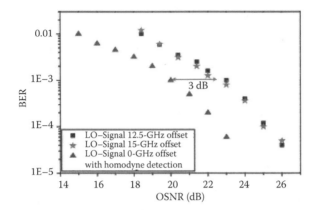

FIGURE 7.33
Back-to-back BER versus OSNR of the heterodyne reception system of PDM-16QAM 112 Gb/s ×8 superchannel with different IF frequency. (Extracted from Z. Dong et al., *Opt. Express*, 21(2), 1773, January 18, 2013. With permission.)

It is further noted that for the equalization based on DSP, a T/2-spaced time-domain FIR filter is used for the compensation of CD. The two complex-valued, 13-tap, T/2-spaced adaptive FIR filters are based on the classic CMA followed by three-stage CMA, to realize multimodulus recovery and polarization demultiplexing.

The back-to-back BER versus the OSNR of the heterodyne reception optical transmission system is shown in Figure 7.33 for PDM-16QAM 128 Gb/s per subchannel with different IF and compared with homodyne detection scheme. It is noted that a 3 dB penalty due to heterodyne detection is suffered by the receiver as compared to the homodyne reception because only one-sided spectrum of the mixed signals can be recovered in the photodetector. Thus, at BER of 4e–3, then the required OSNR is about 23 dB. The transmission distance is 720 km of SSMF incorporating EDFA and non-DCF.

7.5 Concluding Remarks

The total information capacity transmitted over a single-mode fiber has been increased tremendously due to the spectral packing of subchannels by pulse-shaping techniques and DSP processing algorithms as well as soft FEC so as to allow higher-order modulation of QAM to be realized. Thus, digital processing in real-time and digitally based coherent optical transmission systems have been proven to be the most modern transmission systems for the global Internet in the near future. The principal challenges are now lying on the realization of application-specific integrated circuits using microelectronics technology or ultrafast field programmable gate array-based systems.

The concepts of processing in the digital domain have been proven in offline systems and thus more efficient algorithms are required for real-time systems. These remain topical research topics and engineering issues. Higher symbol rates may be possible when wider bandwidth optical modulators and DAC and ADC systems are available, for example, grapheme plasmonic silicon modulators [23] and their integration with microelectronic DAC and ADC DSP systems.

The processing algorithms for QAM vary from level to level but they can be based essentially on the number of circles existing in multilevel QAM as compared to a monocycle constellation of the QPSK scheme. The algorithms developed for QPSK can be extended and modified for higher-level circular constellation.

References

1. T. E. Bell, Communications: Coherent optical communication shows promise, the FCC continues on its path of deregulation, and satellite communications go high-frequency, *IEEE Spectrum*, 23(1), 49–52, 1986.
2. Linear Circuits Inc., From single ended to different input trans-impedance amplifier, http://circuits.linear.com/267.
3. J. S. Weiner, A. Leven, V. Houtsma, Y. Baeyens, Y.-K. Chen, P. Paschke, Y. Yang, J. et al., SiGe differential transimpedance amplifier with 50-GHz bandwidth, *IEEE J. Solid-State Circuits*, 38(9), 1512–1517, 2003.
4. E. Ip and J. M. Kahn, Digital equalization of chromatic dispersion and polarization mode dispersion, *IEEE J. Lightw. Tech.*, 25(8), 2033, August 2007.
5. E. Ip and J. M. Kahn, Compensation of dispersion and nonlinear impairments using digital back propagation, *IEEE J. Lightw. Tech.*, 26(20), 3416, October 15, 2008.
6. I. Fatadin and S. J. Savory, Compensation of frequency offset for 16-QAM optical coherent systems using QPSK partitioning, *IEEE Photonic Tech. Lett.*, 23(17), 1246–1248, 2001.
7. I. Fatadin, D. Ives, and S. J. Savory, Laser linewidth tolerance for 16-QAM coherent optical systems using QPSK partitioning, *IEEE Photonic Tech. Lett.*, 22(9), 631–633, 2010.
8. H. Louchet, K. Kuzmin, and A. Richter, Improved DSP algorithms for coherent 16-QAM transmission, in *Proc. ECOC'08*, paper tu.1.E6, Belgium, September 2008.
9. R. Noe, T. Pfau, M. El-Darawy, and S. Hoffmann, Electronic polarization control algorithms for coherent optical transmission, *IEEE J. Sel. Quant. Elect.*, 16(5), 1193–1199, September 2010.
10. Technical specification of Corning fiber G.652 SSMF, given in L. N. Binh, *Digital Optical Communications*, Chapter 3, Appendix, CRC Press, Boca Raton, 2010.
11. D. Hillerkuss, R. Schmogrow, T. Schellinger, M. Jordan, M. Winter, G. Huber, T. Vallaitis et al., 26 Tbit/s line-rate super-channel transmission utilizing all-optical fast Fourier transform processing, *Nat. Photonics*, 5, 364, June 2011.

12. N. Godard, Passand timing recovery in an all-digital modem receiver, *IEEE Trans. Commun.*, 26, 517–523, May 1978.
13. N. Stojanovic, N. G. Gonzalez, C. Xie, Y. Zhao, B. Mao, J. Qi, and L. N. Binh, Timing recovery in Nyquist coherent optical systems, in *International Conf. Telecommunications Systems*, Serbia, 2012.
14. T. T. Fang and C. F. Liu, Fourth-power law clock recovery with pre-filtering, in *Proc. ICC*, Geneva, Switzerland, 2, 811–815, May 1993.
15. Z. Dong, X. Li, J. Yu, and N. Chi, 128-Gb/s Nyquist-WDM PDM-16QAM generation and transmission over 1200-km SMF-28 with SE of 7.47 b/s/Hz, *IEEE J. Lightw. Tech.*, 30(24), 4000–4006, December 15, 2012.
16. Z. Don et al., 128-Gb/s Nyquist-WDM PDM-16QAM generation and transmission over 1200-km SMF-28 with SE of 7.47 b/s/Hz, *IEEE J. Lightw. Tech.*, 30(24), 4000–4006, December 15, 2012.
17. X. Zhou, L. E. Nelson, R. Magill, R. Isaac, B. Zhu, D. W. Peckham, P. I. Borel, and K. Carlson, PDM-Nyquist-32QAM for 450-Gb/s per-channel WDM transmission on the 50 GHz ITU-T grid, *IEEE J. Lightw. Tech.*, 30(4), 553, February 15, 2012.
18. X. Zhou, L. E. Nelson, R. Isaac, P. Magill, B. Zhu, D. W. Peckham, P. Borel, and K. Carlson, 1200 km transmission of 50 GHz spaced, 5'504-Gb/s PDM-32-64 hybrid QAM using electrical and optical spectral shaping, in *OFC 2012*, CA, USA.
19. G. Bosco, V. Curri, A. Carena, P. Poggiolini, and F. Forghieri, On the performance of Nyquist-WDM terabit superchannels based on PM-BPSK, PM-QPSK, PM-8QAM or PM-16QAM subcarriers, *IEEE J. Lightw. Tech.*, 29(1), 53, January 2011.
20. R. Zhu, K. Xu, Y. Zhang, Y. Li, J. Wu, X. Hong, and J. Lin, QAM coherent subcarrier multiplexing system based on heterodyne detection using intermediate frequency carrier modulation, in *2008 Microwave Photonics*, pp. 165–168, 2008.
21. M. Nakazawa, M. Yoshida, K. Kasai, and J. Hongou, 20 Msymbol/s, 64 and 128 QAM coherent optical transmission over 525 km using heterodyne detection with frequency-stabilized laser, *Electron. Lett.* 42(12), 710–712, 2006.
22. Z. Dong, X. Li, J. Yu, and J. Yu, Generation and transmission of 8 × 112-Gb/s WDM PDM-16QAM on a 25-GHz grid with simplified heterodyne detection, *Opt. Express*, 21(2), 1773, January 28, 2013.
23. M. Liu and X. Zhang, Graphene-based optical modulators, in *Proc. OFC 2012*, CA, USA, paper OTu1I.7.

8

Higher-Order Spectrum Coherent Receivers

This chapter presents the processing of digital signals using higher-order spectral techniques [1–4] to evaluate the performance of a coherent transmission system, especially the bispectrum method in which the signal distribution in the frequency domain is obtained in a plane. The cross correlation as well as the signals themselves can be spacially identified. Thence evaluation of impairments and distortion can be evaluated, and the sources of causes and so on can be determined. Although equalization has not been described here in this chapter, the techniques for equalization using a higher-order spectrum can be found in Refs. [1–3].

8.1 Bispectrum Optical Receivers and Nonlinear Photonic Pre-Processing

In this section, we present the processing of optical signals before the opto-electronic detection in the optical domain in a nonlinear (NL) optical waveguide as an NL signal processing technique for digital optical receiving systems for long-haul optically amplified fiber transmission systems. The algorithm implemented is a high-order spectrum (HOS) technique in which the original signals and two delayed versions are correlated via the fourth wave mixing or third-harmonic conversion process. The optical receivers employing higher-order spectral photonic pre-processor and VLSI (very large-scale integration) electronic system for the electronic decoding and evaluation of the bit error rate of the transmission system are presented. A photonic signal pre-processing system is developed to generate the triple correlation via the third-harmonic conversion in an NL optical waveguide is employed as the photonic pre-processor to generate the essential part of ·a triple correlator. The performance of an optical receiver incorporating the HOS processor is given for long-haul phase modulated fiber transmission.

8.1.1 Introductory Remarks

As stated in previous chapters, tremendous efforts have been initiated for reaching higher transmission bit rates and longer haul for optical fiber communication systems. The bit rate can reach several hundreds of Gb/s and even Tb/s. In this extremely high-speed operational region, the limits of electronic speed processors have been surpassed and optical processing is assumed to play an important role in the optical receiving circuitry. Furthermore, novel processing techniques are required to minimize the bottlenecks of electronic processing and noises and distortion due to the impairment of the transmission medium, the linear and NL distortion effects.

This chapter deals with the photonic processing of optically modulated signals prior to the electronic receiver for long-haul optically amplified transmission systems. NL optical waveguides in planar or channel structures are studied and employed as third-harmonic converters so as to generate a triple product of the original optical waves and its two delayed copies. The triple product is then detected by an opto-electronic receiver. Thence, the detected current would be electronically sampled and digitally processed to obtain the bispectrum of the data sequence and a recovery algorithm is used to recover the data sequence. The generic structures of such high-order spectral optical receivers are shown in Figure 8.1. For the NL photonic processor (Figure 8.1a), the optical signals at the input are delayed and then coupled with an NL photonic device in which the NL conversion process is implemented via the uses of third-harmonic conversion or degenerate four-wave mixing. In the NL digital processor (see Figure 8.1b) the optical signals are detected coherently and then sampled and processed using the NL triple correlation and decoding algorithm.

We propose and simulate this NL optical pre-processor receiver under the MATLAB Simulink platform for differentially coded phase shift keying, the DQPSK modulation scheme.

This section is organized as follows: Section 8.1.3 gives a brief introduction of the triple correlation and bispectrum processing techniques. Section 8.1.4 introduces the simulation platform for the long-haul optically amplified fiber communication systems. Section 8.2.2.1 then gives the implementation of the NL optical processing and its association with an optical receiver with digital signal processing in the electronic domain so as to recover the data sequence. The performance of the transmission system is given with evaluation of the bit-error rate under linear and NL transmission regimes.

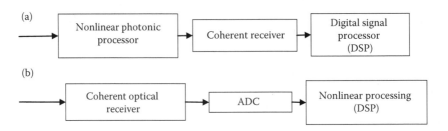

FIGURE 8.1
Generic structure of an HOS optical receiver: (a) photonic pre-processor; (b) NL DSP-based coherent receiver. Note both inphase and quadrature components can be processed as combined signals.

8.1.2 Bispectrum

In signal processing, the power spectrum estimation showing the distribution of power in frequency domain is a useful and popular tool to analyze or characterize a signal or process; however, the phase information between frequency components is suppressed in the power spectrum. Therefore it is necessarily useful to exploit higher-order spectra known as multi-dimensional spectra instead of the power spectrum in some cases, especially in NL processes or systems [6]. Different from the power spectrum, the Fourier transform of the autocorrelation, multi-dimensional spectra are known as Fourier transforms of high-order correlation functions, hence they provide us with not only the magnitude information but also the phase information.

In particular, the two-dimensional spectrum also called bispectrum is by definition the Fourier transform of the triple correlation or the third-order statistics [2]. For a signal $x(t)$, its triple-correlation function C_3 is defined as

$$C_3(\tau_1, \tau_2) = \int x(t)x(t - \tau_1)x(t - \tau_2)dt \tag{8.1}$$

where τ_1, τ_2 are the time-delay variables. Thus, the bispectrum can be estimated through the Fourier transform of C_3 as follows:

$$B_i(f_1, f_2) \equiv F\{C_3\} = \iint C_3(\tau_1, \tau_2)\exp(-2\pi j(f_1\tau_1 + f_2\tau_2))\, d\tau_1 d\tau_2 \tag{8.2}$$

where $F\{\}$ is the Fourier transform, and f_1, f_2 are the frequency variables. From definitions (8.1) and (8.2), both the triple-correlation and the bispectrum are represented in a 3D graph with two variables of time and frequency, respectively. Figure 8.2 shows the regions of power spectrum and bispectrum and their relationship. The cutoff frequencies are determined by intersection

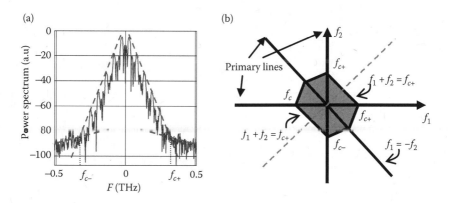

FIGURE 8.2
A description of (a) power spectrum regions and (b) bispectrum regions for explanation.

between the noise and spectral lines of the signal. These frequencies also determine the distinct areas that are basically bounded by a hexagon in bispectrum. The area inside the hexagon shows the relationship between frequency components of the signal only, otherwise the area outside shows the relationship between the signal components and noise. Due to a two-dimension representation in bispectrum, the variation of the signal and the interaction between signal components can be easily identified.

Because of unique features of the bispectrum, it is really useful in characterizing the non-Gaussian or NL processes, and applicable in various fields such as signal processing [1,2], biomedicine, and image reconstruction. Extension of number of representation dimensions makes the bispectrum become more easily and significantly representative of different types of signals and differentiation of various processes, especially NL processes such as doubling and chaos. Hence, the multi-dimensional spectra technique is proposed as a useful tool to analyze the behaviors of signals generated from these systems.

8.1.3 Bispectrum Coherent Optical Receiver

Figure 8.1 shows the structure of a bispectrum optical receiver in which there are three main sections: an all-optical pre-processor, an optoelectronic detection, and amplification, including an ADC to generate sampled values of the triple-correlated product. This section is organized as follows: The next subsection gives an introduction to bispectrum and the associated noise elimination, as well as the benefits of the bispectrum techniques. Then the details of the bispectrum processor are given, followed by some implementation aspects of the bispectrum processor using VLSI.

8.1.4 Triple Correlation and Bispectra

8.1.4.1 Definition

The power spectrum is the Fourier transform of the autocorrelation of a signal. The bispectrum is the Fourier transform of the triple correlation of a signal. Thus, both the phase and amplitude information of the signals are embedded in the triple-correlated product.

While autocorrelation and its frequency domain power spectrum do not contain the phase information of a signal, the triple correlation contains both, due to the definition of the triple correlation,

$$c(\tau_1, \tau_2) = \int S(t)S(t + \tau_1)S(t + \tau_2)\, dt \qquad (8.3)$$

where $S(t)$ is the continuous time-domain signals to be recovered, and τ_1, τ_2 are the delay time intervals. For the special cases where $\tau_1 = 0$ or $\tau_2 = 0$, the triple correlation is proportional to the autocorrelation. This means that the amplitude information is also contained in the triple correlation. The benefit of holding phase and amplitude information is that it has the potential of recovering the signal back from its triple correlation. In practice, the delays τ_1 and τ_2 indicate the path difference between the three optical waveguides. These delay times correspond to the frequency regions in the spectral domain. Thus, different time intervals would determine the frequency lines in the bispectrum.

8.1.4.2 Gaussian Noise Rejection

Given that a deterministic sampled signal $S(n)$ which is the sampled version of the continuous signals $S(t)$, is corrupted by Gaussian noise $w(n)$, with n as the sampled time index. The observed signals can then take the form $Y(n) = S(n) + w(n)$. The poly-spectra of any Gaussian process is zero for any order greater than 2 [5]. The bispectrum is the third-order poly-spectrum and offers significant advantage for signal processing over the second-order poly-spectrum, commonly known as the power spectrum, which is corrupted by Gaussian noise. Theoretically speaking, the bispectral analysis allows us to extract a non-Gaussian signal from the corrupting effects of Gaussian noise.

Thus, for a signal arrived at the optical receiver, the steps to recovery of the amplitude and phase of the lightwave modulated signals are: (i) Estimating the bispectrum of $S(n)$ based on observations of $Y(n)$; (ii) the amplitude and phase bispectra form an estimate of the amplitude and phase distribution in a one-dimensional frequency of the Fourier transform of $S(n)$. These form the constituents of the signal $S(n)$ in the frequency domain; and (iii) thence taking the inverse Fourier transform to recover the original signal $S(n)$. This type of receiver can be termed as the bispectral optical receiver.

8.1.4.3 Encoding of Phase Information

The bispectra contains almost complete information about the original signal (magnitude and phase). If the original signal $x(n)$ is real and finite it can be reconstructed, except for a shift a. Equivalently, the Fourier transform can be determined except for a linear shift factor of $e^{-j2\pi\omega a}$. By determining two adjacent pulses, any differential phase information will be readily available [6]. In other words, the bispectra, hence the triple correlation, contain the phase information of the original signal, allowing it to "pass through" the square law photodiode, which would otherwise destroy this information. The encoded phase information can then be recovered up to a linear phase term, thus necessitating a differential coding scheme.

8.1.4.4 Eliminating Gaussian Noise

For any processes that have zero mean and the symmetrical probability density function (PDF), their third-order cumulants are equaled to zero.

Therefore, in a triple correlation, those symmetrical processes are eliminated. Gaussian noise is assumed to affect signal quality. Mathematically, the third cumulant is defined as

$$c_3(\tau_1, \tau_2) = m_3(\tau_1, \tau_2) - m_1 \begin{bmatrix} m_2(\tau_1) + m_2(\tau_2) \\ + m_2(\tau_1 - \tau_2) + 2(m_1)^3 \end{bmatrix} \tag{8.4}$$

where m_k is the kth order moment of the signal, especially if it is the mean of the signal. Thus, for the zero mean and symmetrical PDF, its third-order cumulant becomes zero [1,2]. Theoretically, considering the signal as $u(t) = s(t) + n(t)$, where $n(t)$ is an additive Gaussian noise, the triple correlation of $u(t)$ will reject Gaussian noise affecting the $s(t)$.

8.1.5 Transmission and Detection

8.1.5.1 Optical Transmission Route and Simulation Platform

Shown in Figure 8.3 is the schematic diagram of the optical transmission link between Melbourne of Victoria, Australia to George Town of Tasmania whose total link length is 700 km, with sections from Melbourne City to the coastal area Gippsland; then an undersea section of more than 300 km crossing the Bass Strait to George Town of Tasmania (undersea or submarine); and then inland transmission to Hobart of Tasmania (coastal area to city connection). Other inland sections in Victoria and Tasmania are structured with optical fibers and lumped optical amplifiers (Er:doped fiber amplifiers–EDFA). Raman distributed optical amplification (ROA) is used by pump sources located at both ends of the Melbourne, Victoria to Hobart, Tasmania link, including the

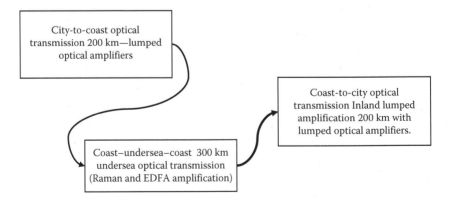

FIGURE 8.3
Schematic of the transmission link, including inland and undersea sections between city and coast (inland beach), then undersea and coast (inland) to city.

300 km undersea section. The undersea section of nearly 300 km consists of only the transmission and dispersion compensating fibers; no active sub-systems are included. This 300 km distance is fairly long, so only Raman distributed gain is used, with pump sources located at both sides of the section and installed inland. Simulink models of the transmission system include the optical transmitter, the transmission line, and the bispectrum optical receiver.

8.1.5.2 Four-Wave Mixing and Bispectrum Receiving

We have also integrated a MATLAB Simulink model of an installed transmission system to investigate the impacts of NL parametric conversion. The spectra of the optical signals before and after this amplification are shown in Figure 8.4, indicating the conversion efficiency. This indicates the performance of the bispectrum optical receiver.

8.1.5.3 Performance

We implement the models for both techniques for binary phase shift keying (BPSK) modulation format for serving as a guideline for phase modulation optical transmission systems using NL pre-processing. We note the following:

1. The arbitrary white Gaussian noise (AWGN) block in the Simulink platform can be set in different operating modes. This block then accepts the signal input and assumes the sampling rate of the input signal, then estimated the noise variance based on Gaussian

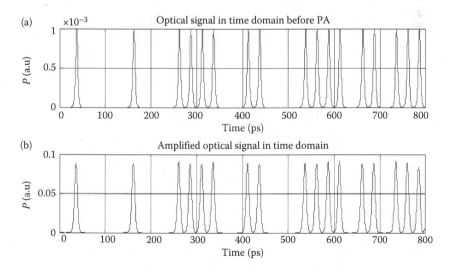

FIGURE 8.4
Time traces of the optical signal (a) before and (b) after the parametric amplifier.

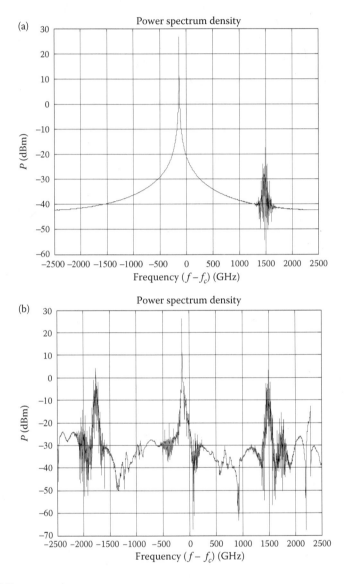

FIGURE 8.5
Corresponding spectra of the optical signal (a) before and (b) after the parametric amplifier.

distribution and the specified SNR. This is then superimposed on
the amplitude of the sampled value. Thus, we believe at that stage
that the noise is contributed evenly across the entire band of the
sampled time (converted to spectral band). The spectra of the optical
signal at the input and output of the nonlinear optical parametric
amplifier are shown in Figure 8.5a and b, respectively. Then in the

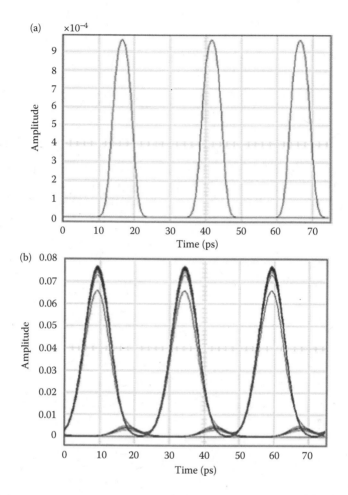

FIGURE 8.6
(a) Input data sequence and (b) detected sequence processed using triple correlation NL photonic processing and recovery scheme bispectrum receiver.

time domain the corresponding waveforms are captured as shown in Figure 8.6a and b, respectively.

2. The ideal curve signal-to-noise ratio (SNR) versus BER plotted in the graph provided is calculated using the commonly used formula in several textbooks on communication theory. This is evaluated based on the geometrical distribution of the phase states and then the noise distribution over those states. That means that all modulation and de-modulation is assumed to be perfect. However, in the digital system simulation the signals must be sampled. This is even more complicated when a carrier is imbedded in the signal, especially when the phase shift keying modulation format is used.

3. We thus re-set up the models of (i) AWGN in a complete BPSK modulation format with both the ideal coherent modulator and demodulator and any necessary filtering required and (ii) AWGN blocks with the coherent modulator and demodulator incorporating the triple correlator and necessary signal processing block. This is done in order to make fair comparison between the two pressing systems.

4. In our former model, the AWGN block was being used incorrectly in that it was being used in "SNR" mode, which applies the noise power over the entire bandwidth of the channel and of course is larger than the data bandwidth, meaning that the amount of noise in the data band was a fraction of the total noise applied. We accept that this was an unfair comparison to the theoretical curve, which is given against E_b/N_0 as defined in Ref. [7].

5. In the current model a fair comparison noise was added to the modulated signal using the AWGN block in E_b/N_0 mode (E_0 is the energy per bit and N_0 is the noises contained within the bit period), with the "symbol period" set to the carrier period. In effect, this set is the "carrier-to-noise" ratio (CNR). Also, the triple correlation receiver was modified slightly from the original, namely the addition of the BP filter and some tweaking of the triple correlation delays. This resulted in the BER curve shown in Figure 8.7. Also, an ideal homodyne receiver model was constructed, with noise added and measured in

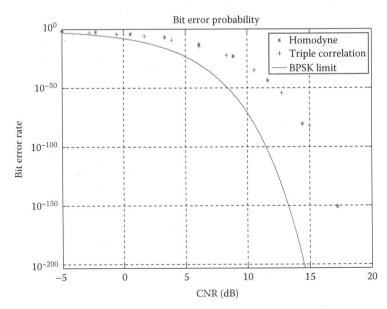

FIGURE 8.7
BER versus carrier–noise ratio for NL triple correlation, ideal BPSK under coherent detection and ideal BPSK limit.

the exact same method as the triple correlation model. This provided a benchmark for comparing the triple correlation receiver.

6. Furthermore, we can compare the simulated BER values with the theoretical limit set by

$$P_b = \frac{1}{2} erfc \sqrt{\frac{E_b}{N_0}} \tag{8.5}$$

by relating the CNR to E_b/N_0 like so

$$\frac{E_b}{N_0} = CNR \frac{B_w}{f_s} \tag{8.6}$$

where channel bandwidth B_w is 1600 Hz set by the sampling rate and f_s is the symbol rate. In our case, a symbol is one carrier period (100 Hz) as we are adding noise to the carrier. These frequencies are set at the normalized level so as to scale to wherever the spectral regions would be of interest. As can be seen in Figure 8.7, the triple correlation receiver matches the performance of the ideal homodyne case and closely approaches the theoretical limit of BPSK (~3 dB at BER of 10^{-10}). As discussed the principle benefit from the triple correlation over the ideal homodyne case will be the characterization of the noise of the channel, which is achieved by analysis of the regions of symmetry in the 2D bispectrum. Finally, we still expect possible performance improvement when symbol identification is performed directly by the triple correlation matrix, as opposed to the traditional method which involves recovering the pulse shape first. It is not possible at this stage to model the effect of the direct method.

7. Equalization of digital sampled signals can be implemented employing algorithms described in Ref. [7].

8.2 NL Photonic Signal Processing Using Higher-Order Spectra

8.2.1 Introductory Remarks

With increasing demand of high capacity, communication networks are facing several challenges, especially in the signal processing at the physical layer at ultra-high speed. When the processing speed is over that of the electronic limit or requires massive parallel and high-speed operations, processing in the optical domain offers significant advantages. All-optical signal processing is a promising technology for future optical communication net-works. An advanced optical network requires a variety of signal processing functions, including optical regeneration, wavelength conversion, optical switching, and signal monitoring. An attractive way to realize these process-ing functions in transparent and high-speed mode is to exploit the third-order nonlinearity in optical waveguides, particularly parametric processes.

Nonlinearity is a fundamental property of optical waveguides, including channel, rib-integrated structures, or circular fibers. The origin of nonlin-earity comes from the third-order NL polarization in optical transmission media. It is responsible for various phenomena such as self-phase modula-tion (SPM), cross-phase modulation (XPM), and four-wave mixing (FWM) effects. In these effects, the parametric FWM process is of special interest because it offers several possibilities for signal processing applications. To implement all-optical signal processing functions, highly NL optical wave-guides are required where the field of the guided waves is concentrated in its core region, hence efficient NL effects. Therefore, the highly NL fibers (HNLF) are commonly employed for this purpose because the NL coefficient of HNLF is about 10-fold higher than that of standard transmission fibers. Indeed, the third-order nonlinearity of conventional fibers is often very small to prevent the degradation of the transmission signal from NL distor-tion. Recently, NL chalcogenite and tellurite glass waveguides have emerged as promising devices for ultra-high-speed photonic processing. Because of their geometries, these waveguides are called planar waveguides. A planar waveguide can confine a high intensity of light within an area comparable to that of the wavelength of light, over a short distance of a few centimeters. Hence they are very compact for signal processing.

8.2.2 FWM and Photonic Processing

8.2.2.1 Bispectral Optical Structures

Figure 8.8 shows the generic and detailed structure of the bispectral optical receiver, respectively, which consists of: (i) An all-optical pre-processor front-end, followed by (ii) a photo-detector and electronic amplifier to transfer the detected electronic current to a voltage level appropriate for sampling by an analog-to-digital converter (ADC), thus the signals at this stage are in sam-pled form; (iii) the sampled triple correlation product is then transformed to the Fourier domain using the FFT. The product at this stage is the row of the matrix of the bispectral amplitude and phase plane (see Figure 8.9). A num-ber of parallel structures may be required if passive delay paths are used. (iv) A recovery algorithm is used to derive the one-dimensional distribution of the amplitude and phase as a function of the frequency, which are the essen-tial parameters required for taking the inverse Fourier transform to recover the time-domain signals.

The physical process of mixing the three waves to generate the fourth wave whose amplitude and phase are proportional to the product of the three

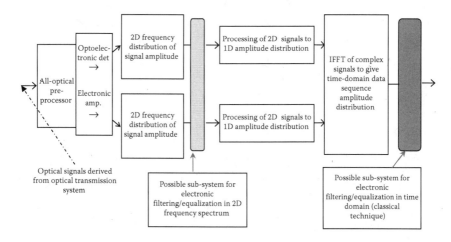

FIGURE 8.8
Generic structure of an optical pre-processing receiver employing bispectrum processing technique.

input waves is well known in literature of NL optics. This process requires: (i) A highly NL medium so as to efficiently convert the energy of the three waves to that of the fourth wave; and (ii) satisfying the phase matching conditions of the three input waves of the same frequency (wavelength) to satisfy the conservation of momentum.

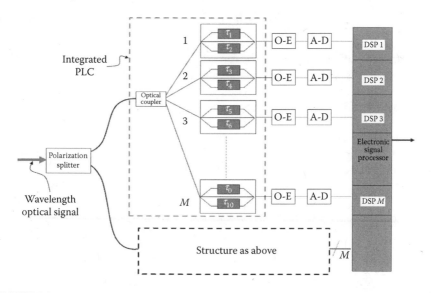

FIGURE 8.9
Parallel structures of photonic pre-processing to generate the triple correlation product in the optical domain.

8.2.2.2 The Phenomena of FWM

The origin of FWM comes from the parametric processes that lie in the NL responses of bound electrons of a material to applied optical fields. More specifically, the polarization induced in the medium is not linear in the applied field, but contains NL terms with a magnitude governed by the NL susceptibilities [8–10]. The first-, second-, and third-order parametric processes can occur due to these NL susceptibilities [χ^1 χ^2 χ^3]. The coefficient χ^3 is responsible for the FWM, which is exploited in this work. Simultaneous with this FWM, there is also a possibility of generating third-harmonic waves by mixing the three waves and parametric amplification. The third-harmonic generation (THG) is normally very small due to the phase mismatching of the guided wave number (the momentum vector) between the fundamental waves and the third-harmonic waves. FWM in a guided wave medium such as single-mode optical fibers has been extensively studied due to its efficient mixing to create the fourth wave [10]. The exploitation of the FWM processes has not yet been extensively exploited in channel optical waveguides. In this section, we demonstrate the use of nonlinear effects in the bispectral processing of optical signals by theoretical and experimental techniques.

The three lightwaves are mixed to generate the polarization vector $\vec{P}$ due to the NL third-order susceptibility, given as

$$\vec{P}_{NL} = \varepsilon_0 \chi^{(3)} \vec{E} \cdot \vec{E} \cdot \vec{E} \tag{8.7}$$

where ε_0 is the permittivity in vacuum; $\vec{E}_1, \vec{E}_2, \vec{E}_3$ are the electric field components of the lightwaves; $\vec{E} = \vec{E}_1 + \vec{E}_2 + \vec{E}_3$ is the total field entering the NL waveguide; and $\chi^{(3)}$ is the third-order susceptibility of the NL medium. P_{NL} is the product of the three total optical fields of the three optical waves here that gives the triple product of the waves required for the bispectrum receiver. It is in the bispectrum receiver that the NL waveguide acts as a multiplier of the three waves, which are considered pump waves in this section. The mathematical analysis of the coupling equations via the wave equation is complicated but straightforward. Let ω_1, ω_2, ω_3, and ω_4 represent the angular frequencies of the four waves of the FWM process that are linearly polarized along the horizontal direction y of the channel waveguides and propagating along the z-direction. The total electric field vector of the four waves is given by

$$\vec{E} = \frac{1}{2}\vec{a}_y \sum_{i=1}^{4} \vec{E}_j e^{j(k_i z - \omega_i t)} + c.c. \tag{8.8}$$

with $\vec{a}_y$ = unit vector along y axis; and *c.c.* = complex conjugate.

The propagation constant can be obtained by $k_i = (n_{eff,i}\omega_i/c)$, with $n_{eff,i}$ the effective index of the *i*th guided waves E_i ($i = 1,...,4$), which can be either TE or TM polarized guided mode propagating along the channel NL optical waveguide. All four waves are assumed to propagate in the same direction. Substituting Equation 8.8 into 8.7, we have

$$\vec{P}_{NL} = \frac{1}{2}\vec{a}_y \sum_{i=1}^{4} P_i e^{j(k_i z - \omega_i t)} + c.c. \tag{8.9}$$

where P_i ($i = 1,2...,4$) consists of a large number of terms involving the product of three electric fields of the optical guided waves, for example, the term P_4 can be expressed as

$$P_4 = \frac{3\varepsilon_0}{4} \chi_{xxxx}^{(3)} \begin{Bmatrix} |E_4|^2 E_4 + 2(|E_1|^2 + |E_2|^2 + |E_3|^2)E_4 \\ +2E_1 E_2 E_3 e^{j\varphi^+} + 2E_1 E_2 E_3^* e^{j\varphi^-} + c.c. \end{Bmatrix}$$

with $\tag{8.10}$

$$\varphi^+ = (k_1 + k_2 + k_3 + k_4)z - (\omega_1 + \omega_2 + \omega_3 + \omega_4)t$$

$$\varphi^- = (k_1 + k_2 - k_3 - k_4)z - (\omega_1 + \omega_2 - \omega_3 - \omega_4)t$$

The first four terms in Equation 8.10 represents the SPM and XPM, effects which are dependent on the intensity of the waves. The remaining terms result in FWM. Thus, the question is which terms are the most effective components resulting from the parametric mixing process. The effectiveness of the parametric coupling depends on the phase matching terms governed by φ^+ and φ^- or a similar quantity.

It is obvious that significant FWM would occur if phase matching is satisfied. This requires the matching of both the frequency as well as the wave vectors as given in Equation 8.10. From Equation 8.10 we can see that the term φ^+ corresponds to the case in which three waves are mixed to result in the fourth wave, which has a frequency three times that of the original wave. This is the THG. However, the matching of the wave vector would not normally be satisfied due to the dispersion effect or the difference in the wave vectors of a channel optical waveguide of the fundamental and third order harmonic are largely different and only minute THG might occur.

The conservation of momentum derived from the four wave vectors requires that:

$$\Delta k = k_1 + k_2 - k_3 - k_4 = \frac{n_{eff,1}\omega_1 + n_{eff,2}\omega_2 - n_{eff,3}\omega_3 - n_{eff,4}\omega_4}{c} = 0 \tag{8.11}$$

The effective refractive indices of the guided modes of the three waves E_1, E_2, and E_3 are the same, as well as their frequencies. This condition is automatically satisfied, provided the NL waveguide is designed such that it supports only a single polarized mode TE or TM and has minimum dispersion difference within the band of the signals.

8.2.3 Third-Order Nonlinearity and Parametric FWM Process

8.2.3.1 NL Wave Equation

In optical waveguides, including optical fibers, the third-order nonlinearity is of special importance because it is responsible for all NL effects. The confinement of lightwaves and their propagation in optical waveguides are generally governed by the NL wave equation (NLE), which can be derived from Maxwell's equations under the coupling of the NL polarization. The NL wave propagation of the NL waveguide in the time-spatial domain in vector form can be expressed as [11] (see also Chapter 2)

$$\nabla^2 \vec{E} - \frac{1}{c^2}\frac{\partial^2 \vec{E}}{\partial t^2} = \mu_0 \left(\frac{\partial^2 \overrightarrow{P_L}}{\partial t^2} + \frac{\partial^2 \overrightarrow{P_{NL}}}{\partial t^2} \right) \tag{8.12}$$

where $\vec{E}$ is the electric field vector of the lightwave, μ_0 is the vacuum permeability assuming a nonmagnetic waveguiding medium, c is the speed of light in vacuum, and $\overrightarrow{P_L}$, $\overrightarrow{P_{NL}}$ are, respectively, the linear and NL polarization vectors, which are formed as

$$\overrightarrow{P_L}(\vec{r},t) = \varepsilon_0 \chi^{(1)} \cdot \vec{E}(\vec{r},t) \tag{8.13}$$

$$\overrightarrow{P_{NL}}(\vec{r},t) = \varepsilon_0 \chi^{(3)} : \vec{E}(\vec{r},t)\vec{E}(\vec{r},t)\vec{E}(\vec{r},t) \tag{8.14}$$

where $\chi^{(3)}$ is the third-order susceptibility. Thus, the linear and NL coupling effects in optical waveguides can be described by Equation 8.12. The second term on the right-hand side (RHS) is responsible for NL processes, including interaction between optical waves through the third-order susceptibility.

In most telecommunication applications, only a complex envelope of optical signal is considered in analysis because bandwidth of the optical signal is much smaller than the optical carrier frequency. To model the evolution of the light propagation in optical waveguides, it requires that Equation 8.12 can be further modified and simplified by some assumptions, which are valid in most telecommunication applications. Hence, the electrical field $\vec{E}$ can be written as

$$\vec{E}(\vec{r},t) = \frac{1}{2}\hat{x}\{F(x,y)A(z,t)\exp[i(kz - \omega t)] + c.c.\} \tag{8.15}$$

where $A(z,t)$ is the slowly varying complex envelope propagating along z in the waveguide and k is the wave number. After some algebra using method of separating variables, the following equation for propagation in optical waveguide is obtained as [11]:

$$\frac{\partial A}{\partial z} + \frac{\alpha}{2}A - i\sum_{n=1}^{\infty}\frac{i^n\beta_n}{n!}\frac{\partial^n A}{\partial t^n} = i\gamma\left(1 + \frac{i}{\omega_0}\frac{\partial}{\partial t}\right) \times A\int_{-\infty}^{\infty} g(t')|A(z, t - t')|^2\, dt' \quad (8.16)$$

where the effect of propagation constant β around ω_0 is Taylor-series expanded, and $g(t)$ is the NL response function including the electronic and nuclear contributions. For the optical pulses wide enough to contain many optical cycles, Equation 8.15 can be simplified as

$$\frac{\partial A}{\partial z} + \frac{\alpha}{2}A + \frac{i\beta_2}{2}\frac{\partial^2 A}{\partial \tau^2} - \frac{\beta_3}{6}\frac{\partial^3 A}{\partial \tau^3} = i\gamma\left[|A|^2 A + \frac{i}{\omega_0}\frac{\partial(|A|^2 A)}{\partial \tau} - T_R A\frac{\partial(|A|^2)}{\partial \tau}\right] \quad (8.17)$$

where a frame of reference moving with the pulse at the group velocity v_g is used by making the transformation $\tau = t - z/v_g \equiv t - \beta_1 z$; A is the total complex envelope of propagation waves; α and β_k are the linear loss and dispersion coefficients, respectively; and $\gamma = \omega_0\, n_2/cA_{eff}$ is the NL coefficient of the guided wave structure and the first moment of the NL response function, defined as

$$T_R \equiv \int_0^{\infty} tg(t')\, dt' \quad (8.18)$$

Equation 8.17 is the basic propagation equation, commonly known as the NL Schrodinger equation (NLSE), which is very useful for investigating the evolution of the amplitude of the optical signal and the phase of the lightwave carrier under the effect of third-order nonlinearity in optical waveguides. The left-hand side of Equation 8.17 contains all linear terms, while all NL terms are contained on the RHS. In this equation, the first term on the RHS is responsible for the intensity-dependent refractive index effects, including FWM.

8.2.3.2 FWM Coupled-Wave Equations

FWM is a parametric process through the third order susceptibility $\chi^{(3)}$. In the FWM process, the superposition and generation of the propagating of the waves with different amplitudes A_k, frequencies ω_k, and wave numbers k_k through the waveguide can be represented as

$$A = \sum_n A_n e^{[j(k_n z - \omega_n \tau)]} \quad \text{with } n = 1, \ldots, 4 \quad (8.19)$$

By ignoring the linear and scattering effects, and with the introduction of Equation 8.19 into Equation 8.17, the NLSE can be separated into coupled differential equations, each of which responsible for one distinct wave in the waveguide

$$\frac{\partial A_1}{\partial z} + \frac{\alpha}{2} A_1 = i\gamma A_1 \left[|A_1|^2 + 2\sum_{n \neq 1} |A_n|^2 \right] + i\gamma\, 2A_3 A_4 A_2^* \exp(-i\Delta k_1 z)$$

$$\frac{\partial A_2}{\partial z} + \frac{\alpha}{2} A_2 = i\gamma A_2 \left[|A_2|^2 + 2\sum_{n \neq 2} |A_n|^2 \right] + i\gamma\, 2A_3 A_4 A_1^* \exp(-i\Delta k_2 z)$$

$$\frac{\partial A_3}{\partial z} + \frac{\alpha}{2} A_3 = i\gamma A_3 \left[|A_3|^2 + 2\sum_{n \neq 3} |A_n|^2 \right] + i\gamma\, 2A_1 A_2 A_4^* \exp(-i\Delta k_3 z) \qquad (8.20)$$

$$\frac{\partial A_4}{\partial z} + \frac{\alpha}{2} A_4 = i\gamma A_4 \left[|A_4|^2 + 2\sum_{n \neq 4} |A_n|^2 \right] + i\gamma\, 2A_1 A_2 A_3^* \exp(-i\Delta k_4 z)$$

where $\Delta k = k_1 + k_2 - k_3 - k_4$ is the wave vector mismatch. The equation system (8.20) thus describes the interaction between different waves in NL waveguides. The interaction, which is represented by the last term in Equation 8.20, can generate new waves. For three waves with different frequencies, a fourth wave can be generated at frequency $\omega_4 = \omega_1 + \omega_2 - \omega_3$. The waves at frequencies ω_1 and ω_2 are called pump waves, whereas the wave at frequency ω_3 is the signal, and the generated wave at ω_4 is called the idler wave, as shown in Figure 8.10a. If all three waves have the same frequency $\omega_1 = \omega_2 = \omega_3$, the interaction is called a degenerate FWM, with the new wave at the same frequency. If only two of the three waves are at the same frequency ($\omega_1 = \omega_2 \neq \omega_3$), the process is called partly degenerate FWM, which is important for some applications like the wavelength converter and parametric amplifier.

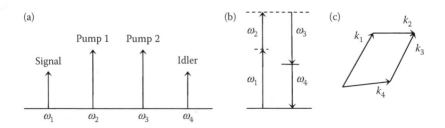

FIGURE 8.10
(a) Position and notation of the distinct waves, (b) diagram of energy conservation, and (c) diagram of momentum conservation in FWM.

8.2.3.3 Phase Matching

In parametric NL processes such as FWM, the energy conservation and momentum conservation must be satisfied to obtain a high efficiency of the energy transfer as shown in Figure 8.10a. The phase matching condition for the new wave requires:

$$\Delta k = k_1 + k_2 - k_3 - k_4 = \frac{1}{c}(n_1\omega_1 + n_2\omega_2 - n_3\omega_3 - n_4\omega_4)$$

$$= 2\pi\left(\frac{n_1}{\lambda_1} + \frac{n_2}{\lambda_2} - \frac{n_3}{\lambda_3} - \frac{n_4}{\lambda_4}\right) \tag{8.21}$$

During propagation in optical waveguides, the relative phase difference $\theta(z)$ between four involved waves is determined by

$$\theta(z) = \Delta k z + \phi_1(z) + \phi_2(z) - \phi_3(z) - \phi_4(z) \tag{8.22}$$

where $\phi_k(z)$ relates to the initial phase and the NL phase shift during propagation. An approximation of the phase matching condition can be given as

$$\frac{\partial\theta}{\partial z} \approx \Delta k + \gamma(P_1 + P_2 - P_3 - P_4) = \kappa \tag{8.23}$$

where P_k is the power of the waves and κ is the phase mismatch parameter. Thus, the FWM process has maximum efficiency for $\kappa = 0$. The mismatch comes from the frequency dependence of the refractive index and the dispersion of optical waveguides. Depending on the dispersion profile of the NL waveguides, it is very important in selection of pump wavelengths to ensure that the phase mismatch parameter is minimized.

Once the fourth wave is generated, the interaction of the four waves along the section of the waveguide continues happening, thus the NLSE must be used to investigate the evolution of the waves.

8.2.3.4 Coupled Equations and Conversion Efficiency

To derive the wave equations to represent the propagation of the three waves and therefore generate the fourth wave, we can resort to the Maxwell equations. It is lengthy to write down all the steps involved in this derivation, so we summarize the standard steps usually employed to derive the wave equations as follows: First, add the NL polarization vector given in Equation 8.7 to the electric field density vector **D**. Then, taking the curl of the first Maxwell equation, use the second equation of the four Maxwell equations and substitute the electric field density vector. Finally, use the fourth equation to come up with the vectorial wave equation.

For the FWM process occurring during the interaction of the three waves along the propagation direction of the NL optical channel waveguide, the evolution of the amplitudes, A_1 to A_4, of the four waves E_1 to E_4 given in Equation 8.20, is given by (only A_1 term is given)

$$\frac{dA_1}{dz} = \frac{jn_2\omega_1}{c}\left[\left(\Gamma_{11}|A_1|^2 + 2\sum_{k \neq 1}\Gamma_{1k}|A_k|^2\right)A_1 + 2\Gamma_{1234}A_2^*A_3A_4 e^{j\Delta kz}\right] \qquad (8.24)$$

where the wave vector mismatch Δk is given in Equation 8.11, and * denotes the complex conjugation. Note that the coefficient n_2 in Equation 8.24 is the NL coefficient related to the NL susceptibility coefficient, and is defined as

$$n_2 = \frac{3}{8n\,\mathrm{Re}(\chi_{xxxx}^3)} \qquad (8.25)$$

8.2.4 Optical Domain Implementation

8.2.4.1 NL Wave Guide

To satisfy the condition of FWM of the waves guided along a rib-waveguide, we can employ chalcogenide glass as the guiding material whose NL refractive index coefficient is about 100,000 times greater than that of silica. The material used in this waveguide is the chalcogenide glass type (e.g., AS_2S_3), or TeO_2. The three waves are guided in this waveguide structure, and their optical fields overlap. The cross section of the waveguide is of the order of $4\,\mu m \times 0.4\,\mu m$. The waveguide cross section can be designed such that the dispersion is "flat" over the spectral range of the input waves, ideally from 1520 to 1565 nm. This can be done by adjusting the thickness of the rib structure.

The fourth wave generated from the FWM waveguide is then detected by the photodetector which acts as an integrating device. Thus, the output of this detector is the triple correlation product in the electronic domain that we are looking for.

If equalization or filtering is required, then these functional blocks can be implemented in the bispectral domain, as shown in Figure 8.8. Figure 8.10 shows the parallel structures of the bispectral receiver so as to obtain all rows of the bispectral matrix. The components of the structure are almost similar, except for the delay time of the optical pre-processor.

In the NL channel waveguide fabricated using TeO_2 (tellurium oxide) on silica, the interaction of the three waves, one original and two delayed beams, happens via the electronic processes with high value NL coefficient material to allow efficient generation of the fourth wave.

When the three waves are co-propagating and satisfy the phase matching condition then the conservation of the momentum of the three waves and the fourth wave is satisfied to produce efficient FWM. Indeed, phase matching can also be satisfied by one forward wave and two backward-propagating waves (a delayed version of the first wave), leading to almost 100% conversion efficiency to generate the fourth wave.

The interaction of the three waves via electronic process and the $\chi^{(3)}$ gives rise to the polarization vector P, which in turn couples with the electric field density of the light waves and then with the NLSE. By solving and modeling this wave equation with the FWM term on the RHS of the equation, one can obtain the wave output (the fourth wave) at the output (the end of the NL waveguide section).

8.2.4.2 Third-Harmonic Conversion

Third-harmonic conversion may happen, but at extremely low efficiency—at least 1000 times less than that of FWM due to the nonmatching of the effective refractive indices of the guided modes at 1550 nm (fundamental wave) and 517 nm (third-harmonic wave).

It is noted that the common term for this process is the matching of the dispersion characteristics, that is, k/omega versus the thickness of the waveguide, with omega as the radial frequency of the waves evaluated at 1550 and 1517 nm.

8.2.4.3 Conservation of Momentum

The conservation of momentum and thus phase matching condition for the FWM is satisfied without much difficulty, as the wavelengths of the three input waves are the same. The optical NL channel waveguide is to be designed such that there is mismatching of the third harmonic conversion and most efficient for FWM. It is considered that single polarized mode, either TE or TM, will be used to achieve efficient FWM. Thus, the dimension of the channel waveguide would be estimated about 0.4 μm (height) $\times$4 μm (width).

8.2.4.4 Estimate of Optical Power Required for FWM

To achieve the most efficient FWM process, the NL coefficient n_2, which is proportional to $\chi^{(3)}$ by a constant ($8n/3$, where n is the refractive index of the medium or the approximate effective refractive index of the guided mode). This NL coefficient is then multiplied by the intensity of the guided waves to give an estimate of the phase change and thence estimation of the efficiency of the FWM. With the cross section estimated in Equation 8.3 and the well confinement of the guided mode, the effective area of the guided waves is very close to the cross section area. Thus, an average power of the guided waves would be about 3–5 mW or about 6 dBm.

With the practical data of the loss of the linear section (section of multimode interference and delay split—similar to array waveguide grating technology) is estimated at 3 dB. Thus, the input power of the three waves required for efficient FWM is about 10 dBm (maximum).

8.2.5 Transmission Models and NL Guided Wave Devices

To model the parametric FWM process between multi-waves, the basic propagation equations described in Section 8.2.3 are used. There are two approaches to simulate the interaction between waves. The first approach, named the separating channel technique, is to use the coupled equations system (8.20), in which the interactions between different waves are obviously modeled by certain coupling terms in each coupled equation. Thus, each optical wave is considered one separate channel, and is represented by a phasor. The coupled equations system is then solved to obtain the solutions of the FWM process. The outputs of the NL waveguide are also represented by separated phasors, hence the desired signal can be extracted without using a filter.

The second or alternating approach is to use the propagation Equation 8.17, which allows us to simulate all evolutionary effects of the optical waves in the NL waveguides. In this technique a total field is used instead of individual waves. The superimposed complex envelope A is represented by only one phasor, which is the summation of individual complex amplitudes of different waves, given as

$$A = \sum_k A_k e^{[j(\omega_k - \omega_0)\tau]} \tag{8.26}$$

where ω_0 is the defined angular central frequency, and A_n and ω_n are the complex envelope and the carrier frequency of individual waves, respectively. Hence, various waves at different frequencies are combined into only a total signal vector that facilitates integration of the NL waveguide model into the Simulink® platform. Equation 8.17 can also be numerically solved by the split-step Fourier method (SSFM). The Simulink block representing the NL waveguide is implemented with an embedded MATLAB program. Because only complex envelopes of the guided waves are considered in the simulation, each of the different optical waves is shifted by a frequency difference between the central frequency and the frequency of the wave, to allocate the wave in the frequency band of the total field. Then the summation of individual waves, which is equivalent to the combination process at the optical coupler, is performed prior to entering the block of NL waveguides, as depicted in Figure 8.12. The output of the NL waveguide will be selected by an optical bandpass filter (BPF). In this way, the model of the NL waveguide can be easily connected to other Simulink® blocks, which are available in the platform for simulation of optical fiber communication systems [3].

8.2.6 System Applications of Third-Order Parametric Nonlinearity in Optical Signal Processing

In this section, a range of signal processing applications are demonstrated through simulations that use the NL waveguide to model the wave mixing process.

8.2.6.1 Parametric Amplifiers

One of important applications of the $\chi^{(3)}$ nonlinearity is parametric amplification. The optical parametric amplifiers (OPA) offer a wide gain bandwidth, high differential gain and optional wavelength conversion, and operation at any wavelength [12]. These important features of OPA are obtained because the parametric gain process does not rely on energy transitions between energy states, but it is based on highly efficient FWM in which two photons at one or two pump wavelengths interact with a signal photon. The 4th photon, the idler, is formed with a phase such that the phase difference between the pump photons and the signal and idler photons satisfies the phase matching condition (8.21). A generic Simulink model for evaluating the parametric amplification is shown in Figure 8.11, then Figure 8.12 depicts a more detailed model for a fiber-based parametric amplifier.

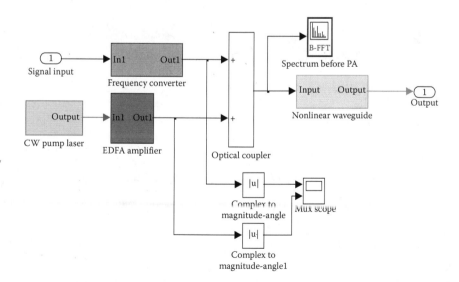

FIGURE 8.11
Typical Simulink setup of the parametric amplifier using the model of NL waveguide.

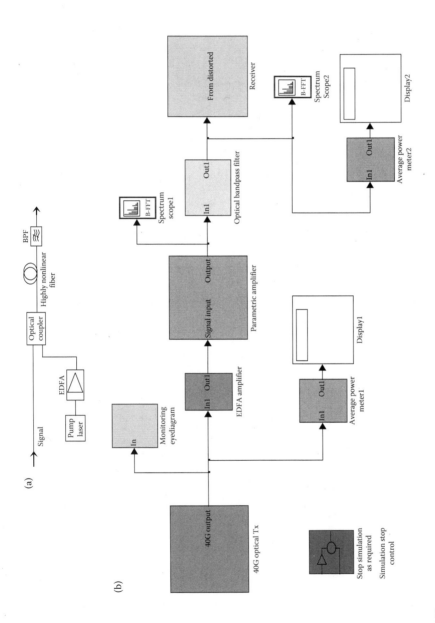

FIGURE 8.12

(a) A typical setup of an optical parametric amplifier; (b) Simulink model of optical parametric amplifier.

For a parametric amplifier using one pump source, from the coupled Equations 8.20 with $A_1 = A_2 = A_p$, $A_3 = A_s$, and $A_4 = A_i$, it is possible to derive three coupled equations for a complex field amplitude of the three waves $A_{p,s,i}$:

$$\frac{\partial A_p}{\partial z} = -\frac{\alpha}{2} A_p + i\gamma A_p \left[|A_p|^2 + 2\left(|A_s|^2 + |A_i|^2 \right) \right] + i2\gamma A_s A_i A_p^* \exp(-i\Delta kz)$$

$$\frac{\partial A_s}{\partial z} = -\frac{\alpha}{2} A_s + i\gamma A_s \left[|A_s|^2 + 2\left(|A_p|^2 + |A_i|^2 \right) \right] + i\gamma A_p^2 A_i^* \exp(-i\Delta kz) \qquad (8.27)$$

$$\frac{\partial A_i}{\partial z} = -\frac{\alpha}{2} A_i + i\gamma A_i \left[|A_i|^2 + 2\left(|A_s|^2 + |A_p|^2 \right) \right] + i\gamma A_p^2 A_s^* \exp(-i\Delta kz)$$

The analytical solution of these coupled equations determines the gain of the amplifier [10]

$$G_s(L) = \frac{|A_s(L)|^2}{|A_s(0)|^2} = 1 + \left[\frac{\gamma P_p}{g} \sinh(gL) \right]^2 \qquad (8.28)$$

where L is the length of the highly NL fiber/waveguide, P_p is the pump power, and g is the parametric gain coefficient:

$$g^2 = -\Delta k \left(\frac{\Delta k}{4} + \gamma P_p \right) \qquad (8.29)$$

The phase mismatch Δk can be approximated by extending the propagation constant in a Taylor series around ω_0:

$$\Delta k = -\frac{2\pi c}{\lambda_0^2} \frac{dD}{d\lambda} (\lambda_p - \lambda_0)(\lambda_p - \lambda_s)^2 \qquad (8.30)$$

Here, $dD/d\lambda$ is the slope of the dispersion factor $D(\lambda)$ evaluated at the zero dispersion of the guided wave component, that is, at the optical wavelength $\lambda_k = 2\pi c/\omega_k$.

Figure 8.13b shows the Simulink setup of the 40 Gb/s RZ transmission system using a parametric amplifier. The setup contains a 40 Gb/s optical RZ transmitter, an optical receiver for monitoring, a parametric amplifier block, and a BPF that filters the desired signal from the total field output of the amplifier. Details of the parametric amplifier block can be seen in Figure 8.12. The block setup of the parametric amplifier consists of a continuous wave (CW) pump laser source, an optical coupler to combine the signal and the pump, and a highly NL fiber block containing the embedded MATLAB model for NL propagation. The important simulation parameters of the system are listed in Table 8.1.

(a)

(b)

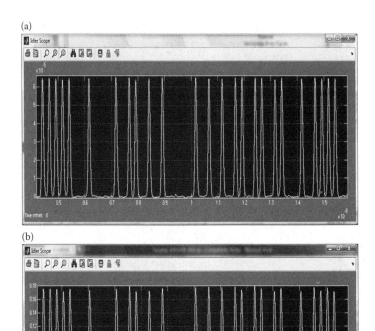

FIGURE 8.13
Time traces of the 40 Gb/s signal before (a) and after (b) the parametric amplifier.

TABLE 8.1

Critical Parameters of the Parametric Amplifier in a 40 Gb/s Transmission System

---

RZ 40 Gb/s Signal
$\lambda_0 = 1559$ nm, $\lambda_s = \{1531.12, 1537.4, 1543.73, 1550.12\}$ nm,
$P_s = 1$ mW (peak), $B_r = 40$ Gb/s

Parametric Amplifier
L pump source: $P_p = 100$ mW (after EDFA), $\lambda_p = 1560.07$ nm
HNLF: $L_f = 200$ m, $D = 0.02$ ps/km/nm, $S = 0.03$ ps/nm^2/km, $\alpha = 0.5$ dB/km, $A_{eff} = 12$ μm^2,
$\gamma = 13$ 1/W/km
BPF: $\Delta\lambda_{BPF} = 0.64$ nm, $\lambda_i = \{1587.91, 1581.21, 1574.58, 1567.98\}$ nm

---

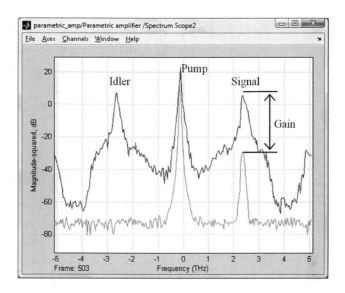

FIGURE 8.14
Optical spectra at the input (gray) and the output (black) of the OPA.

Figure 8.14 shows the signals before and after the amplifier in time domain. The time trace indicates the amplitude fluctuation of the amplified signal as a noisy source from the wave mixing process. Their corresponding spectra are shown in Figure 8.15. The noise floor of the output spectrum of the amplifier shows the gain profile of OPA. Simulated dependence of OPA gain on the wavelength difference between the signal and the pump is shown

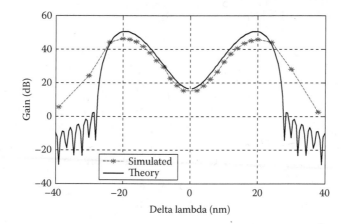

FIGURE 8.15
Calculated and simulated gain of the OPA at $P_p = 30$ dBm.

(a)   (b)

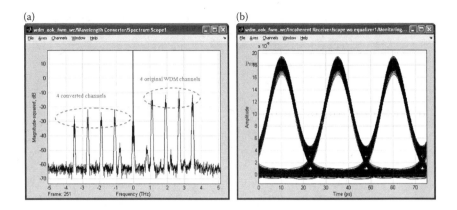

FIGURE 8.16
(a) The wavelength conversion of 4 WDM channels; (b) eye diagram of the converted 40 Gb/s signal after BPF.

in Figure 8.16, together with theoretical gain using Equation 8.15. The plot shows an agreement between theoretical and simulated results. The peak gain is achieved at the phase-matched condition, where the linear phase mismatch is compensated by the NL phase shift.

8.2.6.2 Wavelength Conversion and NL Phase Conjugation

Beside the signal amplification in a parametric amplifier, the idler is generated after the wave mixing process. Therefore, this process can also be applied to wavelength conversion. Due to very fast response of the third-order nonlinearity in optical waveguides, the wavelength conversion based on this effect is transparent to the modulation format and the bit rate of signals. For a flat wideband converter, which is a key device in wavelength-division multiplexing (WDM) networks, a short length HNLF with a low dispersion slope is required by design. By a suitable selection of the pump wavelength, the wavelength converter can be optimized to obtain a bandwidth of 200 nm [7]. Therefore, the wavelength conversion between bands such as C and L can be performed in WDM networks. Figure 8.16 shows an example of the wavelength conversion for four WDM channels at the C-band. The important parameters of the wavelength converter are shown in Table 8.1. The WDM signals are converted into the L-band with the conversion efficiency of −12 dB.

Another important application with the same setup is the NL phase conjugation (NPC). A phase-conjugated replica of the signal wave can be generated by the FWM process. From Equation 8.8, the idler wave is approximately given in case of degenerate FWM. For simplification: $E_i \sim A_p^2 A_s^* e^{-j\Delta kz}$ or $E_i \sim r A_s^* e^{[j(-kz-\omega\tau)]}$ with the signal wave $E_s \sim A_s e^{[j(kz-\omega\tau)]}$. Thus, the idler field is the complex conjugate of the signal field. In appropriate conditions, optical distortions can be compensated by using NPC, and optical pulses propagating in the fiber link can be recovered. The basic principle of distortion compensation with NPC refers to spectral inversion. When an optical pulse propagates in an optical fiber, its shape will be spread in time and distorted by the group velocity dispersion. The phase-conjugated replica of the pulse is generated in the middle point of the transmission link by the NL effect. On the other hand, the pulse is spectrally inverted, where spectral components in the lower frequency range are shifted to the higher frequency range and vice versa. If the pulse propagates in the second part of the link with the same manner in the first part, it is inversely distorted again and can cancel the distortion in the first part to recover the pulse shape at the end of the transmission link. With using NPC for distortion compensation, a 40–50% increase in transmission distance compared to a conventional transmission link can be obtained [8–11]. Figure 8.17 shows the setup of a long-haul 40 Gb/s transmission system demonstrating the distortion compensation using NPC. The fiber transmission link of the system is divided into two sections by an NPC based on parametric amplifier. Each section consists of five spans, with a 100 km standard single mode fiber (SSMF) in each span. Figure 8.18a shows the eye diagram of the signal after propagating through the first fiber section. After the parametric amplifier at the midpoint of the link, the idler signal, a phase conjugated replica of the original signal, is filtered for transmission into next section. The signal in the second section suffers the same dispersion as in the first section. At the output of the transmission system the optical signal is regenerated as shown in Figure 8.18b. Due to change in wavelength of the signal in NPC, a tunable dispersion compensator can be required to compensate the residual dispersion after transmission in real systems.

8.2.6.3 High-Speed Optical Switching

When the pump is an intensity-modulated signal instead of a CW signal, the gain of the OPA is also modulated due to its exponential dependence on the pump power in a phase-matched condition. The width of the gain profile in the time domain is inversely proportional to the product of the gain slope (S_p), or the NL coefficient and the length of the NL waveguide (L). Therefore, an OPA with high gain or a large S_pL operates as an optical switch with an ultra-high bandwidth, which is very important in some signal processing applications such as pulse compression or short-pulse generation. A Simulink setup for a 40 GHz short-pulse generator is built with the configuration shown in Figure 8.19. In this setup, the input signal is a CW source with low power and the pump is amplitude-modulated by a Mach–Zehnder intensity modulator (MZIM), driven by a RF sinusoidal wave at 40 GHz. The waveform of the modulated pump is shown in Figure 8.20a. Important parameters of the FWM-based short-pulse generator are shown in Table 8.2. Figure 8.20b shows the generated short-pulse sequence with the pulse width of 2.6 ps at the signal wavelength after the optical BPF.

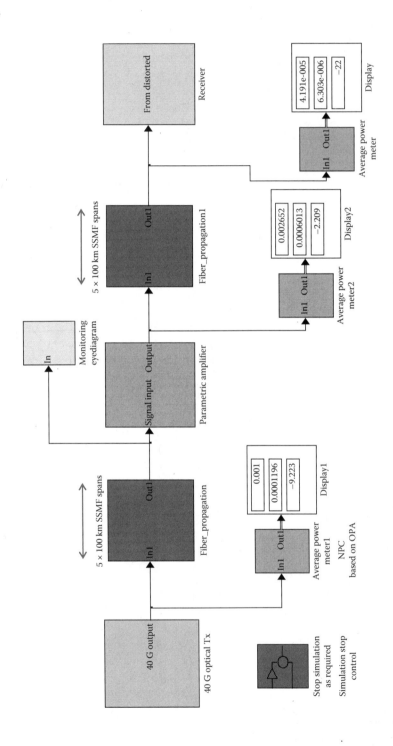

FIGURE 8.17
Simulink setup of a long-haul 40 Gb/s transmission system using NPC for distortion compensation.

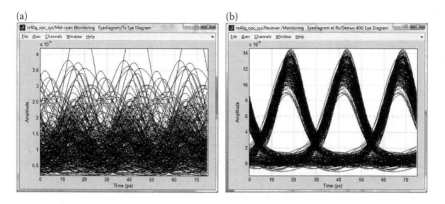

FIGURE 8.18

Eye diagrams of the 40 Gb/s signal at the end (a) of the first section and (b) of the transmission link.

Another important application of the optical switch based on FWM process is the demultiplexer, a key component in ultra-high-speed optical time division multiplexing (OTDM) systems. OTDM is a key technology for Tb/s Ethernet transmission, which can meet the increasing demand of traffic in future optical networks. A typical scheme of an OTDM demultiplexer is that the pump is a mode-locked laser (MLL) to generate short pulses for control, as shown in Figure 8.21a. The working principle of the FWM-based demultiplexing is described as follows: The control pulses generated from an MLL at tributary rate are pumped and co-propagated with the OTDM signal through the NL waveguide. The mixing process between the control pulses and the OTDM signal during propagation through the NL waveguide converts the desired tributary channel to a new idler wavelength. Then the demultiplexed signal at the idler wavelength is extracted by a bandpass filter before going to a receiver, as shown in Figure 8.21a.

Using HNLF is relatively popular in structures of an OTDM demultiplexer [13]. However, its stability, especially the walk-off problem, is still a serious obstacle. Recently, planar NL waveguides have emerged as promising devices for ultra-high-speed photonic processing [14,15]. These NL waveguides offer a lot of advantages, such as no free-carrier absorption, stability at room temperature, no requirement of quasi-phase matching, and possibility of dispersion engineering. With the same operational principle, planar waveguide-based OTDM demultiplexers are very compact and suitable for photonic integrated solutions. Figure 8.21b shows the Simulink setup of the FWM-based demultiplexer of the on–off keying (OOK) 40 Gb/s signal from the 160 Gb/s OTDM signal using a highly NL waveguide instead of HNLF. Important parameters of the OTDM system in Table 8.3 are used in the simulation. Figure 8.22a shows the spectrum at the output of the NL waveguide. Then the demultiplexed signal is extracted by the BPF as shown in Figure 8.22b. Figure 8.23 shows the time traces of the 160 Gb/s

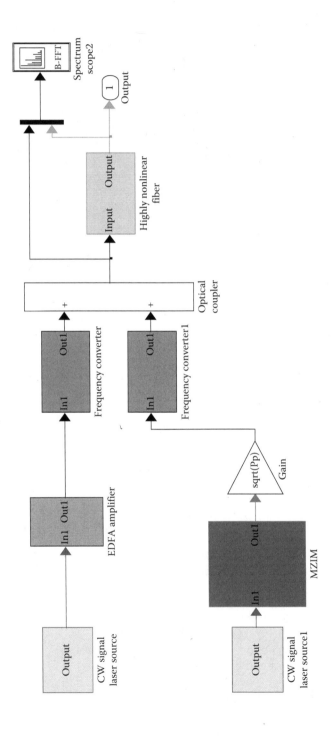

FIGURE 8.19
Simulink setup of the 40 GHz short-pulse generator.

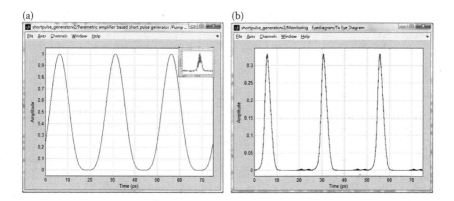

FIGURE 8.20

Time traces of (a) the sinusoidal amplitude modulated pump, and (b) the generated short-pulse sequence. (Inset: The pulse spectrum.)

OTDM signal, the control signal, and the 40 Gb/s demultiplexed signal, respectively. The red dots in Figure 8.23a indicate the timeslots of the desired tributary signal in the OTDM signal. The developed model of the OTDM demultiplexer can be applied not only to the conventional OOK format but also to advanced modulation formats such as DQPSK, which increases the data load of the OTDM system without increase in bandwidth of the signal. By using available blocks developed for a DQPSK system [16], a Simulink model of the DQPSK-OTDM system is also set up for demonstration. The bit rate of the OTDM system is doubled to 320 Gb/s with the same pulse repetition rate. Figure 8.24 shows the simulated performance of the demultiplexer in both 160 Gb/s OOK- and 320 Gb/s DQPSK-OTDM systems. In the case of the DQPSK-OTDM signal, the BER curve shows a low error floor, which may be a result of the influence of NL effects on phase-modulated signals in the waveguide.

TABLE 8.2

Important Parameters of the FWM Based on an OTDM Demultiplexer Using a NL Waveguide

OTDM Transmitter

MLL: $P_0 = 1$ mW, $T_p = 2.5$ ps, $f_m = 40$ GHz

Modulation formats: OOK and DQPSK; OTDM multiplexer: 4×40 Gsymbols/s

FWM-Based Demultiplexer

Pumped control: $P_p = 500$ mW, $T_p = 2.5$ ps, $f_m = 40$ GHz, $\lambda_p = 1556.55$ nm

Input signal: $P_s = 10$ mW (after EDFA), $\lambda_s = 1548.51$ nm

Waveguide: $L_w = 7$ cm, $D_w = 28$ ps/km/nm, $S_w = 0.003$ ps/nm²/km,

$\alpha = 0.5$ dB/cm, $\gamma = 10^4$ 1/W/km

BPF: $\Delta\lambda_{BPF} = 0.64$ nm

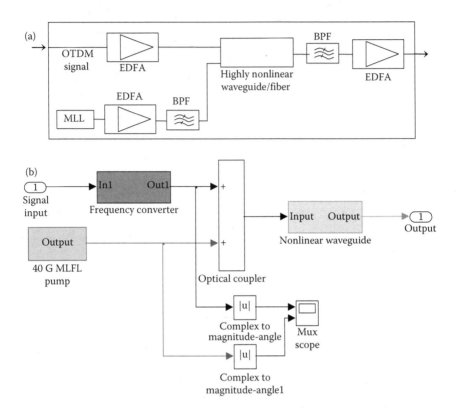

FIGURE 8.21
(a) A typical setup of the FWM-based OTDM demultiplexer, (b) Simulink model of the OTDM demultiplexer.

TABLE 8.3

Parameters of the 40 GHz Short Pulse Generator

Signal: $P_s = 0.7$ mW, $\lambda_s = 1535$ nm, $\lambda_0 = 1559$ nm
Pump source: $P_p = 1$ W (peak), $\lambda_p = 1560.07$ nm, $f_m = 40$ GHz
HNLF: $L_f = 500$ m, $D = 0.02$ ps/km/nm, $S = 0.03$ ps/nm²/km, $\alpha = 0.5$ dB/km, $A_{eff} = 12$ μm², $\gamma = 13$ 1/W/km
BPF: $\Delta\lambda_{BPF} = 3.2$ nm

8.2.6.4 Triple Correlation

One of the promising applications exploiting the $\chi^{(3)}$ nonlinearity is implementation of triple correlation in optical domain. Triple correlation is a higher-order correlation technique. Its Fourier transform, called bispectrum, is very important in signal processing, especially in signal recovery [3, 17]. The triple correlation of a signal $s(t)$ can be defined as

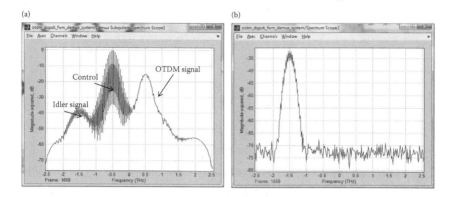

FIGURE 8.22
Spectra at the outputs of (a) the NL waveguide and (b) BPF.

$$C^3(\tau_1, \tau_2) = \int s(t)s(t - \tau_1)s(t - \tau_2)dt \qquad (8.31)$$

where τ_1, τ_2 are time-delay variables. To implement the triple correlation in the optical domain, the product of three signals, including different delayed versions of the original signal, need to be generated and then detected by an

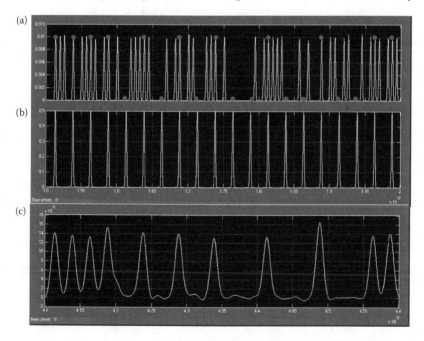

FIGURE 8.23
Time traces of (a) the 160 Gb/s OTDM signal, (b) the control signal, and (c) the 40 Gb/s demultiplexed signal.

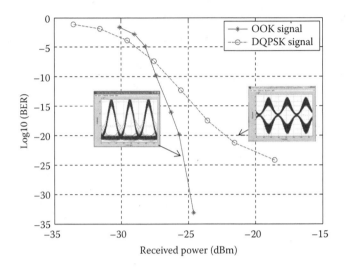

FIGURE 8.24
Simulated performance of the demultiplexed signals for 160 Gb/s OOK- and 320 Gb/s DQPSK-OTDM systems. (Insets: Eye diagrams at the receiver.)

optical photodiode to perform the integral operation. From the representation of the NL polarization vector (see Equation 8.14), this triple-product can be generated by the $\chi^{(3)}$ nonlinearity. One way to generate the triple correlation is based on THG, where the generated new wave containing the triple product is at the frequency of three times the original carrier frequency. Thus, if the signal wavelength is in the 1550 nm band, the new wave needs to be detected at around 517 nm. The triple-optical autocorrelation based on single-stage THG has been demonstrated in direct optical pulse shape measurement [18]. However, it is hard to obtain high efficiency in the wave-mixing process this way, due to the difficulty of phase matching between three signals. Moreover, the triple-product wave is in 517 nm, where wideband photo-detectors are not available for high-speed communication applications. Therefore, a possible alternative to generating the triple product is based on other NL interactions such as FWM. From Equation 8.20, the fourth wave is proportional to the product of three waves $A_4 \sim A_1 A_2 A_3^* e^{-j\Delta kz}$. If A_1 and A_2 are the delayed versions of the signal A_3, the mixing of three waves results in the fourth wave A_4, which is obviously, the triple product of three signals. As mentioned in Section 8.2, all three waves can take the same frequency; however, these waves should propagate into different directions to possibly distinguish the newly generated wave in a diverse propagation direction, which requires a strict arrangement of the signals in spatial domain. An alternative way we propose is to convert the three signals into different frequencies (ω_1, ω_2, and ω_3). Then the triple-product wave can be extracted at the frequency $\omega_4 = \omega_1 + \omega_2 - \omega_3$, which is still in the 1550 nm band.

Figure 8.25a shows the Simulink model for the triple correlation based on FWM in the NL waveguide. The structural block consists of two variable delay lines to generate delayed versions of the original signal as shown in Figure 8.25, and frequency converters to convert the signal into different three waves before combining at the optical coupler to launch into the NL waveguide. Then the fourth wave signal generated by FWM is extracted by the passband filter. To verify the triple-product based on FWM, another model shown in Figure 8.25b is used to estimate the triple product by using Equation 8.31 for comparison. The integration of the generated triple-product

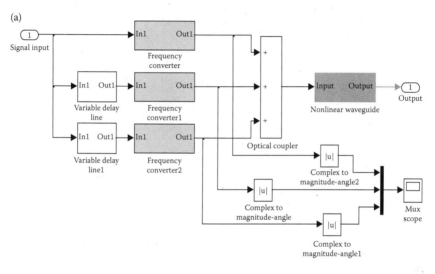

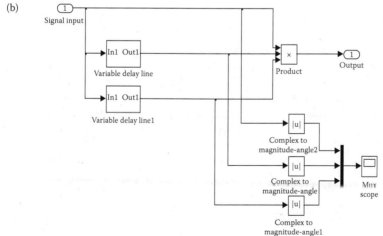

FIGURE 8.25
(a) Simulink setup of the FWM-based triple-product generation, and (b) Simulink setup of the theory-based triple-product generation.

TABLE 8.4

Important Parameters of the FWM-Based OTDM
Demultiplexer Using a NL Waveguide

Signal Generator

Single-pulse: $P_0 = 100$ mW, $T_p = 2.5$ ps, $f_m = 10$ GHz

Dual-pulse: $P_1 = 100$ mW, $P_2 = 2/3P_1$, $T_p = 2.5$ ps, $f_m = 10$ GHz

FWM-Based Triple-Product Generator

Original signal: $\lambda_{s1} = 1550$ nm, $\lambda_{s1} = 1552.52$ nm

Delayed τ_1 signal: $\lambda_{s2} = 1552.52$ nm, Delayed τ_2 signal: $\lambda_{s3} = 1554.13$ nm

Waveguide: $L_w = 7$ cm, $D_w = 28$ ps/km/nm, $S_w = 0.003$ ps/nm²/km

$\alpha = 0.5$ dB/cm, $\gamma = 10^4$ 1/W/km

BPF: $\Delta\lambda_{BPF} = 0.64$ nm

signal is then performed by the photo-detector in the optical receiver to estimate the triple correlation of the signal. A repetitive signal, as depicted in
Figure 8.26 (modeling) and Figure 8.27 (close up scale), which is a dual-pulse
sequence with unequal amplitude, is generated for investigation. Important
parameters of the setup are shown in Table 8.4. This table shows the waveform of the dual-pulse signal and the spectrum at the output of the NL waveguide. The wavelength spacing between three waves is unequal, to reduce
the noise from other mixing processes. The triple-product waveforms estimated by theory and FWM process are shown in Table 8.4. In the case of
the estimation based on FWM, the triple-product signal is contaminated
by the noise generated from other mixing processes, as indicated in Table
8.4. Figure 8.28 shows the triple correlations of the signal after processing at
the receiver, in both cases based on theory and FWM. The triple correlation
is represented by a 3D plot, which is displayed by the image. The x and y

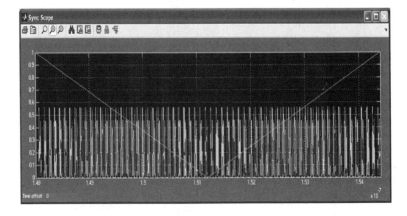

FIGURE 8.26
The variation in time domain of the time-delay signals (low level) and the original signal (high
level).

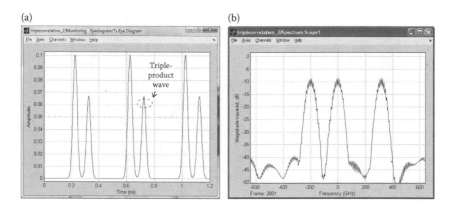

FIGURE 8.27
(a) Time trace of the dual-pulse sequence for investigation. (b) Spectrum at the output of the NL waveguide.

axes of the image represent the time-delay variables (τ_1 and τ_2) in terms of samples, with a step-size of $T_m/32$, where T_m is the pulse period. The intensity of the triple correlation is represented by colors, with scale specified by the color bar. Although the FWM-based triple correlation result is noisy, the triple-correlation pattern is still distinguishable as compared to the theory. A simpler signal pattern of single pulse has also been investigated, as shown

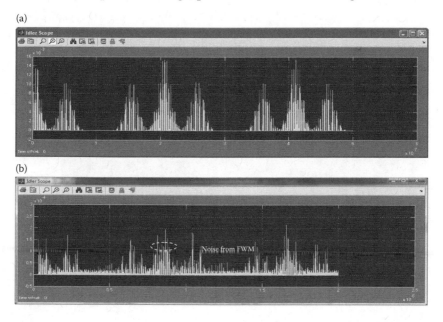

FIGURE 8.28
Generated triple-product waves in the time domain of the dual-pulse signal based on (a) theory, and (b) FWM in NL waveguide.

(a) (b)

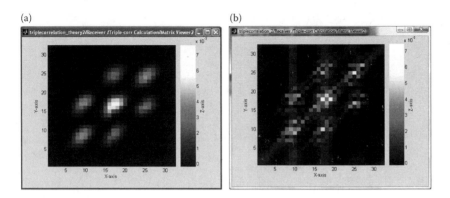

FIGURE 8.29
Triple correlation of the dual-pulse signal based on (a) theoretical estimation, and (b) FWM in the NL waveguide.

in Figures 8.29 and 8.30. Figure 8.31 depicts a simplified schematic of a signal processor using bispectrum technique integrated with a coherent receiver.

8.2.6.5 Remarks

This section demonstrates the employment of an NL optical waveguide and the associated NL effects such as parametric amplification, four-wave mixing, and THG for the implementation of the triple correlation and thence the bispectrum creation and signal recovery techniques to reconstruct the data sequence transmitted over a long-haul optically amplified fiber transmission link.

(a) (b)

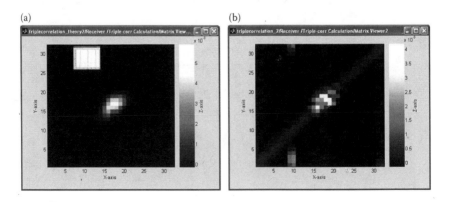

FIGURE 8.30
Triple correlation of the single-pulse signal based on (a) theoretical estimation, and (b) FWM in the NL waveguide (Inset: the single-pulse pattern.)

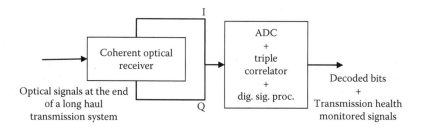

FIGURE 8.31
Schematic diagram of a high-order spectral optical receiver and electronic processing.

8.2.7 NL Photonic Pre-Processing in Coherent Reception Systems

This section looks at the uses of NL effects and applications in modern optical communications networks in which 100 Gb/s optical Ethernet is expected to be employed.

Table 8.5 gives a principal parameters for the parametric amplifiers employed in the 40 Gb/s transmission system. Further, Table 8.6 laid out all parameters for the 40 Gb/s systems incorporating the NPL. Typical performance of a photonic signal pre-processor employing no linear four wave mixing is given and that of an advanced processing of such received signals in the electronic domain processed by a digital triple correlation system. At least 10 dB improvement is achieved on the receiver sensitivity.

Regarding the NL effects, the nonlinearity of the optical fibers hinders and limits the maximum level of the total average power of all the multiplexed channels for maximizing the transmission distance. This is due to the change of the refractive index of the guided medium as a function of the intensity of the guided waves. This in turn creates the phase changes and hence different group delays, thence distortion. Furthermore, other associated NL effects such as the four-wave mixing, Raman scattering, Brillouin

TABLE 8.5

Critical Parameters of the Parametric Amplifier in a 40 Gb/s System

RZ 40 Gb/s Transmitter
$\lambda_s = 1520$ nm $- 1600$ nm, $\lambda_0 = 1559$ nm
Modulation: RZ-ASK, $P_s = 0.01$ mW (peak), $B_r = 40$ Gb/s

Parametric Amplifier
Pump source: $P_p = 1$ W (after EDFA), $\lambda_p = 1560.07$ nm
HNLF: $L_f = 500$ m, $D = 0.02$ ps/km/nm, $S = 0.09$ ps/nm^2/km, $\alpha = 0.5$ dB/km, $A_{eff} = 12$ μm^2, $\gamma = 13$ 1/W/km
BPF: $\Delta\lambda_{BPF} = 0.64$ nm

Receiver
Bandwidth $B_e = 28$ GHz, $i_{eq} = 20$ pA/Hz$^{1/2}$, $i_d = 10$ nA

TABLE 8.6

Critical Parameters of the Long-Haul Transmission System Using NPC for Distortion Compensation

RZ 40 Gb/s Transmitter
$\lambda_s = 1547$ nm, $\lambda_0 = 1559$ nm
Modulation: RZ-OOK, $P_s = 1$ mW (peak), $B_r = 40$ Gb/s

Fiber Transmission Link
SMF: $L_{SMF} = 100$ km, $D_{SMF} = 17$ ps/nm/km, $\alpha = 0.2$ dB/km
EDFA: Gain = 20 dB, NF = 5 dB;
Number of spans: 10 (5 in each section), $L_{link} = 1000$ km

NPC Based on OPA
Pump source: $P_p = 1$ W (after EDFA), $\lambda_p = 1560.07$ nm
HNLF: $L_f = 500$ m, $D = 0.02$ ps/km/nm, $S = 0.09$ ps/nm²/km, $\alpha = 0.5$ dB/km,
  $A_{eff} = 12$ µm², $\gamma = 13$ 1/W/km
BPF: $\Delta\lambda_{BPF} = 0.64$ nm

Receiver
Bandwidth $B_e = 28$ GHz, $i_{eq} = 20$ pA/Hz$^{1/2}$, $i_d = 10$ nA

scattering, and inter-modulation have also created jittering and distortion of the received pulse sequences after a long transmission distance.

However, recently we have been able to use to our advantages these NL optical effects as pre-processing elements before the optical receiver to improve its sensitivity. A higher-order spectrum technique is employed with the triple correlation implemented in the optical domain via the use of the degenerate four-wave mixing effects in a high NL optical waveguide. However, this may add additional optical elements and filtering in the processor, and hence complicate the receiver structure. We can overcome this by bringing this NL higher-order spectrum processing to after the opto-electronic conversion, and in the digital processing domain after a coherent receiving and electronic amplification sub-system.

In this section we illustrate some uses of NL effects and NL processing algorithms for improving the sensitivity of optical receivers employing NL processing at the front end of the photo-detector and NL processing algorithms in the electronic domain.

The spectral distribution of the FWM and the simulated spectral conversion can be achieved. There is a degeneracy of the frequencies of the waves so that efficient conversion can be achieved by satisfying the conservation of momentum. The detected phase states and bispectral properties are depicted in Figure 8.32, in which the phases can be distinguished based on the diagonal spectral lines. Under noisy conditions these spectral distributions can be observed in Figure 8.33.

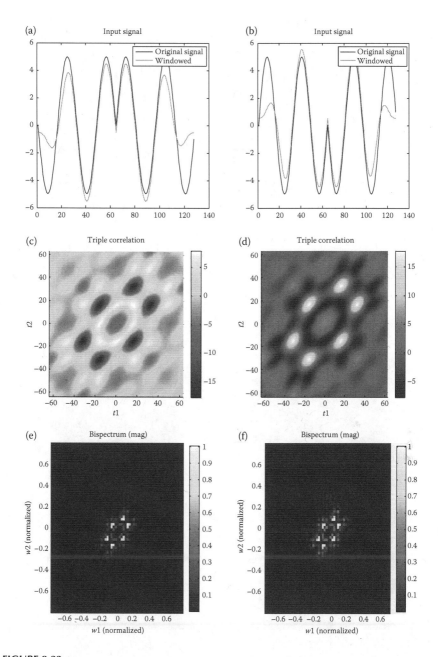

FIGURE 8.32
Input waveform with phase changes at the transitions (a and b), triple correlation, and bispectrum, (c through h) of both phase and amplitude.

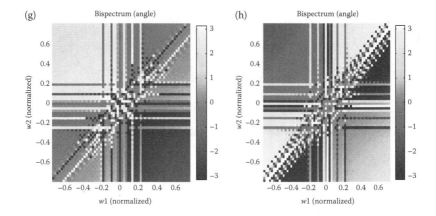

FIGURE 8.32 (continued)
Input waveform with phase changes at the transitions (a and b), triple correlation, and bispectrum, (c through h) of both phase and amplitude.

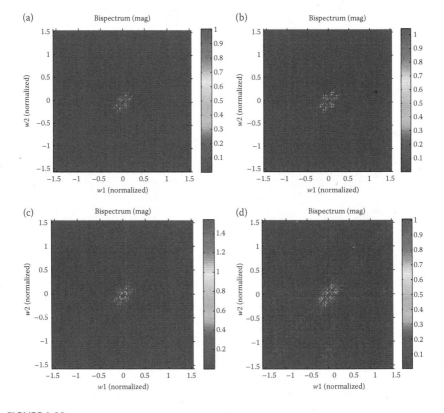

FIGURE 8.33
Effect of Gaussian noise on the bispectrum (a and c). Amplitude distribution in two-dimensional (b and d) phase spectral distribution.

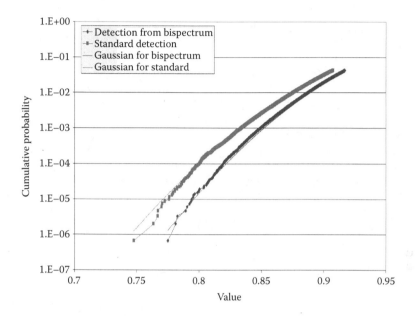

FIGURE 8.34
Error estimation version detection level of the HOS processor.

In contrast to the optical processing described above, an NL processing technique using HOS can be implemented in the electronic domain. This is implemented after the ADC, which samples the incoming electronic signals produced by the coherent optical receiver as shown in Figure 8.34. The operation of a third-order spectrum analysis is based on the combined interference of three signals (in this case, the complex signals produced at the output of the ADC, two of which are the delayed version of the original. Thence the amplitude and phase distribution of the complex signals are obtained in three-dimensional graphs, which allow us to determine the signal and noise power and the phase distribution. These distributions allow us to perform several functions necessary for evaluation of the performance of optical transmission systems. Simultaneously, the processed signals allow us to monitor the health of the transmission systems, such as the effects due to NL effects, the distortion due to chromatic dispersion of the fiber transmission lines, and the noises contributed by in line amplifiers, among others. A typical curve that compares the performance of this innovative processing with convention detection techniques is shown in Figure 8.34. If we aim to achieve an error rate at a bit error rate of 1e–9 at the output of the receivers employing conventional techniques and our high-order spectral receiver employing digital signal processing techniques, then at least 1000 times lower in the sensitivity of the receiver would be required. This is equivalent to at least one unit improvement on the quality factor of the eye opening. This is in turn equivalent to about 10 dB in the SNR.

These results are very exciting for network and system operators, as significant improvement of the receiver sensitivity can be achieved. This allows significant flexibility in the operation and management of the transmission systems and networks. Simultaneously, the monitored signals produced by the high-order spectral techniques can be used to determine the distortion and noises of the transmission line and thus the management of the tuning of the operating parameters of the transmitter, the number of wavelength channels, and the receiver or inline optical amplifiers.

The effect of additive white Gaussian noise on the bispectrum magnitude is shown in Figure 8.33, and the sequence of figures is indicated in Figure 8.32. The uncorrupted bispectrum magnitude is shown in Figure 8.32a while (b), (c), and (d) where generated using SNRs of 10, 3, and 0 dB, respectively. This provides a method of monitoring the integrity of a channel and illustrates another attractive attribute of the bispectrum. It is noted that the bispectrum phase is more sensitive to Gaussian white noise than is the magnitude and quickly becomes indistinguishable below 6 dB.

Although the triple correlation algorithm employing the NL processing can be implemented in either optical preprocessing domain or electronic processing, the processor would involve both optical hardware and algorithm processors. From the point of view of practical systems, it needs to deliver to the market at the right time for systems and networks operating in the Tera-bits/s speed. One can thus be facing the following dilemma: (i) the optical pre-processing demanding an efficient NL optical waveguide that must be an integrated structure, whereby efficient coupling and interaction can be achieved. If not, then the gain of about 3 dB in SNR would be defeated by this loss. Furthermore, the integration of the linear optical waveguiding section and an NL optical waveguide is not matched due to the difference in waveguide structures for both regions. For linear waveguide structure to be efficient for coupling with circular optical fibers, silica on silicon would be best suited due to the small refractive index difference and the technology of burying such waveguides to form an imbedded structure with an optical spot size that would match that of a single guided mode fiber. This silica on silicon would not match with an efficient NL waveguide made by As_2S_3 on silicon. (ii) On the other hand, if electronic processing is employed, then it requires an ultrafast ADC and then fast electronic signal processors. Currently, a 56 GSamples/s ADC is available from Fujitsu, as shown in Figure 8.35. It is noted that the data output samples of the ADC are structured in parallel forms with the referenced clock rate of 1.75 GHz. Thus, all processing of the digital samples must be in parallel form and thus parallel processing algorithms must be structured in parallel. This is the most challenging problem that we must overcome in the near future.

An application-specific integrated circuit (ASIC) must also be designed for this processor. Hard decisions must also be made regarding the fitting of such an ASIC and the associated optical and opto-electronic components into

(a)

(b)

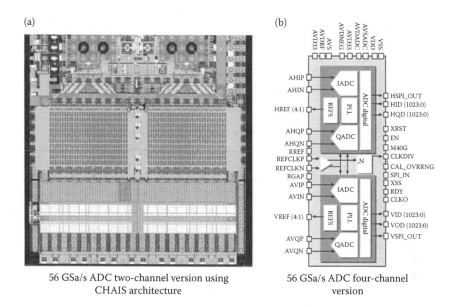

56 GSa/s ADC two-channel version using CHAIS architecture

56 GSa/s ADC four-channel version

FIGURE 8.35
Plane view of the Fujitsu ADC operating at 56 GSa/s. (a) Integrated view and (b) operational schematic.

international standard compatible size. Thus, all designs and components must meet this requirement.

8.2.8 Remarks

In this section we have demonstrated a range of signal processing applications exploiting the parametric process in NL waveguides. A brief mathematical description of the parametric process through third-order nonlinearity has been reviewed. A NL waveguide has been proposed to simulate interaction of multi waves in optical waveguides, including optical fibers. Based on the developed Simulink modeling platform, a range of signal-processing applications exploiting parametric FWM process has been investigated through simulation. With a CW pump source, the applications such as parametric amplifier, wavelength converter, and optical phase conjugator have been implemented for demonstration. The ultra-high-speed optical switching can be implemented by using an intensity modulated pump to apply in the short-pulse generator and the OTDM demultiplexer. Moreover, the FWM process has been proposed to estimate the triple-correlation, which is very important in signal processing. The simulation results showed the possibility of the FWM-based triple correlation using the NL waveguide with different pulse patterns. Although the triple correlation is contaminated by noise from other

FWM processes, it is distinguishable. The wavelength positions as well as the power of three delayed signals need to be optimized to obtain the best results.

Furthermore, we have also addressed the important issues of nonlinearity and its uses in optical transmission systems, such as the management of networks if the signals indicating the health of the transmission system are available. There is no doubt that the NL phenomena play several important roles in the distortion effects of signals transmitted, but also allow us to employ to improve the transmission quality of the signals. This has been briefly described in this chapter by using four-wave mixing effects and NL signal processing using high-order spectral analysis and processing in the electronic domain. These ultra-high-speed optical pre-processing and/or electronic triple-correlation and bispectrum receivers are the first systems using NL processing for 100–400 Gb/s and Tbps optical Internet.

The degree of complexity of ADC and DAC, as well as high-speed DSP, will allow higher-complex processing algorithms to be embedded for real-time processing. Thus, higher-order spectral processing techniques will be potentially employed for coherent optical communication with higher-dimensional processing.

References

1. J. M. Mendel, Tutorial on high-order statistic (Spectra) in signal processing and system theory: Theoretical results and some applications, *Proceedings of the IEEE*, 79(3), 1991.
2. P. A. Delaney, A bibliography of higher-order spectra and cumulants, *IEEE Signal Processing Magazine*, 11(3), 61–70, July 1994.
3. G. Sundaramoorthy et al., Bispectral reconstruction of signals in noise: Amplitude reconstruction issues, *Acoustics, Speech and Signal Processing, IEEE Transactions on*, 38(7), 1297–1306, July 1990.
4. H. Bartelt, A. W. Lohmann, and B. Wirnitzer, Phase and amplitude recovery from bispectra, *Applied Optics*, 23, 3121–3129, 1984.
5. C. L. Nikias et al., Signal processing with higher-order spectra, *IEEE Signal Processing Magazine*, 10(3), 10–37, July 1993.
6. T.M. Liu et al., Triple-optical autocorrelation for direct optical pulse-shape measurement, *Applied Physics Letters*, 81(8), 1402–1404, 2002.
7. J.M. Mendel, Tutorial on high-order statistic (Spectra) in signal processing and system theory: Theoretical results and some applications, *Proceedings of the IEEE*, 79(3), 1991.
8. D. R. Brillinger, Introduction to polyspectra, *Annals of Mathematical Statistics*, 36, 1351–1374, 1965.
9. G. Sundaramoorthy, M.R. Raghuveer, and S.A. Dianat, Bispectral reconstruction of signals in noise: Amplitude reconstruction issues, *IEEE Transactions on Acoustics, Speech and Signal Processing*, 38(7), 1297–1306, July 1990.

10. H. Bartelt, A. W. Lohmann, and B. Wirnitzer, Phase and amplitude recovery from bispectra, *Applied Optics* 23, 3121–3129, 1984.
11. G. P. Agarwal, *Nonlinear Fiber Optics*, Academics, 2004.
12. M. C. Ho et al., 200-nm-bandwidth fiber optical amplifier combining parametric and Raman gain, *IEEE Journal of Lightweight Technology*, 19, 977–981, July 2001.
13. H.C. H. Mulvad et al., 1.28 Tbit/s single-polarization serial OOK optical data generation and demultiplexing, *Electronics Letters*, 45(5), 280–281, Feb 2009.
14. M.D. Pelusi et al., Applications of highly-nonlinear chalcogenide glass devices tailored for high-speed all-optical signal processing, *IEEE Journal of Selected Topics Quantum Electronics*, 14, 529–539, 2008.
15. T.D. Vo et al., Photonic chip based 1.28 Tbaud transmitter optimization and receiver OTDM demultiplexing, *Proceedings of OFC 2010*, San Diego, 2010, Paper PDPC5.
16. L.N. Binh, *Optical Fiber Communications Systems: Theory and Practice with Matlab and Simulink Models*, CRC Press, 2010.
17. C.L. Nikias and J.M. Mendel, Signal processing with higher-order spectra, *IEEE Signal Processing Magazine*, 10(3), 10–37, 1993.
18. T.M. Liu et al., Triple-optical autocorrelation for direct optical pulse-shape measurement, *Applied Physics Letters*, 81(8), 1402–1404, 2002.

Index

A

ADC, *see* Analog-to-digital converter (ADC)

Advanced specific integrated circuit (ASIC), 151

AGC, *see* Automatic gain control (AGC)

All-optical signal processing, 419–420

Amplified spontaneous emission (ASE), 80, 350

Amplitude shift keying (ASK), 11, 13, 122, 377

format generation, 140–141

Analog-to-digital converter (ADC), 7, 20, 301, 369, 420, 454, 455

APD, *see* Avalanche photodiode (APD)

Application-specific integrated circuit (ASIC), 340, 454

Arbitrary white Gaussian noise (AWGN), 415

ASE, *see* Amplified spontaneous emission (ASE)

ASIC, *see* Advanced specific integrated circuit (ASIC); Application-specific integrated circuit (ASIC)

ASK, *see* Amplitude shift keying (ASK)

ASK coherent system, 187; *see also* Optical heterodyne detection

amplitude-demodulated envelope, 187

BER, 188, 189

envelope detection, 188–189

equivalent current model, 189

IF frequency, 187

PDF, 188

SNR, 188

synchronous detection, 189

Attenuation coefficient, 52–53

Autocorrelation reception process, 183

Automatic gain control (AGC), 182

Avalanche photodiode (APD), 180

AWGN, *see* Arbitrary white Gaussian noise (AWGN)

B

B2B, *see* Back-to-back (B2B)

Back propagation (BP), 42, 111; *see also* Nonlinear fiber transfer functions

drawbacks of, 111

equalized constellations, 112

implementation, 113

with low complexity, 114

multi-span equivalent transmission system, 113

QPSK constellation and phase rotation, 112

receiver model incorporating, 112

total residual dispersion, 114

Back-to-back (B2B), 228, 372

Bandpass filter (BPF), 185, 430

Bending loss, 50

BER, *see* Bit error rate (BER)

Binary phase shift keying (BPSK), 141

modulation, 415

Bipolar junction transistor (BJT); *see also* Coherent balanced receiver and noise suppression; Electronic amplifiers

advantages, 233, 234

disadvantages, 234

noise-equivalent and small-signal model, 250

optimum emitter resistance, 252

spectral density of noises, 249

total equivalent noise power, 252

types of, 238

Bispectrum, 411–412, 442; *see also* Four-wave mixing (FWM); Triple correlation

coherent optical receiver, 412

estimation, 411

fourier transforms of high-order correlation functions, 411

multi-dimensional spectra, 411

optical pre-processing receiver, 421

as third-order poly-spectrum, 413

Bispectrum optical receivers, 409; *see also* Bispectrum; Nonlinear photonic signal processing
    autocorrelation, 412
    power spectrum, 411–412
    transmission system performance, 410
    triple product of original optical waves, 410
Bit error rate (BER), 68, 110, 145, 181, 352, 369
BPF, *see* Bandpass filter (BPF)
BPSK, *see* Binary phase shift keying (BPSK)
Brillouin scattering, 45; *see also* Stimulated Brillouin scattering (SBS)

C

Carrier phase estimation, 339; *see also* Carrier recovery (CR)
    coherent receiver, 341
    forward phase estimation QPSK optical coherent receivers, 341–342
    *M*th power phase estimation, 340
    optical delay differential detection, 341
    phase noise and nonlinear effects correction, 340–341
Carrier recovery (CR), 317; *see also* Carrier phase estimation
    algorithm and demonstration of, 345
    DSP structure in PDM-coherent receivers
    EMCORE external cavity laser measured oscillation, 344
    FO estimation, 343
    FO oscillations and Q-penalties, 343
    frequency estimation, 346
    high-precision, 348
    modified algorithms for, 345
    in polarization division multiplexed receivers, 342
    Q gain, 347
    10G hybrid off-line experiment, 347
    V&V CPR penalties, 344, 345

Carrier-suppressed return to zero (CSRZ), 131, 140; *see also* Return-to-zero optical pulses
    format generation, 141
    pulse, 141
    QPSK transmitter, 371, 372
    spectra of, 162, 176
Carrier-to-noise ratio (CNR), 418
CCFE, *see* Coarse carrier frequency estimation (CCFE)
CCFE-R, *see* Coarse carrier frequency estimation and recovery (CCFE-R)
CD, *see* Chromatic dispersion (CD)
CHAIS, *see* Charged mode interleaved sampler (CHAIS)
Chalcogenide glass, 428
Charged mode interleaved sampler (CHAIS), 228
Chromatic dispersion (CD), 54, 110, 301, 371
Chromatic dispersion factor, total, 59
Circular optical waveguides, 25; *see also* Optical fibers
    refractive index, 28
CM, *see* Constant modulus (CM)
CMA, *see* Constant modulus algorithm (CMA); Constant modulus amplitude (CMA)
CNR, *see* Carrier-to-noise ratio (CNR)
Coarse carrier frequency estimation (CCFE), 345
    residual FO, 346
Coarse carrier frequency estimation and recovery (CCFE-R), 345
    residual FO, 346
Coherent balanced detector optical receiver, 231
Coherent balanced receiver and noise suppression, 231, 248–249; *see also* Bipolar junction transistor (BJT); Coherent detection; Coherent optical receivers; Electronic amplifiers
    amplifier step response, 244
    analytical noise expressions, 233–235
    arrangement of coherent optical communications systems, 236

current from PDs, 232
current generator, 235
electronic current difference, 233
equivalent input noise current,
236–238
equivalent noise current, 249
excess noise cancellation
technique, 246
excess noise measurement, 247–248
fields at coupler output, 232
first-stage pole effect on electronic
amplifier root locus, 240
high-frequency root locus, 240
noise components, 233
noise equations, 249–252
noise generators, 235
noise measurement and
suppression, 242, 244–245
noise power, 251
noise spectral density, 235, 251
PD current, 233
poles of singularities, 242
pole-zero pattern, 238–242, 243
preamplifier, 237
preamplifier design, 239
quantum shot noise, 235
requirement for quantum limit,
245–246
rise-time and 3 dB bandwidth,
242–244
rise time measurement in electrical
domain, 244
shot noise power spectral density,
235
small equivalent circuits, 234
Coherent detection, 20, 180, 182,
228; *see also* Coherent
balanced receiver and noise
suppression; Coherent optical
receivers; DSP-assisted
coherent detection; Electronic
amplifiers; Optical heterodyne
detection; Optical homodyne
detection; Optical intradyne
detection
advantages, 184
challenges in, 202
coherent optical transmission, 183
heterodyne detection, 183

homodyne reception, 183
LO, 183
ML phase estimator, 203
nonlinear phase noises, 203
processes and DSP, 305
self-coherent detection and
electronic DSP, 201–203
spectrum of, 201
synchronous detection, 183
synchronous processes, 305–306
types, 185
in WDM systems, 184
Coherent optical communications, 180;
see also Coherent detection
algorithms for, 304; *see also* Linear
equalization; Nonlinear
equalizer
arrangement of, 236
incorporating NLE, 320
ML phase estimation for, 334
simplified schematic of, 306
Coherent optical receivers, 90, 179,
228, 410; *see also* Coherent
balanced receiver and noise
suppression; Coherent
detection; Optical phase lock
loop (OPLL)
advantages, 184, 202
using balanced detection
technique, 341
bispectrum, 412
carrier phase tracking, 202
components, 181
disadvantages, 184
DSP processing functionalities flow,
211
equivalent circuit of electronic
preamplifier, 182
in-line fiber optical amplifier, 182
QPSK, 342
structure of, 182
synchronous coherent receiver, 210
Combed optical source, 298
Constant modulus (CM), 316
Constant modulus algorithm
(CMA), 314; *see also* Linear
equalization (LE)
carrier phase recovery plus, 317
CM criterion, 316

Constant modulus algorithm (*Continued*)
  DD processing, 318
  equivalent model for baseband
      equalization, 315
  operation with carrier phase
      recovery, 317–318
  output of equalized signals, 315
  phase-blind nature of, 317
  phase updating rule, 317
  received signal before processing,
      314
  tap update with CMA cost function,
      317
Constant modulus amplitude (CMA),
      71, 400
Continuous phase FSK (CPFSK), 147
Continuous waves (CW), 121, 433
Coupled equations system, 430
CPFSK, *see* Continuous phase FSK
      (CPFSK)
CR, *see* Carrier recovery (CR)
Cross-phase modulation (XPM), 420; *see
      also* Nonlinear optical effects
  effect, 43, 44
  illustration of, 45
  phase shift of ith channel effect, 44
  power dependence of RI, 78
CSRZ, *see* Carrier-suppressed return to
      zero (CSRZ)
Cut-off wavelength, 40
CW, *see* Continuous waves (CW)

D

DAC, *see* Digital-to-analog conversion
      (DAC)
Data phase modulator, 143
DC drift, 128
DCF, *see* Dispersion compensating fiber
      (DCF)
DCM, *see* Dispersion compensating
      module (DCM)
DD, *see* Decision-directed (DD);
      Directly modulated (DD)
Decision-directed (DD), 384
  processing, 318
Decision feedback equalizers (DFEs),
      318; *see also* Nonlinear
      equalizer

incorporating linear and nonlinear
      filter, 322
  with linear FFE and decision device,
      320
Demultiplexer, *see* Optical time
      division multiplexing
      demultiplexer (OTDM
      demultiplexer)
Device under test (DUT), 62
DFEs, *see* Decision feedback equalizers
      (DFEs)
DFT, *see* Discrete Fourier transform
      (DFT)
DGD, *see* Differential group delay
      (DGD)
Differential detection, 190; *see also*
      Optical heterodyne detection
  BER, 191
  process modification, 190
  SNR, 191
  square noise current, 191
Differential-dispersion parameter, 63
Differential group delay (DGD), 97, 349,
      398
Differential nonlinearity (DNL), 221
Differential phase compensation
      (DPC), 317
Differential phase detection, 195;
      see also Optical homodyne
      detection; Differential phase
      shift keying (DPSK)
  BER analytical approximation and
      numerical evaluation, 198
  of continuous phase FSK
      modulation format, 197
  CPFSK differential detection
      configuration, 199
  detected signal phase, 199
  frequency to voltage conversion
      relationship, 199
  heterodyne signal current, 195
  linewidth neglecting criteria, 198
  with local oscillator, 195–197
  modulation index parameter, 200
  PDF of quantum phase noise, 196
  probability of error, 196, 197
  receiver power penalty
      dependence, 200
  under self-coherence, 197–200

standard deviation from central
frequency, 196
Differential phase shift keying (DPSK),
6, 122, 137; *see also* Differential
phase detection; External
modulators
background, 137–138
bias point and RF driving signals,
139
encoded differential data, 142
format generation, 141
optical transmitter, 138
pre-coder, 138
RZ-DPSK transmitter, 142
signal constellation diagram of, 138,
142
spectral properties of, 164
Differential quadrature phase shift
keying (DQPSK), 141, 371
format generation, 143
offset, 146
phase component recovery, 334
signal constellation diagram of, 143
Differential Viterbi (DiffV), 343
DiffV, *see* Differential Viterbi (DiffV)
Digital-based optical transmitter, 90
Digital frequency estimation (FDE), 343
Digital frequency recovery (FDR), 343
Digital modulation formats, 11; *see also*
M-ary QAM; Nyquist pulse-
shaping techniques; Optical
fiber communication
lightwave carrier inphase
and quadrature phase
constellations, 14
lightwave carrier modulation, 11
modulation formats, 11–13
Nyquist filter for spectral
equalization, 12
optical signal, 11
output spectrum of Nyquist filtered
QPSK signal, 12
partial response, 13–15
pulse shaping and modulations, 13
spectra of pulse sequences, 12
Digital signal processing (DSP), 1,
90, 102, 105, 301, 369, 410;
see also Linear equalization
(LE); Maximum a posteriori

technique (MAP technique);
Maximum likelihood detection
(MLSD); Maximum likelihood
estimation equalizer-MSK
optical transmission system
performance; Nonlinear
equalizer (NLE); Nonlinear
fiber transfer functions; Viterbi
algorithm
advantages, 106
algorithms, 2
-based coherent optical
communication systems, 202
based coherent optical transmission
systems, 369, 406–407; *see also*
Superchannel transmission
-system; Quadrature phase
shift keying (QPSK)
-based coherent receiver, 10, 304
-based receiver, 100
electronic, 330
equalizers at transmitter, 329–331
feedback linear filter, 332
functionalities of, 370
nonlinear processor, 331
optical digital receiver and, 22
optical receiver based on, 100
in PDM-coherent receivers, 343
at receiver end, 301–302
sequence flow in, 302–303
shared equalization, 329, 332–333
signal mixing and LO laser, 301
subsystem processing, 304
at transmitter, 301
z-transform at nonlinear distortion
equalizer output, 330
Digital signal processing algorithms
(DSP algo), 77
Digital-to-analog conversion (DAC), 6,
158, 301, 386; *see also* External
modulators
B2B performance evaluation, 173
for DSP-based modulation and
transmitter, 168
frequency transfer characteristics
of, 175
Fujitsu DAC, 168–170, 172
I and Q components generation,
171–173

Digital-to-analog conversion (*Continued*)
  noise spectral characteristics of, 175
  NTT InP-based, 169–170
  Nyquist PDM-QPSK transmitter
    experimental setup, 173
  pressing steps, 171
  16 QAM 4-level generated signals, 171
  structure, 170
  test signals and converted analog
    signals, 171
  28Gbaud RF signals spectrum and
    eye diagram, 174
Direct modulation techniques, 121
Directly modulated (DD), 248
Discrete Fourier transform (DFT), 157
Discrete phase modulation NRZ
        format generation, 141; *see also*
        Minimum shift keying (MSK);
        Modulation format generation
  continuous phase modulation
        PM-NRZ formats, 146–147
  differential phase shift keying,
        141–142
  differential quadrature phase shift
        keying, 143
  linear and nonlinear MSK, 147–151
  M-Ary ADPSK generation, 144–146
  non return-to-zero differential
        phase shift keying, 143
  NRZ-DPSK photonics transmitter,
        144
  return-to-zero differential phase
        shift keying, 143
  RZ-DPSK photonics transmitter, 144
Dispersion compensated fiber, 88
Dispersion compensating fiber (DCF),
        87, 99, 208, 392
Dispersion compensating module
        (DCM), 107
Dispersion-shifted fibers (DSF), 49, 88
Dispersion slope, 63
Distributed feedback (DFB), 10, 54, 202
DNL, *see* Differential nonlinearity
        (DNL)
Double sideband (DSB), 285
DPC, *see* Differential phase
        compensation (DPC)
DQPSK, *see* Differential quadrature
        phase shift keying (DQPSK)

Driving voltage, 125, 127
DSB, *see* Double sideband (DSB)
DSF, *see* Dispersion-shifted fibers (DSF)
DSP algo, *see* Digital signal processing
        algorithms (DSP algo)
DSP-assisted coherent detection,
        208, 228; *see also* Coherent
        detection; Digital signal
        processing (DSP); Effective
        number of bits (ENOB)
  ADC and DAC operating speed
        evolution, 219
  ADC operation principles, 230
  ADC subsystems, 230, 231
  coherent reception, 211, 217
  DAC structures, 229
  digital processing system, 217, 228
  DSP-based reception systems, 209
  DSP processing functionalities
        flow, 211
  equivalent current model at optical
        balanced receiver input, 212
  linewidth resolution progress, 218
  parameters in sensitivity
        analysis, 212
  polarization and phase diversity
        receivers, 209
  receiver sensitivity analysis,
        216–217
  semiconductor manufacturing
        with resolution of line
        resolution, 219
  sensitivity analysis, 211–215
  shot-noise-limited receiver
        sensitivity analysis, 215
  synchronous coherent receiver, 210
DuoB, *see* Duobinary (DuoB)
Duobinary (DuoB), 15
DUT, *see* Device under test (DUT)

E

EA effect, *see* Electro-absorption effect
        (EA effect)
EAM, *see* Electro-absorption modulator
        (EAM)
ECL, *see* External cavity laser (ECL)
EDFA, *see* Erbium doped fiber amplifier
        (EDFA)

Effective number of bits (ENOB), 218;
    see also DSP-assisted coherent
    detection
  of ADC, 223
  B2B performance, 226
  best-fit INL, 221–222
  comprehensive effects of AGC
    clipping and, 227
  corresponding to Gaussian noise
    levels, 225
  of digitalized system, 220
  DNL in ADC transfer curve, 221
  end-point INL, 222–223
  frequency response of commercial
    real-time DSA, 224
  impact on transmission
    performance, 226–228
  signal-dependent, 225
  SINAD, 220
Eigenvalue equation of wave equation,
    31
Electrical spectrum analyzer (ESA), 248
Electrically time division-multiplexed
    transponders (ETDM
    transponders), 156
Electro-absorption (EA effect), *see* Stark
    effect
Electro-absorption effect (EA effect), 122
Electroabsorption modulated laser
    (EML), 123
Electro-absorption modulator (EAM),
    122, 394, 396; *see also* External
    modulators
  combined with CW laser source, 123
  drawback of, 124
  driving voltage for, 123
  integrated with distributed feedback
    laser, 123
  package EAM of Huawei Center, 123
  power vs. applied voltage of, 124
  semiconductor intensity modulator
    insertion loss, 122
Electronic amplifiers, 203; *see also*
    Bipolar junction transistor
    (BJT); Coherent balanced
    receiver and noise
    suppression; Coherent
    detection; Trans-impedance
    amplifiers (TIA)

amplifier noise, 206–208
design circuit for balanced optical
    amplifier, 239
differential amplifiers, 206
equivalent electronic noise source,
    207
first-stage pole effect on electronic
    amplifier root locus, 240
optoelectronic receiver, 204
p-i-n detector, 204
quantum shot noises, 207
shot noises, 207
SNR at electronic amplifier output,
    208
thermal noises, 207
three-stage electronic preamplifier,
    237
transimpedance amplification type,
    204
transistor selection, 233
Electro-optic effect, 125
Electro-optic modulator (EOM), 124; *see
    also* External modulators
  bias control, 128–129
  chirp-free optical modulators,
    129–130
  crystal cuts of $LiNbO_3$ integrated
    structures, 129
  DC drift, 128
  electro-optic phase modulation,
    126
  integrated optical modulator bias
    control arrangement, 129
  intensity modulation, 126
  intensity modulators, 125–127
  Mach–Zehnder modulator, 128, 130
  phase modulators, 125
  phasor representation and transfer
    characteristics, 127–128
  photonic modulator structures,
    130–131
  polarization modulator, 130, 131
  typical operational parameters, 131
Electro-optic phase modulator
    (E-OPM), 151
EML, *see* Electroabsorption modulated
    laser (EML)
ENOB, *see* Effective number of bits
    (ENOB)

EOM, *see* Electro-opticmodulator
          (EOM)
E-OPM, *see* Electro-optic phase
          modulator (E-OPM)
Equivalent-step index profile technique
          (ESI profile technique), 41
Erbium doped fiber amplifier (EDFA),
          1, 109, 255, 414
Error vector magnitude (EVM), 85
ESA, *see* Electrical spectrum analyzer
          (ESA)
ESI profile technique, *see* Equivalent-
          step index profile technique
          (ESI profile technique)
ETDM transponders, *see* Electrically
          time division-multiplexed
          transponders (ETDM
          transponders)
Ethernet technology, 179
EVM, *see* Error vector magnitude
          (EVM)
External cavity laser (ECL), 54, 371
External modulation, 121; *see also*
          External modulators
     and advanced modulation formats,
          122
     electro-optic phase modulation, 126
     intensity modulation, 126
     techniques, 121
     types, 122
External modulators, 121, 173–176; *see
          also* Differential phase shift
          keying (DPSK); Digital-to-
          analog conversion (DAC);
          Electro-absorption modulator
          (EAM); Electro-optic
          modulator (EOM); External
          modulation; I–Q integrated
          modulators; Modulation
          format generation; Photonic
          MSK transmitter; Return-to-
          zero optical pulses
     intensity manipulation, 122

F

FBCFE-R, *see* Feedback carrier
          frequency estimation and
          recovery (FBCFE-R)

FBG, *see* Fiber Bragg gratings (FBG)
FDE, *see* Digital frequency estimation
          (FDE)
FDR, *see* Digital frequency recovery
          (FDR)
FEC, *see* Forward error coding (FEC)
Feedback carrier frequency estimation
          and recovery (FBCFE-R), 345
Feedforward (FF), 310
Feed-forward carrier phase estimation
          (FFCPE), 342
Feedforward equalizer (FFE), 310; *see
          also* Linear equalization (LE)
     electrical SNR, 312
     mean square noise, 312
     noisy detection error, 312
     output noise signals of, 311
     probability noise density
          function, 312
Feedfoward carrier phase recovery
          (FFCPR), 343
FF, *see* Feedforward (FF)
FF transversal filter, 320
FFCPE, *see* Feed-forward carrier phase
          estimation (FFCPE)
FFCPR, *see* Feedfoward carrier phase
          recovery (FFCPR)
FFE, *see* Feedforward equalizer (FFE)
Fiber Bragg gratings (FBG), 6
Fiber dispersion factor, 60, 61
Fiber nonlinearity, *see* Nonlinear
          optical effects
Field-effect transistor (FET); *see also*
          Electronic amplifiers
     advantages, 234
Finite impulse response (FIR), 69, 388,
          400
     fractionally spaced FIR filter, 309
Finite state machine (FSM), 326; *see also*
          Trellis structure
     optical fiber as, 328
     state trellis, 327
FIR, *see* Finite impulse response (FIR)
Forward error coding (FEC), 373
4-ary quadrature amplitude
          modulation (4 QAM) signal,
          404
4PPD, *see* Fourth-power law PD (4PPD)
Fourth-power law PD (4PPD), 399

Four-wave mixing (FWM), 420;
  see also Nonlinear optical
  effects; Optical domain
  implementation; Optical
  time division multiplexing
  demultiplexer (OTDM
  demultiplexer); Third-order
  nonlinearity
  based OTDM demultiplexer, 442
  bispectral optical structures, 420–421
  conservation of momentum, 423–424
  coupled-wave equations, 425–427
  energy conservation diagram, 426
  idler wave, 426
  momentum conservation diagram,
    426
  on optical channels, 48
  optical pre-processing receiver, 421
  phase matching, 48–49
  phase matching and, 423
  phenomena, 422
  photonic pre-processing, 421
  polarization vector, 422
  propagation constant, 422
  pump waves, 426
  SPM and XPM, 423
  total electric field vector, 422
Franz and Keldysh effect, 122
Frequency shift keying (FSK), 11,
    13, 122
  coherent system, 191–192
FSK, *see* Frequency shift keying (FSK)
FSM, *see* Finite state machine (FSM)
Fujitsu PDM-IQ modulator, 167
Full-width-half-mark (FWHM), 54, 134
FWHM, *see* Full-width-half-mark
    (FWHM)
FWM, *see* Four-wave mixing (FWM)

G

GbE, *see* Gigabit Ethernet (GbE)
Gigabit Ethernet (GbE), 179
Global submarine cable systems, 3
Gordon–Mollenauer effect, 203, *see*
    Nonlinear phase noise
Group delay factor, 89
Group velocity delay (GVD), 349
Group velocity dispersion (GVD), 54

in frequency domain transfer
    function, 65
optical pulse distortion, 437
of single-mode optical fiber, 106
Guiding lightwaves, 3; *see also* Optical
    fiber communication
  advanced transmission, 5, 7–8
  attenuation factor, 4, 5
  cable networks, 4
  coherent detection with DSP, 6
  DAC, 6
  optical amplification, 6
  optical fibers, 3
  refractive index, 8
  SSMF, 8
  transmission link dispersion
    management, 6
GVD, *see* Group velocity delay (GVD);
    Group velocity dispersion
    (GVD)

H

Half-band filters, 332
Higher-order spectrum coherent
    receivers, *see* Nonlinear
    photonic signal processing
Highly NL fibers (HNLF), 420
High-order spectrum (HOS), 409
High-resolution spectrometer
    (HRS), 273
HNLF, *see* Highly NL fibers (HNLF)
Hopping effect, 81
Horizontal polarized (H-pol), 369
HOS, *see* High-order spectrum (HOS)
H-pol, *see* Horizontal polarized (H-pol)
HRS, *see* High-resolution spectrometer
    (HRS)

I

I, *see* Inphase (I)
IF, *see* Intermediate frequency (IF)
IFFT, *see* Inverse fast Fourier transform
    (IFFT)
IM, *see* Intensity modulation (IM)
IM optical waves, *see* Intensity
    modulated optical waves (IM
    optical waves)

IMDD, *see* Intensity-modulation/direct-detection (IMDD)
Impulse sequence, 306
Infra-red (IR), 50
INL, *see* Integral nonlinearity (INL)
Inphase (I), 369
Integral nonlinearity (INL), 221
Intensity modulated optical waves (IM optical waves), 92
Intensity modulation (IM), 89, 126
Intensity-modulation/direct-detection (IMDD), 179
Interferometric intensity modulator, 127
    to carve pulses, 142
    electrical to optical transfer curve, 127
Intermediate frequency (IF), 183, 404
Intermodal delay effects, 54
International Telecommunication Union (ITU), 131
Intersymbol interference (ISI), 15, 157, 305, 387
Intra-modal dispersion, *see* Group velocity dispersion (GVD)
Inverse fast Fourier transform (IFFT), 157
I–Q integrated modulators, 164; *see also* External modulators
    advantages, 165
    alternate structure of, 165
    amplitude modulation of lightwave path, 165
    and electronic digital multiplexing, 167–168
    Fujitsu PDM-IQ modulator, 167
    inphase and quadrature phase optical modulators, 164–167
    QAM transmitter schemes, 166
    16 QAM modulation scheme constellation, 165
    superstructures, 166
    time division multiplexing, 168
IR, *see* Infra-red (IR)
ISI, *see* Intersymbol interference (ISI)
ITU, *see* International Telecommunication Union (ITU)

J

JFF-CPE-R, *see* Joint H-V FF carrier phase estimation and recovery (JFF-CPE-R)
Joint H-V FF carrier phase estimation and recovery (JFF-CPE-R), 345

L

Laser sources, 286
LE, *see* Linear equalization (LE)
Least-mean-square (LMS), 400
Least significant bits (LSB), 221
LE-ZF filter, *see* Linear equalizer zero-forcing filter (LE-ZF filter)
Lightwave carrier modulation, 11
Likelihood function, 333
Linear distortion impairments, 302
Linear equalization (LE), 305; *see also* Constant modulus algorithm; Feedforward equalizer (FFE); Nonlinear equalizer (NLE); Zero-forcing linear equalization (ZF-LE)
    additive Gaussian noise tolerance, 310–312
    basic assumptions, 306–307
    detection processes, 305
    equalization with minimizing MSE, 312–314
    feedback transversal filter, 310
    impulse sequence, 306
    linear feedback transversal equalizers, 311
    narrowband filter receiver integrated with, 349
    and NLE, 321
    z-transform, 310
Linear equalizer zero-forcing filter (LE-ZF filter), 308
Linearly polarized (LP), 371
    mode, 41
LMS, *see* Least-mean-square (LMS)
LMS equalizer (LMSE), 400
LMSE, *see* LMS equalizer (LMSE)
LN modulator, 125–127
    in integrated optical modulator, 129
LO, *see* Local oscillator (LO)

LO frequency offset (LOFO), 200, 282, 299, 301, 400

Local oscillator (LO), 24, 179, 255, 301, 371

LOFO, *see* LO frequency offset (LOFO)

Lorentzian linewidth formula, 336

Low pass filter (LPF), 185, 260; *see also* Optical phase lock loop (OPLL); Phase lock loop (PLL)
  Bode plot, 270
  digital, 256–257, 266, 268–270
  direct form II digital filter topology, 269
  fixed-point arithmetic, 266, 268
  FPGA board implementing digital PLL, 272
  FPGA implementation, 272
  interface board, 270–272
  locking state indication, 272–273
  magnitude response, 260
  OPLL hardware details, 273
  optical power budget, 273
  phase error and ratio of time constants, 283
  quantization errors for filter coefficients, 268
  realization using analog circuit, 260
  state diagram of, 267
  transfer function of, 260, 268

Lower sidebands (LSB), 151, 349

LP, *see* Linearly polarized (LP)

LSB, *see* Least significant bits (LSB)

M

Mach–Zehnder in-phase/quadrature modulator (QMZ-IQ modulator), 166

Mach–Zehnder intensity modulators (MZIM), 90, 437

Mach–Zehnder interferometric modulator (MZIM), 9, 62, 175; *see also* Electro-optic modulator (EOM)
  power transfer function, 128
  structure of, 122, 130

Mach–Zehnder modulators (MZMs), 399

Manakov-PMD equations, 106

MAP technique, *see* Maximum a posteriori technique (MAP technique)

M-Ary amplitude differential phase shift keying (M-Ary ADPSK)
  driving voltages for signal constellation, 145
  format generation, 144
  signal constellation, 145
  waveforms of NRZ and RZ pulses, 146

M-ary phase-shift keying (M-ary PSK), 202
  to increase spectral efficiency, 202
  ML phase estimator, 203
  structures of, 183

M-ary PSK, *see* M-ary phase-shift keying (M-ary PSK)

M-ary QAM, 12; *see also* Digital modulation formats
  BER vs. SNR for multi-level, 14
  digital modulation formats, 13
  to increase spectral efficiency, 202

Material dispersion, *see* Group velocity dispersion (GVD)

Material dispersion factor, 57

Material dispersion parameter, *see* Material dispersion factor

Maximum a posteriori technique (MAP technique), 333; *see also* Digital signal processing (DSP)
  complex phase reference vector, 335
  DWDM optical DQPSK system, 337
  estimates, 334–339
  likelihood function, 333
  method, 333
  ML DSP phase estimation, 339
  ML estimate, 333
  ML receiving processor, 335–336
  Monte Carlo simulated BER, 338
  Mth-power phase estimation scheme, 337
  output signal from photocurrents, 335
  for phase estimation, 333
  phase noise difference, 336
  posterior distribution of θ, 333
  Q-factor, 339

Maximum likelihood (ML), 203, 321
    estimate, 333
    phase estimator, 203
    receiving processor, 335–336
Maximum likelihood detection
        (MLSD), 324; *see also*
        Maximum likelihood
        estimation (MLSE)
Maximum likelihood estimation
        (MLSE), 106, 303, 324,
        381; *see also* Maximum
        likelihood detection (MLSD);
        Maximum likelihood
        estimation equalizer-MSK
        optical transmission system
        performance
    equalizer as FSM, 326
    nonlinear, 325–326
Maximum likelihood estimation
        equalizer-MSK optical
        transmission system
        performance, 348; *see also*
        Minimum shift keying (MSK);
        MLSE equalizer; Viterbi
        algorithm; Viterbi-MLSE
        equalizers
    BER vs. required OSNR, 355
    40 Gb/s transmission systems,
        351–352
    key simulation parameters, 352
    MLSE scheme performance, 351
    OFDR-based 10 Gb/s MSK optical
        transmission, 356
    quadratic phase channel limitations,
        364–365
    simulation set-up for performance
        evaluation, 351
    10 Gb/s optical MSK signals
        transmission, 352–355
Maximum transmission limit, 64
Maxwellian distribution, 63
    of PMD random process, 64
MCM, *see* Multi-carrier multiplexing
        (MCM)
Mean square error (MSE), 377
MFD, *see* Mode field diameter (MFD)
Microbending loss, 51
MIMO, *see* Multiple input multiple
        output (MIMO)

Minimum shift keying (MSK), 11,
        122, 192, 303, 348; *see also*
        Frequency shift keying
        (FSK); Maximum likelihood
        estimation equalizer-MSK
        optical transmission system
        performancee; Photonic MSK
        transmitter
    base-band equivalent optical MSK
        signal, 150
    characteristics of, 148, 151
    CPFSK signal, 147
    in-phase component, 148
    linear and nonlinear MSK, 147
    logic gates construction, 149
    as offset DQPSK, 150–151
    optical MSK transmission system, 350
    optical power spectra of, 163
    phase trellis for, 148
    pre-coding logic equations, 149
    quadrature component, 148
    signal, 147
    as special case of CPFSK, 150
    spectra of, 160, 176
    spectral properties of, 164
    state diagram for, 148
    transmitter, 348
    truth table, 149
ML, *see* Maximum likelihood (ML)
MLL, *see* Mode-locked laser (MLL)
MLSD, *see* Maximum likelihood
        detection (MLSD)
MLSE, *see* Maximum likelihood
        estimation (MLSE)
MLSE equalizer; *see also* Maximum
        likelihood estimation
        equalizer-MSK optical
        transmission system
        performance
    configuration, 348–349
    effectiveness investigation, 359
    noise distribution, 350
    performance comparison, 361
    performance vs. normalized values,
        360
    receiver integrated with equalizers,
        349
    with reduced-state template
        matching, 351

1024-state, 358
with Viterbi algorithm, 349–351, 352
Mode fiber, 32
Mode field diameter (MFD), 41
Mode-locked laser (MLL), 439
Modulation, 140
  index parameter, 200
  instability, 80
  spectrum of optical gain, 81
  techniques, 2
Modulation format generation, 140;
    see also Discrete phase
    modulation NRZ format
    generation; External
    modulators
  ASK formats, 140–141
  baseband NRZ and RZ line coding,
    140
  carrier-suppressed RZ formats, 141
  NRZ photonics transmitter, 140
  RZ photonics transmitter, 140
  spectra of, 159–164, 176
  typical parameters of optical
    intensity modulators for, 161
M-PSK, *see* Multiple phase shift keying
    (M-PSK)
MQW, *see* Multi-quantum well (MQW)
MSE, *see* Mean square error (MSE)
MSK, *see* Minimum shift keying (MSK)
Mth-power phase estimation scheme,
    337
Multi-carrier multiplexing (MCM), 156
Multi-dimensional spectra technique,
    412
Multi-level modulation formats, 105
Multilevel phase shift keying, 255; *see
    also* M-ary phase-shift keying
    (M-ary PSK)
Multiple input multiple output
    (MIMO), 303
  techniques, 26
Multiple phase shift keying (M-PSK), 343
Multiplexer (mux), 371
Multi-quantum well (MQW), 122
Multi-span equivalent transmission
    system, 113
mux, *see* Multiplexer (mux)
MZ delay interferometer (MZDI), 191,
    144

MZDI, *see* MZ delay interferometer
    (MZDI)
MZIM, *see* Mach–Zehnder intensity
    modulators (MZIM); Mach–
    Zehnder interferometric
    modulator (MZIM)

N

NF, *see* Noise figure (NF)
90°Optical hybrid, 283, 284
NLE, *see* Nonlinear equalizer (NLE);
    Nonlinear wave equation
    (NLE)
NLP, *see* Nonlinear phase (NLP)
NLPN, *see* Nonlinear phase noise
    (NLPN)
NLSE, *see* Nonlinear Schroedinger
    equation (NLSE)
Noise
  components, 233
  generators, 235
  power, 252
Noise figure (NF), 84
Nonlinear
  SPM coefficient, 89
  SPM effects, 301
Nonlinear (NL), 409
  photonic processor, 410
Nonlinear equalizer (NLE), 319; *see
    also* Digital signal processing
    (DSP); Linear equalization
    (LE); Zero-forcing nonlinear
    equalization (ZF-NLE)
  combining FFE and DFE
    equalization slicer, 333
  DD cancellation of ISI, 319–321
  equalization with minimizing MSE,
    324
  equalized sampled signals, 319
  error probability, 321
  of factorized channel response, 323
  FF transversal filter, 320
  LE and, 321
  probability density function, 321
  transmission system, 319, 320
  at transmitter, 331
  z-transform at distortion equalizer
    output, 330

Nonlinear fiber transfer functions, 99;
        see also Back propagation (BP);
        Digital signal processing;
        Optical fibers; Volterra series
        transfer functions (VSTF)
    cascades of linear and, 101–103
    coherent receiver with ADCs, 105
    digital processing unit, 105
    DSP-based receiver, 100
    DSP techniques, 106
    electronic compensation, 104–106,
        108–110
    nonlinear phase effects, 101
    nonlinearity suppression, 105
    QPSK received signal constellation,
        102
    signals through optical fibers, 102
    simulation system setup, 109
Nonlinear optical effects, 42, 78; *see
        also* Cross-phase modulation
        (XPM); Four-wave mixing
        (FWM); Optical fibers; Self-
        phase modulation effect (SPM
        effect); Stimulated Brillouin
        scattering (SBS); Stimulated
        Raman scattering (SRS)
    of mode hopping effects, 81
    phase noises, 86–87
    self-phase modulation effects, 42
    spectral broadening, 80
    stimulated scattering effects, 45
    transfer function, 82
    types, 42
    wave propagation equation, 80
Nonlinear phase (NLP), 111
Nonlinear phase conjugation (NPC), 436
    demultiplexer, 439
    distortion compensation, 437, 438
    regenerated optical signal, 437, 439
Nonlinear phase noise (NLPN), 86
    variance, 86–87
Nonlinear photonic pre-processing, 449
    efficient NL optical waveguide
        requirement, 454
    Gaussian noise on bispectrum, 452,
        454
    higher-order spectrum processing,
        450
    using HOS, 453

Long-Haul transmission
        parameters, 450
    optical effects as pre-processing
        elements, 450
    parametric amplifier
        parameters, 449
    phase states and bispectral
        properties, 450, 451, 452
    spectral distribution and
        conversion, 450
    uncorrupted bispectrum
        magnitude, 452, 454
Nonlinear photonic signal processing,
        419, 448, 455–456; *see also* Four-
        wave mixing (FWM); Optical
        domain implementation;
        Third-order nonlinearity;
        Triple correlation
    BER curve, 418
    coherent optical receiver, 410
    detected sequence, 417
    input data sequence, 417
    photonic processor, 410
    propagation equation, 430
    separating channel technique, 430
    transmission models and wave
        devices, 430
Nonlinear Schroedinger equation
        (NLSE), 23, 45, 89, 95, 381, 425
Nonlinear wave equation (NLE), 424
Nonlinearity, 420
Non return-to-zero differential phase-
        shift keying (NRZ-DPSK), 138
Nonreturn-to-zero (NRZ), 15, 140, 377
    -DPSK photonics transmitter, 144
    photonics transmitter, 140
    spectra of, 162, 176
Nonzero dispersion-shifted fibers
        (NZ-DSF), 49, 88
Normalized propagation constant, 58
NPC, *see* Nonlinear phase conjugation
        (NPC)
NRZ, *see* Nonreturn-to-zero (NRZ)
NRZ-DPSK, *see* Non return-to-zero
        differential phase-shift keying
        (NRZ-DPSK)
Nyquist limit, 202
Nyquist pulse and spectra, 386
    excess bandwidth, 386

frequency-domain representation,
386
frequency response, 386, 387, 388
impulse response, 386, 387, 388
raised-cosine filter, 386
symbol values, 387
Nyquist pulse-shaping techniques, 15;
see also Digital modulation
formats
filter design and transmission
system performance, 18
frequency response, 16, 17
impulse and frequency response, 19
impulse response, 17
raised-cosine filter, 15, 18
super-channel Nyquist spectrum, 16
Nyquist-WDM system, 15; *see also*
Nyquist pulse-shaping
techniques
NZ-DSF, *see* Nonzero dispersion-
shifted fibers (NZ-DSF)

O

O/E, *see* Optical/electric converter
(O/E)
OFDM, *see* Orthogonal frequency
division modulation (OFDM);
Orthogonal frequency
division multiplexing (OFDM)
OFFT, *see* Optical fast Fourier
transform (OFFT)
112G QPSK coherent transmission
systems, 371
BER vs. SNR, 373, 374
CSRZ QPSK transmitter, 371–372
I-Q signals, 372
noise analysis at receiver, 373–374
OSNR vs. BER, 373
polarization splitter, 372
receiver types, 372
symbol rate, 371
On–off keying (OOK), 344, 439
OOK, *see* On–off keying (OOK)
OPA, *see* Optical parametric amplifiers
(OPA)
OPL, *see* Optical phase locking (OPL)
OPSK, *see* Optical phase shift keying
(OPSK)

Optical amplification, 25
Optical coherent detection, *see*
Coherent detection
Optical communication system; *see also*
Optical fiber communication
DWDM optical DQPSK system, 337
with EDFAs, 337
modulation for, 122
Optical delay differential detection, 341
Optical demodulation, 18; *see also*
Optical fiber communication
Optical digital receiver, 22
employing coherent detection and
DSP, 23
Optical dispersion, 26
Optical domain implementation; *see
also* Four-wave mixing (FWM)
conservation of momentum, 429
NL wave guide, 428–429
optical power required for FWM,
429–430
polarization vector, 429
third-harmonic conversion, 429
Optical/electric converter (O/E), 273
Optical fast Fourier transform (OFFT),
394
Optical fiber coherent detection, 258
Optical fiber communication, 1; *see also*
Digital modulation formats;
Guiding lightwaves; Weakly
guiding fibers
coherent and incoherent receiving
techniques, 9–10
coherent detection, 20, 23
coherent reception techniques, 10
coherent solution, 10
over decades, 7
development in, 25–26
digital processing in advanced,
10–11
direct modulation, 10
dispersion-managed, 21
DSP-based coherent receiver, 10
electromagnetic spectrum of
waves, 22
and equivalent transfer
functions, 77
fiber span concatenation, 84
index profiles of modern fibers, 42

Optical fiber communication (*Continued*)
  lightwave region for silica based, 22
  mode fiber, 32
  modulation formats, 8–9
  optical demodulation, 18
  optical digital receiver and DSP, 22
  optical receiver function, 18
  $\pi/2$ hybrid coupler, 36
  progress in, 1
  single-mode, 5
  spatial mode multiplexing, 33
  transmission and receiving for, 3
  transmission capacity, 2
Optical fibers, 3, 25, 114–115; *see*
        also Circular optical
        waveguides; Nonlinear
        fiber transfer functions;
        Nonlinear optical effects;
        Signal distortion in optical
        fibers; Single-mode fiber
        (SMF); Split-step-Fourier
        method; Step-index fiber;
        Weakly guiding fibers
  absorption losses, 49–50
  attenuation coefficient, 52–53
  bending loss, 50
  cut-off wavelength, 40
  dispersion compensated fiber, 88
  etched cross-section, 27
  fiber attenuation spectrum, 51
  fiber cross-section and step index
        profile, 27
  Gaussian profile, 29
  geometrical structures and index
        profile, 26
  graded-index profile, 28
  microbending loss, 51
  optical dispersion, 26
  power-law index profile, 28
  properties, 26
  Raleigh scattering loss, 50
  refractive index profile, 27–29
  Sellmeier's coefficients, 58
  signal attenuation in, 49
  single-mode, 25
  special dispersion, 87–88
  splice loss, 52
  step-index fiber, 27
  waveguide losses, 50–52

Optical heterodyne detection, 185,
        195; *see also* ASK coherent
        system; Coherent detection;
        Differential detection
  asynchronous detection, 186
  configuration of, 185
  electronic signal power, 187
  FSK coherent system, 191–192
  LO, 186
  optical signal electric field, 186
  OSNR, 187
  photodetection current, 186
  PSK coherent system, 189–190, 222
  shot noise, 187
  spectrum of, 201
  square-law detection, 185–186
Optical homodyne detection, 192;
        see also Coherent detection;
        Differential phase detection
  BER, 194
  detection and OPLL, 193–194
  linewidth influences, 195
  matching received signal, 194
  noise power, 194
  PDF of IF, 195
  probability of error, 195
  quantum limit detection, 194–195
  receiver transfer function, 194
  shot-noise power, 194
  signal power, 194
  SNR, 194
  spectrum of, 201
  structure of, 192
Optical intradyne detection, 200; *see*
        also Coherent detection
  advantages of, 200
  spectrum of, 201
Optical Kerr effect, 42–43, 80; *see also*
        Nonlinear optical effects
Optical modulator, 90
Optical OOK transmitter, 131
Optical parametric amplifiers (OPA),
        431, 432; *see also* Non linear
        phase conjugation (NPC)
  critical parameters of, 434
  gain of, 435, 437
  optical spectra at input and
        output, 435
  simulink model of, 432

time traces, 434
wavelength difference between
    signal and pump, 436–437
Optical phase lock loop (OPLL), 180,
    255, 298–299, 393; *see also*
    Coherent optical receivers;
    Low pass filter (LPF); Phase
    lock loop (PLL)
advantages of, 255
attractiveness of, 256
beat signal monitoring, 291–292
beating signal, 289
carrier signal and LO frequency,
    282, 294
closed-loop optical RF, 290–291
comb source generation, 296
combed optical source, 298
configuration settings of
    components, 287
digital implementation, 256
digital LPF, 257
double sideband LO carrier
    spectrum, 282
error function, 283, 284
experimental development of, 284
experimental setup, 275, 278
functional requirements, 265
generic comb generator, 298
high-resolution optical spectrum
    analysis, 293
improvements, 256
issues/problems to be resolved, 295
issues with, 264
laser beating experiments, 288, 289
laser sources, 286
locking of lasers, 291
locking state, 278–281
LO frequency, 289, 294
LO modulation, 285
LO spectrum components and
    carrier signal, 293
loop filter design, 289–290
LPF input, 275
MATLAB Simulink model, 281
90°optical hybrid, 283, 284
noise observed after ADC stage, 279
noise sources, 278
nonfunctional requirements,
    265–266

O/E and digital filter output under
    locking state, 279
OPLL test-bed setup, 287
optical coherent detection and,
    258
optical receiver noise, 292
performances, 274
phase error and time constant,
    293–295
RF beat signal, 288
simulation, 274
subcarrier modulated, 263
super LO- and DSP-based
    superchannel coherent
    receiver system, 297
super LO using OPLL and comb
    generation, 297
for superchannel coherent receiver,
    296–298
transmitted data stream and
    demodulated data, 285
VCO drive voltage, 275
voltage shots, 291, 292
Optical phase locking (OPL), 255, 340
Optical phase shift keying (OPSK),
    137
Optical signal
    detection, 179; *see also* Coherent
    detection
    pre-processing receiver, 421
    traveling, 54
Optical signal spectra, 415
    input and output of nonlinear
        optical parametric amplifier,
        416
    waveforms in time domain, 417
Optical signal-to-noise ratio (OSNR), 87,
    105, 145, 187, 372
Optical time division multiplexing
    demultiplexer (OTDM
    demultiplexer), 439
    extracted demultiplexed signal,
        439, 443
    FWM-based, 442
    FWM parameters, 441
    HNLF walk-off problem, 439
    for modulation formats, 441
    parameters of, 446
    principle of, 439

Optical time division multiplexing
    demultiplexer (*Continued*)
  simulated performance of
    demultiplexed signals, 444
  spectrum at output of NL
    waveguide, 439, 443
  time traces, 441, 443
Optical transmission link, 414
Optical waveguide, 420; *see also*
    Nonlinearity
Optoelectronic receiver, 204
Orthogonal frequency division
    modulation (OFDM), 122
Orthogonal frequency division
    multiplexing (OFDM), 66, 156,
    157
  optical FFT/IFFT based, 159, 160
  serial data sequence in, 159
  signal generation and recovery
    principles, 158
  system arrangement of, 157
Oscilloscope, 288
OSNR, *see* Optical signal-to-noise ratio
    (OSNR)
OTDM demultiplexer, *see* Optical
    time division multiplexing
    demultiplexer (OTDM
    demultiplexer)

P

Partial signal technique, 15
PD, *see* Phase detector (PD);
    Photodetector (PD);
    Photodiode (PD); Pin diode
    (PD)
PDF, *see* Probability density function
    (PDF)
PDM, *see* Polarization division
    multiplexed (PDM)
PDM-16 QAM, *see* Polarization division
    multiplexed channels
    (PDM-QPSK)
PDM-QPSK, *see* Polarization division
    multiplexed channels
    (PDM-QPSK)
PE, *see* Phase estimation (PE)
Phase detector (PD), 258, 259; *see also*
    Phase lock loop (PLL)

output signal, 259
  phase error signal, 260
Phase estimation (PE), 180
Phase lock loop (PLL), 183, 255; *see also*
    Low pass filter (LPF); Optical
    phase lock loop (OPLL); Phase
    detector (PD)
  closed-loop transfer function of, 262
  components, 258
  FPGA-based, 265
  frequency lock condition, 264
  general theory, 258
  generic PLL blocking state
    diagram, 259
  loop filter, 260
  natural frequency of, 257
  noise sources, 278
  phase error response, 262
  response to phase input and phase
    error, 263
  second-order, 261
  voltage-controlled oscillator, 261
Phase modulators, 125; *see also* Electro-
    optic modulator (EOM)
  optical transmission system, 415
Phase rotation (PM), 89
Phase shift keying (PSK), 11, 13, 140
  coherent system, 189–190
  optical heterodyne and differential
    detection for, 190, 222
Phase updating rule, 317
Photodetector (PD), 372
Photodiode (PD), 179
Photonic MSK transmitter, 151; *see
    also* External modulators;
    Minimum shift keying
    (MSK); Orthogonal frequency
    division multiplexing (OFDM)
  band-limited phase-shaped optical
    MSK, 154
  modulation format spectra, 159–164
  multi-carrier FDM signal
    arrangement, 157
  multi-carrier multiplexing optical
    modulators, 156–159
  optical MSK signal sequence time-
    domain phase trellis, 153
  optical MSK transmitter, 152, 153–155
  optical RZ-MSK, 156

phase continuity of signals, 152
phase trellis of MSK transmitted
    signals, 155
single-side band optical modulators,
    155–156
Photonic transmitter, 121, 175
Photorefractive effect, 42; *see also*
    Nonlinear optical effects
Piezoelectric transducer (PZT), 256
Pin diode (PD), 180
$\pi/2$ Hybrid coupler, 36
Planar lightwave circuits (PLCs), 166
Planar optical waveguides, 25
PLCs, *see* Planar lightwave circuits
    (PLCs)
PM, *see* Phase rotation (PM)
PMD, *see* Polarization mode dispersion
    (PMD)
Polarization division multiplexed
    (PDM), 369, 384
Polarization division multiplexed
    channels (PDM-QPSK), 369
    optical transmission system, 372
Polarization mode dispersion (PMD),
    33, 179, 301, 371
    conceptual model of, 63
    effect in digital optical
        communication, 64
    fiber, 64
    first-order, 97
    maximum transmission limit,
        64–65
    Maxwellian distribution, 63, 64
    modeling of, 97–98
Polarization modulator, 130, 131
Polarization shift keying (PolSK), 12
PolSK, *see* Polarization shift keying
    (PolSK)
Power spectrum, 412, 413
    Gaussian noise on, 413
    regions, 411–412
Probability density function (PDF), 188,
    413
Propagation equation, 430
PSK, *see* Phase shift keying (PSK)
Pulse carver, 140, 142; *see also*
    Interferometric intensity
    modulator
PZT, *see* Piezoelectric transducer (PZT)

Q

Q, *see* Quadrature (Q)
QAM, *see* Quadrature amplitude
    modulation (QAM)
$Q_{m.n}$ number format, 266; *see also* Low
    pass filter (LPF)
QMZ-IQ modulator, *see* Mach–Zehnder
    in-phase/quadrature
    modulator (QMZ-IQ
    modulator)
QPSK, *see* Quadrature phase shift
    keying (QPSK)
Quadrature (Q), 369
Quadrature amplitude modulation
    (QAM), 9, 90, 122, 202
Quadrature phase shift keying (QPSK),
    369, 371; *see also* 112 G QPSK
    coherent transmission
    systems;
Quantum-confined Stark effect, 124
Quantum shot noises, 207

R

Raleigh scattering, 45
    loss, 50
Raman amplifiers, 6; *see also* Optical
    fiber communication
Raman distributed optical
    amplification (ROA), 414
Raman scattering, 45; *see also*
    Stimulated Raman scattering
    (SRS)
Received signal (Rx), 256
Recirculating frequency shifting (RFS),
    386
Reconfigurable optical add-drop
    module (ROADMs), 381
Refractive index, 57
    of silica, 55, 57
Return to zero (RZ), 15, 131, 377
    photonics transmitter, 140
    spectra of, 162, 176
Return-to-zero optical pulses, 131; *see
    also* External modulators
    bias point and RF driving
        signals, 133
    CSRZ pulse generation, 135

Return-to-zero optical pulses (*Continued*)
　CSRZ pulse shape, 131–132
　CSRZ pulses phasor representation,
　　135, 136
　DPSK optical transmitter with RZ
　　pulse carver, 139
　duty cycle of, 135
　modulation stages, 134
　optical OOK transmitter, 131
　phasor representation, 134
　pulse carver, 131, 133
　RZ-ASK transmitter, 131, 132
　RZ pulse electric field waveforms, 132
　RZ33 pulse generation, 136
　RZ33 pulses phasor representation,
　　136–137
RFS, *see* Recirculating frequency
　　shifting (RFS)
Rhode Schwartz model, 247
RHS, *see* Right-hand side (RHS)
Right-hand side (RHS), 424
ROA, *see* Raman distributed optical
　　amplification (ROA)
ROADMs, *see* Reconfigurable
　　optical add-drop module
　　(ROADMs)
ROF, *see* Roll-off factor (ROF)
Roll-off factor (ROF), 386
Root-raise cosine (RRC), 19
RRC, *see* Root-raise cosine (RRC)
Rx, *see* Received signal (Rx)
RZ, *see* Return to zero (RZ)
RZ-ASK transmitter, 131, 132
RZ-DPSK transmitter, 142
　photonic, 144

S

SBS, *see* Stimulated Brillouin scattering
　　(SBS)
Scattering processes, 45
SC-OPLL, *see* Subcarrier optical phase
　　locked loop (SC-OPLL)
Second-order poly-spectrum, *see* Power
　　spectrum
Self-heterodyne detection, *see*
　　Autocorrelation reception
　　process
Self-phase modulation (SPM), 301, 420

Self-phase modulation effect (SPM
　　effect), 43; *see also* Nonlinear
　　optical effects
　accumulated nonlinear phase
　　changes, 43–44
　induced nonlinear phase noise,
　　see Nonlinear phase noise
　　(NLPN)
　and intra-channel nonlinear effects,
　　81–86
　maximal launched power, 83
　maximum input power, 44
　modified propagation constant, 43
　and modulation instability, 80–81
　and nonlinear phase shift, 79
　nonlinear power penalty, 82
　power dependence of RI, 78
Sellmeier's dispersion formula, 57
Semiconductor optical amplifier
　　(SOA), 122
Shannon's channel capacity theorem, 66
Shot noises, 207
Signal distortion in optical fibers, 53;
　　see also Polarization mode
　　dispersion (PMD)
　chromatic dispersion factor, 59
　complex envelope, 56
　differential-dispersion parameter, 63
　dispersion slope, 63
　fiber dispersion factor, 60, 61
　group delay, 54, 60
　group velocity, 53
　higher-order dispersion, 62–63
　intermodal delay effects, 54
　intermodal dispersion, 54
　material dispersion, 55–58
　normalized propagation
　　constant, 58
　pulse dispersion per unit length, 57
　pulse spreading, 60
　refractive index, 57
　refractive index of silica, 55
　time signal and spectrum, 55
　V-parameters, 60
　waveguide dispersion, 58
　waveguide dispersion factor,
　　60–62
　wave phase velocity, 53
Signal processing functions, 420

Signal-to-noise and distortion ratio (SINAD), 220
Signal to noise ratio (SNR), 145, 309, 417
SINAD, *see* Signal-to-noise and distortion ratio (SINAD)
Single-mode fiber (SMF), 1, 31; *see also* Optical fibers
  amplitude evolution phasor, 91
  bit rate scaling and transmission distance, 68
  carrier chirping effects, 70
  chirp effects, 66–67, 68
  complex magnitude in phase and quadrature parts, 92
  complex power amplitude, 94
  digital-based optical transmitter and coherent reception, 90
  dispersive pulse sequences, 69
  eye diagram of time signals, 78
  frequency response of, 69
  frequency transfer response phase, 68
  frequency vs. amplitude, 94
  Gaussian pulse, 71–72
  group delay factor, 89
  IM optical waves complex amplitude, 92
  linear and nonlinear fiber properties in, 79
  linear transfer function, 65–72, 75
  microbending loss of, 51
  modulating frequency vs. complex power, 93
  modulation implementation, 66
  NLSE, 89
  nonlinear fiber transfer function, 72–77
  nonlinear SPM coefficient, 89
  optical modulator, 90
  overall fiber transfer function, 94
  pulse response, 71
  rectangular pulse transmission through, 67
  response to Gaussian pulse, 71
  Shannon's channel capacity theorem, 66
  simplified linear and nonlinear operating region, 88–94

  step response of, 70
  system step response, 65–66
  transfer function of, 65, 68
  transmission bit rate and dispersion factor, 77
  Volterra series transfer function, 72, 76
  wave propagation inside, 72
Single-sideband (SSB), 94, 284
  modulation, 89
  optical modulators, 155–156
Single-side subcarrier modulation OPLL (SS-SC-OPLL), 265
16 QAM systems, 381
  all-optical 10.8 Tbit/s OFDM, 379
  carrier phase recovery, 371, 384
  cavity structure of laser with back mirror vibration, 377
  CD and PMD equalization, 377, 378
  coherent receiver, 342
  constellation distortions, 383–385
  equalized constellations of QPSK modulation, 379
  frequency offset estimation, 381–382
  I-Q imbalance estimation results, 374–375
  OSNR vs. BER for integrated coherent receivers, 376
  phase distortion and BP compensation, 377–381
  phase errors vs. launched power, 380
  physical phenomena effects on constellation of, 383–384
  refined carrier phase recovery, 382–383
  skew estimation, 375–377
  symbol phase estimation, 381
SMF, *see* Single-mode fiber (SMF)
SNR, *see* Signal to noise ratio (SNR)
SOA, *see* Semiconductor optical amplifier (SOA)
SOP, *see* State-of-polarization (SOP)
Spectrum analyzer, 244
Splice loss, 52
Split-step-Fourier method (SSFM), 84, 95, 430, 111; *see also* Optical fibers
  accuracy improvisation, 96

Split-step-Fourier method (*Continued*)
  computational time optimization, 98
  fiber length, 95
  NLSE, 95
  operators, 96
  optimization, 98
  polarization mode dispersion
      modeling, 97–98
  symmetrical, 95–97
  waveform discontinuity, 98–99
  windowing effect of FFT, 98–99
SPM, *see* Self-phase modulation
      (SPM)
SPM effect, *see* Self-phase modulation
      effect (SPM effect)
SPM-induced nonlinear phase noise,
      see Nonlinear phase noise
Spontaneous scattering, 46
Spot-size r_0, 41
Square-law
  detection, 185–186; *see also* Optical
      heterodyne detection
  photodetectors, 217
SRS, *see* Stimulated Raman scattering
      (SRS)
SSB, *see* Single-sideband (SSB)
SSMF, *see* Standard single-mode optical
      fiber (SSMF)
SS-SC-OPLL, *see* Single-side
      subcarrier modulation OPLL
      (SS-SC-OPLL)
Standard single-mode optical fiber
      (SSMF), 8, 40, 342, 437
  total chromatic dispersion
      factor of, 59
Standard Telephone Cables (STC), 1
Stark effect, 122, *see* Electro-absorption
      effect (EA effect)
  quantum-confined, 124
State-of-polarization (SOP), 398
State trellis, 329; *see also* Trellis
      structure
STC, *see* Standard Telephone Cables
      (STC)
Step-index fiber, 27; *see also* Optical
      fibers; Weakly guiding fibers
  eigenvalue equation of wave
      equation, 31
  equivalent-step-index description, 41

field spatial function, 30
  in Gaussian approximation, 37
  spot-size r_0 approximation, 41
  wave equation solutions, 30–31
Stimulated Brillouin scattering (SBS),
      46, 79; *see also* Nonlinear
      optical effects
  coupled equations, 46
  threshold power for generation of,
      46–47
Stimulated Raman scattering (SRS),
      46, 47, 79; *see also* Nonlinear
      optical effects
  coupled equations, 47
  spectrum of Raman gain, 47
  threshold for stimulated Raman
      gain, 47–48
Subcarrier optical phase locked loop
      (SC-OPLL), 256, 257; *see also*
      Optical phase lock loop
      (OPLL)
Superchannel transmission system,
      385, 392; *see also* Nyquist pulse
      and spectra
  architecture of, 394
  B2B BER vs. OSNR, 406
  BER performance, 401, 402
  CD tolerance, 389
  comb-generated subcarriers, 392
  DSP-based coherent receiver, 389,
      392–394
  DSP-based heterodyne coherent
      reception systems, 403–406
  frequency shift, 392
  Godard phase detector algorithm,
      398, 399
  inferiority of SNR sensitivity, 404
  modulation format, 389
  Nyquist channel spectra, 400
  Nyquist filter for spectral
      equalization, 389
  Nyquist filtered QPSK signal, 389
  Nyquist QPSK, 386
  optical FFT receiver assessment, 397
  optical fourier transform-based
      structure, 394–395
  percentage overloading, 397
  PMD tolerance, 389
  processing of, 395

requirements, 388–389
RFS comb generator, 393
16 QAM superchannel transmission, 399–401
SOP rotation speed, 389
spectra of superchannels, 395, 397
spectral efficiency, 389
subcarriers with modulation, 395
system specifications of various transmission structures, 389, 390–391
32 QAM Nyquist transmission systems, 401
timing recovery in Nyquist QAM channel, 398
transmission distance, 388
transmission reach distance variation, 403
transmission SSMF length, 401
Superchannels, 296–298
Synchronous detection, 189

T

Thermal noises, 207
THG, *see* Third-harmonic generation (THG)
Third-harmonic generation (THG), 422
Third-order nonlinearity, 424; *see also* Four-wave mixing (FWM); Optical parametric amplifiers (OPA)
coupled equations and conversion efficiency, 427–428
high-speed optical switching, 437
NL coefficient, 428
NL wave propagation, 424
NLE, 424
NLSE, 425
parametric amplifiers, 431
phase matching, 427
phase mismatch, 433
propagation in optical waveguide, 425
short-pulse generator, 437, 440, 441
three coupled equations, 433
triple correlation, 442
waveform of modulated pump, 437, 441

wavelength conversion and NL phase conjugation, 436
Third-order statistics, *see* Triple correlation—Fourier transform of
TIA, *see* Trans-impedance amplifiers (TIA)
Trans-impedance amplifiers (TIA), 203, 372; *see also* Electronic amplifiers
with differential feedback paths, 205
inputs and outputs, 205–206
wideband, 205
Transmission and detection, 414; *see also* Optical signal spectra; Nonlinear photonic signal processing
AWGN block, 418
BER value comparison, 419
carrier period, 419
fair comparison noise, 418
four-wave mixing and bispectrum receiving, 415
impacts of NL parametric conversion, 415
optical transmission link, 414–415
performance, 415
signals sampling, 417
time traces of optical signal, 415
Transmission models and NL guided wave devices, 430
Transversal equalizers, 310; *see also* Linear equalization (LE)
linear feedback, 311
Trellis structure, 326; *see also* Finite state machine (FSM); Viterbi algorithm
construction of state, 328–329
state trellis, 329
Triple correlation, 412, 442; *see also* Bispectrum; Four-wave mixing (FWM); Nonlinear photonic signal processing
additive Gaussian noise, 414
dual-pulse sequence with unequal amplitude, 446, 447, 448
eliminating Gaussian noise, 413
Fourier transform of, 411
function, 411

Triple correlation *(Continued)*
  Gaussian noise rejection, 413
  to generate, 444
  in optical domain, 443
  phase and amplitude information, 412
  phase information encoding, 413
  of signal, 442–443
  Simulink model, 445
  single-pulse signal, 448
  symmetrical processes in, 414
  third cumulant, 414
  triple-product, 424, 444, 447
Triple-optical autocorrelation, 444
Triple-product, 424, 444
  waves, 447

U

UI, *see* Unit interval (UI)
Ultraviolet (UV), 50
Unit interval (UI), 398
Upper sidebands (USB), 151, 349
USB, *see* Upper sidebands (USB)
UV, *see* Ultraviolet (UV)

V

VCO, *see* Voltage control oscillator (VCO)
Vertical polarized channel (V-pol), 369
Very large-scale integration (VLSI), 409
Viterbi algorithm, 327; *see also* Maximum likelihood estimation equalizer-MSK optical transmission system performance; Trellis structure
  branch metric calculation in, 350
  detector, 324
  MLSE equalizer with, 349–351, 352
  operation, 325
  phases in, 327–328
Viterbi-MLSE equalizers; *see also* Maximum likelihood estimation equalizer-MSK optical transmission system performance
  equalizers for PMD mitigation, 359–364

OSNR penalties of MLSE performance, 363
performance, 353, 355–359, 361, 362
required OSNR for MLSE performance, 363
required OSNR vs. residual dispersion in, 353
Viterbi-Viterbi (V-V), 342
VLSI, *see* Very large-scale integration (VLSI)
Voltage control oscillator (VCO), 190, 257, 261
Volterra series model, 72
Volterra series transfer functions (VSTF), 83, 86; *see also* Nonlinear fiber transfer functions
  amplitude frequency domain, 103
  and electronic compensation, 103–104
  electronic nonlinearity compensation scheme, 104–106
  fiber span representation by, 104
  fiber transfer functions, 107
  first-and third-order, 106–107
  inverse of, 106–108
  Manakov-PMD equations, 106
  third-order, 107, 111
V-pol, *see* Vertical polarized channel (V-pol)
VSTF, *see* Volterra series transfer functions (VSTF)
V-V, *see* Viterbi-Viterbi (V-V)

W

Wave vector mismatch, 426
Waveguide dispersion, 54, *see* Group velocity dispersion (GVD)
Waveguide dispersion factor, 60
  alternative expression for, 61–62
  vs. wavelength, 62
Waveguide dispersion parameter, *see* Waveguide dispersion factor
Wavelength-division multiplexed (WDM), 401, 436
  channels, 44
Wavelength selective switch (WSS), 401

WDM, *see* Wavelength-division
    multiplexed (WDM)
Weakling guiding, 25; *see also* Optical
    fiber communication
Weakly guiding fibers; *see also* Optical
    fibers; Step-index fiber
  axial power density, 40
  cut-off properties, 38–40
  fundamental mode, 29, 31
  Gaussian approximation, 36–38
  guided mode, 31, 33, 34, 35
  intensity profiles, 33
  LP01 mode intensity distribution, 39
  $\pi/2$ hybrid coupler, 36
  power distribution, 40–41
  properties of fundamental mode
    of, 36
  scalar wave equation, 29
  single and few mode conditions,
    31–36
  spot-size r_0 approximation, 41
  total power, 40
  V parameter, 39
  wave equation solutions, 30–31
Window length (WL), 344
Windowing effect, 98
WL, *see* Window length (WL)
WSS, *see* Wavelength selective switch
    (WSS)

X

XPM, *see* Cross-phase modulation
    (XPM)

Z

Zero-forcing linear equalization
    (ZF-LE), 307; *see also* Linear
    equalization (LE)
  for fiber as transmission channel,
    308–309
  fractionally spaced FIR filter, 309
  frequency response, 308
  impulse response of filter, 308
  transfer function, 308
  transmission system, 307
Zero-forcing nonlinear
    equalization (ZF-NLE), 321;
    see also Nonlinear equalizer
    (NLE)
  error probability, 323
  impulse response, 322
  operational principles, 322
  transfer function, 323
ZF-LE, *see* Zero-forcing linear
    equalization (ZF-LE)
ZF-NLE, *see* Zero-forcing nonlinear
    equalization (ZF-NLE)